HABITechno 3 International Conference 2017

Ecoregion as a Verb of Settlement Technology and Development

IOP Conference Series: Earth and Environmental Science
Volume 152

Bandung, Indonesia
11 November 2017

ISBN: 978-1-5108-6476-4
ISSN: 1755-1307

Printed from e-media with permission by:

Curran Associates, Inc.
57 Morehouse Lane
Red Hook, NY 12571

Some format issues inherent in the e-media version may also appear in this print version.

Printed by Curran Associates, Inc. (2018)

For permission requests, please contact the Institute of Physics
at the address below.

Institute of Physics
Dirac House, Temple Back
Bristol BS1 6BE UK

Phone: 44 1 17 929 7481
Fax: 44 1 17 920 0979

techtracking@iop.org

Additional copies of this publication are available from:

Curran Associates, Inc.
57 Morehouse Lane
Red Hook, NY 12571 USA
Phone: 845-758-0400
Fax: 845-758-2633
Email: curran@proceedings.com
Web: www.proceedings.com

Table of contents

Volume 152

HABITechno 3 International Conference

11 November 2017, Bandung, Indonesia

Accepted papers received: 2 May 2018 Published online: 11 June 2018

Preface

Papers

IOP Conf. Series: Earth and Environmental Science **152** (2018) 011001 doi:10.1088/1755-1315/152/1/011001

PREFACE

The School of Architecture, Planning and Policy Development, Institute of Technology Bandung, is pleased to present the Proceeding of HABITechno International Seminar, held in the Institute of Technology Bandung campus, Bandung, 11 November, 2017. This is the third HABITechno seminar, and we plan to hold the seminar every two year.

From 90's to recent time, discursion of housing and settlement development had been transformed in line with the growth of discussion in social and ecological issue. One of awareness towards this phenomenon is shown with development and urbanization theme which had been brought by HABITAT III in 2016 as commitments of change towards sustainable city development based on integration of social, economic, and environmental dimension. In general those commitments are: (1) Sustainable city development for achieving social inclusion and alleviating poverty; (2) Sustainable and inclusive city welfare for all; (3) Environmentally sustainable and resilient city development.

Selected theme for this third Habitechno is 'ECOREGION AS A VERB OF SETTLEMENT TECHNOLOGY AND DEVELOPMENT". This theme was selected because of numerous aspects in technology and environment, particularly in sustainable development in housing and settlement development which hadn't been explicitly delivered. This conference attempts to integrate technology into design, housing procurement and settlement improvement. Result from this conference is expected could be an input for housing and settlement development guidance in the future.

Ecoregion is defined as geographical region which has climate, soil, water, flora and fauna similarity and pattern of human interaction with nature which shows integration between nature system and living environment. Through ecoregion, recent development problem which is too economic growth oriented and hadn't put environment into an account could be solved. Balancing economic growth and sustainable environment is needed in order to eliminate environmental damage caused by development acceleration. This balance is important for settlement development because settlement has the biggest percentage of space in urban area particularly, so that it could create a bigger impact to the environment.

Integration and comprehensive viewpoint is needed to achieve balance as well as ecosystem. Recent development always based on administrative and creates environmental problem (pollution, damage, etc) interregional and less coordination interregional. Meanwhile, ecoregion based planning is not following administrative boundary but region characteristic boundary. Through integrated ecoregion approach, each built environment development, housing and settlement in particular, will put consideration of interregional functional relation and landscape characteristic with its ecological function into an important aspect in order to accelerate achieving the aim of sustainable development

In order to achieve innovative solutions for solving housing and settlement problem, role of many parties such as government, practitioner, industry, and academician is needed. Therefore, beside as a discussion forum, HABITechno has aim to build connection between those four parties which could participate in this international conference

In this publication, papers are compiled to provide an insight for reflection and sharing the result of researches and best practices.

IOP Conf. Series: Earth and Environmental Science **152** (2018) 011001 doi:10.1088/1755-1315/152/1/011001

We would like to express our thanks and appreciation to all presenters, participants, the members of scientific committee, the members of advisory board, and sponsors as well as our colleagues for their contributions in this seminar.

We hope that this seminar can be fruitful to all involved, and give intellectually engaging knowledge to the participants.

We look forward to meeting you all again in our next HABITechno seminar.

The Committee
HABITechno International Seminar

IOP Conf. Series: Earth and Environmental Science **152** (2018) 011002 doi:10.1088/1755-1315/152/1/011002

Peer review statement

All papers published in this volume of *IOP Conference Series: Earth and Environmental Science* have been peer reviewed through processes administered by the proceedings Editors. Reviews were conducted by expert referees to the professional and scientific standards expected of a proceedings journal published by IOP Publishing.

Public Flat in Indonesia, Their Role in Highly Densed City: Legal Aspect Review and Prototype Assessment

Beta Paramita[1], Rendy Perdana Khidmat[2], Hiroatsu Fukuda[3]

[1] Architecture Study Program, Universitas Pendidikan Indonesia, Jl. Dr. Setiabudhi 229, Bandung 40152.
[2] Architectural Engineering Study Program, ITERA, Jalan Terusan Ryacudu, Way Hui, Jati Agung, Kabupaten Lampung Selatan, Lampung 35365
[3] Architecture Department, The University of Kitakyushu, 1-1 Hibikino, Wakamatsu-ku, Kitakyushu 808135

betaparamita@upi.edu

Abstract. An affordable vertical house, which is known as *rusun* in Indonesian term becomes the utopia of a compact city. The *rusun* to be the main actor contributing to the creation of liveable comfort in dense areas by relocating squatters and slum inhabitants to better living places. Technical guidelines and building standards play a major role in a building's environmental impact. These standards will be reviewed based on their impact on building performance. There are two prototypes of *rusun* referenced in 05/PRT/M/2007, and the assessment of these prototypes has been done by using ENVi-met. The meteorological data for the ENVI-met configuration input is adapted to the same weather and climate conditions taken from the data measured on June 13th, 2012. The analysis of this assessment will focus on building orientation, building length, building height, and also the microclimate which is impacted by these prototypes. Based on the assessment, it is recommended that the layout of outdoor space on the north side should provide as much green coverage as possible, either for shadowing or for the benefit of its albedo. The simulation shows that building ratio of 1:3 produces higher wind speed, which also controls air temperature and humidity.

1. Introduction

High population densities increase exposure to the effects of microclimates. Also, high population densities are vulnerable to climate change, as they can potentially localize climatic effects, such as increased local temperatures, urban heat island effects and high levels of outdoor temperatures. Therefore, density has become one of a variety of factors that influence the sustainability of urban form. Furthermore, density has an effect of compactness, the need to travel, and the feasibility of public transport in reducing emissions. In a broader sense, sustainable cities are a matter of density. Based on the data from BPS Indonesia, there are twelve cities in Indonesia with the population of more than 10,000 people / km2 [1] not only causing demographic problems for the social economic but also giving burdens to the urban living environment. The excess of population in big cities in Indonesia creates slum and squatter pockets nearby the downtowns. Thus, the program of 1000 towers was set by the Indonesian Government to replace the over-crowded settlement by providing vertical housing, which in Indonesian term is called *rusunawa* (rent) and *rusunami* (owned).

IOP Conf. Series: Earth and Environmental Science **152** (2018) 012001 doi:10.1088/1755-1315/152/1/012001

The national government runs the program of *rusun*. Previously, two ministries take responsibility for *rusun* construction that is the ministry of public works and the ministry of public housing. Starting from 2014, these two ministries have been joined as the ministry of public works and housing (PUPR). The demand of *rusun* as affordable housing in Indonesia is rising year by year. The amount of backlog based on the concept of residential in 2014 is 7,4 million unit house. Therefore, the development and construction of *rusun* will be further carried out in the following years to handle the high density in the slum areas, which is essential for city planning and design. The legal aspect is then a primary strategy for *rusun* planning and design. Technical guidelines and building standards play a major role in a building's environmental impact, and they are also related to building performance, a factor correlated with urban built environments. Therefore, the actual and strategic positions of legal aspects are the driving force in examining the technical guidelines and building standards of *rusun*.

2. Legal Aspect of *Rusun*

Legal aspects discussed here refer to the result of a legal product when Kemenpera and PU were yet in the same ministry. The hierarchy of legal aspects of *rusun* includes technical requirements, guidelines, and standards, shown in figure 1. The focus of the discussion is the content of the guidelines. Thus, the hierarchy of the legal aspect will not be necessary to discuss. Specific discussions of the technical guidelines in this paper can be seen in the following:

a. 60/PRT/1992 about Technical Requirements of *rusun* construction [2]
b. 05/PRT/M/2007 about Technical Guidelines of high-rise *rusun* [3]

Figure 1 Legal Aspect of Rusun
Source : UU, Permen and SNI compilation

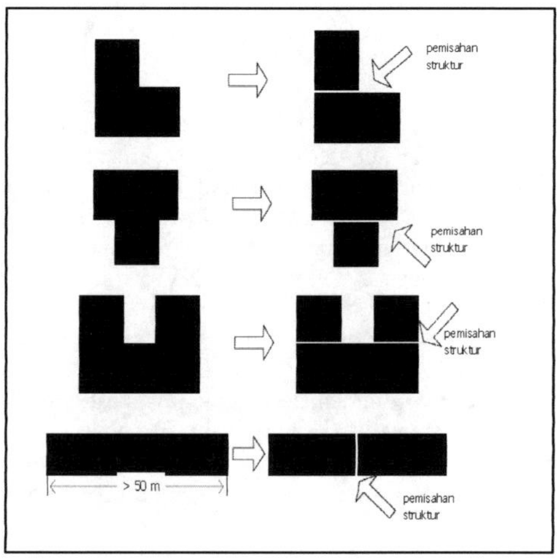

Figure 2 Building Planfor *Rusun*
Source: Legal Aspect of Rusun 05/PRT/M/2007

An explanation of the above legal aspects is as follows:
60/PRT/1992 CHAPTER VII: DENSITY AND LAYOUT OF BUILDING and the detail are as follows:
<u>*Article 48*</u>
(1),(2) Public flats with five floors and the maximum density of the dwellers is 1,736 persons and the BCR is 25% and FAR is 1.25

(4) The use of land is the ratio of total land area for building flats, which is 50% and for infrastructure is 20%

Article 50

(1) The distance between buildings required for fire safety, daylighting and air circulation05/PRT/M/2007
CHAPTER III: THE REQUIREMENTS OF BUILDING FORM AND CONFIGURATION

III.1. (4) d. The building distances (clearances) are set from the ground floor as far as 4m, with an augmentation of 0.5m for each additional floor or building level until it reaches the furthest distance of 12.5 m.

III.2. (1) b. Building blocks with T, L or U shapes which have the building length of more than 50m need to make a dilatation to prevent damages caused by an earthquake or subsidence as seen in Figure 2.

Building plans with centric shapes such as squares, polygons or circles are better than elongated shapes to prevent earthquake-related damages.

As described above, there is an absence of building forms and configurations regarding passive design needed to accommodate the urban microclimate aspect. Furthermore, the low availability of detailed dimensions and the lack of specific explanations make this guideline open to multiple interpretations. Since this guideline is so general, developers will not comply with these rules which in turn put the law enforcement of building regulations in a difficult position

Additionally, an important part of this legal aspect is the *rusun* prototypes, which have two main building plan types. These include symmetric parallel (SP) and symmetric cross (SC), and they can be used for eight, fifteen and twenty floors. To accommodate subsequent discussions, the assessment of these flats will then be referred to as the prototype of *Rusun*.

2.1. Legal Aspects Review

The legal aspect here refers to 60/PRT/1992 CHAPTER VII, article 48 and 50; 05/PRT/M/2007 CHAPTER III, verses III.1 and III.2, which are included in the Indonesian National Standard (SNI) for *rusun* construction. The realization of legal aspect aforementioned seems to overlap each other, and also there are some inconsistencies in the statements which are described in Table 1.

Table 1 Legal Aspect Content and Realization

Legal Aspect	Realization of Prototype
60/PRT/1992 : Public flats with five floors and the maximum density of the dwellers is 1,736 persons, the BCR is 25% and FAR is 1.25 Max uses of land for building is 50%; environmental infrastructure 20% and environmental facilities 30%	**05/PRT/2007** Directly provide 8, 15, and 20 floors in the area of 5,000m^2 8 floors SPModel:BCR 25%, FAR 2.0, density 1,024persons. SCModel BCR 37%, FAR 2.96, density 1,120persons 15 floors SPModel: BCR 25%, FAR 3.85, density 1,920 persons. SCModel: BCR 37%, FAR 5.55, density 2,040persons 20 floors SPModel: BCR 25%, FAR 6.42, density 2,560 persons. SCModel: BCR 37%, FAR 7.4, density 2,720persons
05/PRT/2007 : Building distance from the ground floor as far as 4m, and the farthest is 12.5m SNI 03-2846-1992 : [4] 8floors → BCR 17.5%, FAR 1.375 and density 1,909 persons The number of floors allowed are 11-	**05/PRT/2007** The distance between the building is not synchronous with the availability of prototype, which is only for a single building. In case applied in the building groups, it will have an impact on the availability of daylighting and possibility of wind turbulence 8floors tower → the distance between building 12.5m so H/W = 30/12.5 = 2.4

12 floors	15floors tower → the distance between building 12.5m so H/W = 51/12.5 = 4.08 20floors tower → the distance between building 12.5m so H/W = 66/12.5 = 5.28

3. Assessment of *Rusun*'s Prototypes

The assessment of building forms and configurations for the two models of the prototype was done by using ENVI-met. The meteorological data for the ENVI-met configuration input was adapted with the weather and climate conditions taken from the data measured on June 13th, 2012. Since the prototype only provides a single building for the site, the analysis of this assessment focuses on building orientation, building length, building height, also the microclimate which is impacted by the prototypes. There are two prototypes of *rusun* referenced in 05/PRT/M/2007. The first model is a symmetric parallel model (SP), as shown in figure 3a and figure 3b. The second model is a symmetric cross model (SC), as shown in figure 3c. and figure 3d

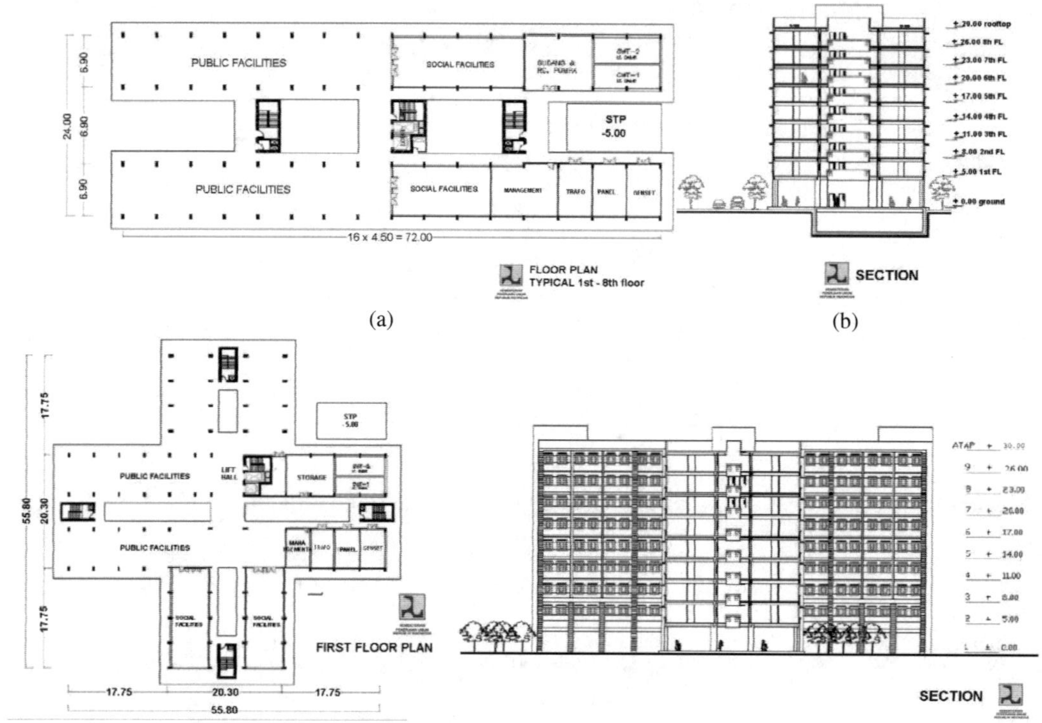

Figure 3(a). Plan of Symmetric Parallel (SP) Model; (b). Section of Symmetric Parallel (SP) Model; (c) Plan of Symmetric Cross (SC) model and (d) Section of Symmetric Cross (SC) Model
Source: Legal Aspect of Rusun 05/PRT/M/2007

3.1 Microclimate Assesment

ENVI-met is a three-dimensional microclimate model designed to analyze the small scale interaction between urban design and the microclimate. The model combines the calculation of fluid dynamic parameters such as wind flow or turbulence with the thermodynamic processes taking place on the ground surface, at walls and roofs or at the plants. With a typical resolution between 0.5m and 10m, the model is

able to simulate even complicated geometric forms, such as terraces, balconies or complex quarters. [5][6]
The model includes the simulation of:
1. flow around and between buildings;
2. exchange processes of heat and vapour at the ground surface and at walls;
3. turbulence;
4. exchange at vegetation and vegetation parameters.

This simulation will help architects and planners predict the quality of urban areas in the complexity of urban microclimate, understand the better living area or give the optimum decision before the construction is held

Table 2. General Condition for Simulation

Location	Bandung, Indonesia -6.54S; 107.37E		Indoor 296.34 K Heat transmission
Climate type	Hot humid climate		Walls: 0.41 W/m^{2}K
Simulation day	13th June, 2013 24hours	Building	Roofs: 0.157 W/m^{2}K Albedo
Meteorological Data	Initial Temperature 298.35 K		Walls: 0.2 Roofs: 0.3
	Specific Humidity: 13 gWater/kg air RH 63% Indoor 296.34 K Wind speed 2.29m/s Direction 321NW	Heat Human Balance	Walking speed: 1.4 m/s Energy Exchg : 116W/m^2 Clo : 0.7

As mention above, there are two models, which are SP and SC. These two models have tilted from the west-east 20 degrees toward north. Accordingly, there is SP_20 for the SP Model titled 20 degrees, and SC_20 for the SC Model titled 20 degrees. Therefore, in this simulation, the total number of building models is four models. The general condition for simulation, like the geographic location and meteorological condition, is described in table 2. The model of SP Model can be seen in Figure. 4(a) and the SP_20 Model is shown in Figure 4(b). Also, the SC Model can be seen in Figure 4(c) and the SC_20 Model is shown in Figure 4(d).

In this study, the assessment is limited to prototypes of only eight floors, as fifteen and twenty floor prototypes are not included in the study. The consideration of the impropriety of fifteen and twenty floor prototypes has explained in Table 1.

The building form and massing, including their different height to width ratio, distances between building and green open space are the most significant aspects to influence the wind speed and the penetration of radiation within the urban area. Here, the interaction between urban form and microclimate from ENVI-met simulation has been shown by air temperature (Ta), humidity (RH), velocity (v) and the most important part is the mean radian temperature (Tmrt) value, as an expression of radiation flux and parameter with high variability in urban area, at least by the modification of the global radiation. Later, table 3 shows microclimate condition of *Rusun*'s prototype and the diagram of this assessment shown at table 4.

Figure 5 as a chart from table 3 presents the microclimate for each prototype. These results indicate that the **south west** orientation of this model reached the lowest temperature (Ta), but did not reach low values in other meteorological parameters. The **north west** orientation has the highest Ta, but the other parameters (Tmrt, Tg and RH) show the lowest value. By tilting the building orientation of west to east to N 20°, the SC Model does not seem to significantly reduce the gap of maximum meteorological parameters.

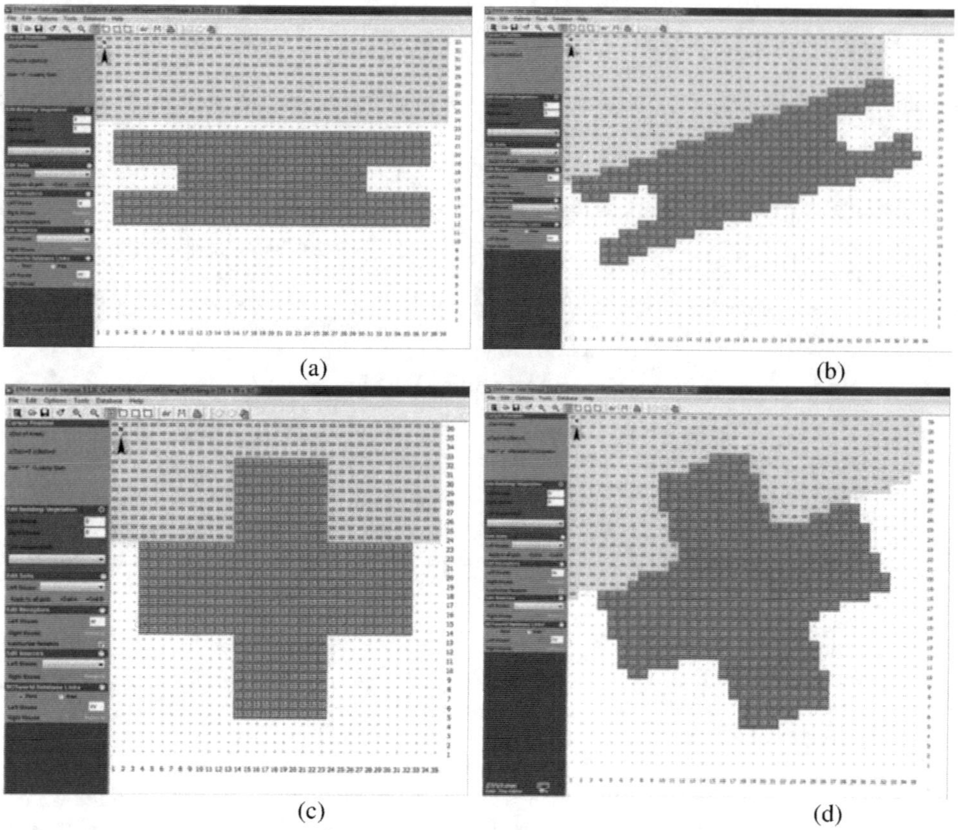

Figure 4. Area input for ENVI-met a. SP Model b. SP-20 model; c. SC Model d. SC-20 Model

Table 3. Microclimate assessment of *Rusun*'s Prototype

Symmetric Parallel Model (SP) Fig. 4a					
Orientation	Ta	Tmrt	Tg	RH	v
N	30.1	76.37	65.35	53	1.74
S	28.48	25.96	26.55	59	1.64
W	29.31	52.91	47.89	56	1.3
E	28.83	29.88	29.82	57.3	0.07
Symmetric Parallel Model titled 20° (SP_20)					
orientation	Ta	Tmrt	Tg	RH	v
N	28.31	63.13	57.37	80.2	0.7
S	28.03	54.09	49.78	76.32	0.7
W	27.95	50.8	46.01	80.08	1.25
E	28.27	50.78	45.85	77.66	1.4
Symmetric Cross Model (SC)					
orientation	Ta	Tmrt	Tg	RH	v
NE	29.5	61.77	55.05	59.5	1.23
SW	29.78	75.25	65.4	55.43	1.36
NW	30.15	62.46	56.61	55.36	0.87
SE	28.99	77.05	70.72	57.41	0.41
Symmetric Cross Model titled 20° (SC_20)					
orientation	Ta	Tmrt	Tg	RH	v

NE	27.31	65.04	57.39	87.92	1.15
SW	26.27	73.33	63.29	85.2	1.31
NW	27.42	61.5	55.48	65.72	0.82
SE	27.32	75.1	68.61	81.74	0.44

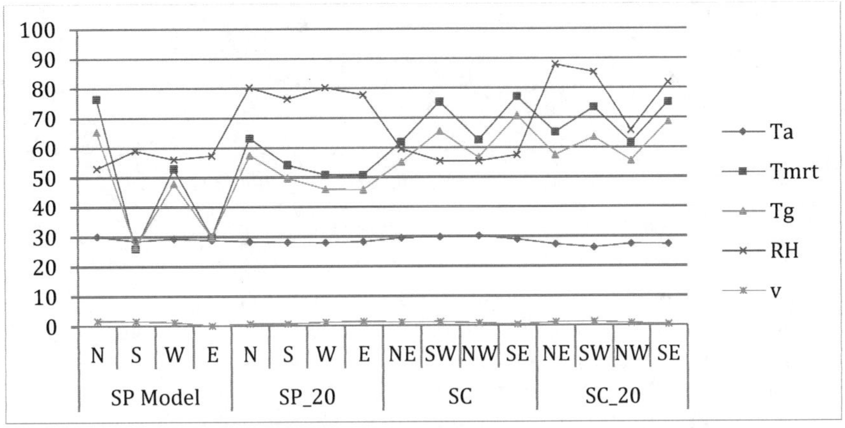

Figure 5. Microclimate condition of *Rusun*'s Prototype (from table 3)

Referring to the previous research [3], building form and massing have an important effect on the microclimate through modifying ground surface temperature with vegetation ratio which is proven to offer lower isolation and bring the optimum outdoor thermal comfort.

The result of ENVI-met simulation of building prototypes: SP; SC; SP_20; SC_20 from 06.00 am to 07.00 pm are shown in figure 5 as a chart from table 3. This value presents the Tmrt value for each prototype, in which all the prototypes reached the highest Tmrt value at 2pm. The SC_20 Model reached the lowest Tmrt value at 50.97 °C, meanwhile the SC Model reached the highest Tmrt value at 52.7 °C.

This simulation provides the information that four of the building prototypes, building form and massing, including their orientation, length, and height give similar responses to the microclimate situation.

3.2 Living Quality based on human perception of outdoorthermal comfort.

The living quality is determined by many aspects. However, it cannot be denied that meteorology and atmospheric conditions have a significant impact on the quality of urban environments, which depend on their physical aspects. Thus, it is a logical correlation that air temperature, humidity, wind speed and insolation within the urban space strongly impact on human health.

Table 4 Tmrt value for four models

Model	Diurnal Mean Radiant temperature (Tmrt in °C)													
	06.00	07.00	08.00	09.00	10.00	11.00	12.00	13.00	14.00	15.00	16.00	17.00	18.00	19.00
SP	15.73	28.30	43.20	47.76	49.54	49.89	50.29	51.08	51.50	50.30	45.20	25.47	17.67	16.97
SC	15.74	28.72	43.10	48.45	50.35	51.00	51.55	52.29	52.67	50.76	45.62	27.75	18.04	17.41
SP_20	15.73	28.28	42.94	47.77	49.62	50.09	50.58	51.46	51.93	50.34	44.90	26.18	17.99	17.35
SC_20	15.74	28.22	42.85	47.81	49.72	50.12	50.31	50.88	50.97	49.40	44.34	24.66	17.15	16.61

As previously described, with respect to the idea of microclimate improvement in the frame of building groups, a passive design gives better outdoor thermal comfort.

Table 5. Thermal sensation classification for Taiwan and Western/Middle Europe

Thermal sensation	PET in Taiwan (°C)	PET range for Western/Middle europe (°C)
Very cold	<14	<4
Cold	14-18	4- 8
Cool	18-22	8-13
Slighty cool	22-26	13-18
Neutral	26-30	18-23
Slighty warm	30-34	23-29
Warm	34-38	29-35
Hot	38-42	35-41
Very hot	<42	<41

In this research, the assessment focuses on the human perceptions of outdoors, which in this step is merely based on measurement and simulation, not through interviews or personal contacts. Studies on thermal comfort of humans in urban areas require meteorological parameters, such as air temperature, humidity, air velocity, short and long wave fluxes. The quality of urban environment as a living area then will be stated as PET (Physiological Equivalent Temperature), which is an index of human well-being or comfortability based on the thermal environment. The PET can make estimations using a free software package called RayMan. In their study, Lin, et.al [7] looked at 1644 interviews in Taiwan and developed PET ranges for Taiwan in their hot and humid climate. This is shown in table 4. The PET indices based on microclimate data from ENVI-met simulation which is shown in table 5. From those tables, it shown that rusun prototypes offer the outdoor thermal comfort in the perception of "warm", which is stated in the range of 34.7°C – 37.4 °C. As mentioned above, the prototypes give similar responses to the microclimate. This also means that the outdoor thermal comfort gives the similar value in PET indices.

4. Conclusion

In order to compile comprehensive and complete building regulations, a holistic and complicated approach needs to be utilized. Thus, it must be acknowledged that this research has a huge limitation regarding the implementation of the legal aspects above. Nevertheless, the important finding of this study can be considered as an improvement of the current regulatory recommendations. This is true especially regarding the building form and configuration. The most important facets of the study's results are as follows:

Table 6. Outdoor thermal perception for *Rusun*'s model

Model	Tmrt max	Ta (	Va (m/s)	RH (%)	PET	Perception
SP	51.50	27.50	1.86	83.13	34.7	Warm
SC	52.67	29.70	1.76	56.75	37.4	Warm
SP_20	51.93	28.28	1.80	77.18	35.8	Warm
SC_20	50.97	27.55	1.47	84.28	35.4	Warm

Based on the four models, it was determined that the northern face receives more insolation compared to other directions such as the west, east or south. Therefore, the layout of outdoor space on the north side should provide as much green coverage as possible, either for shadowing or for the benefit of its albedo.

The building ratio of 1:3 produces higher wind speed, which also controls air temperature and humidity. However, this is difficult to apply due to the limited availability of land.

The given prototypes of eight, fifteen and twenty floors provide the possibility of opening it up to *rusun* construction unwell planned. Instead of creating bounds with this prototype, it would be better to carefully measure and accommodate the number of people who live in a certain area.

The strategy of building group passive design is summarized as follows: the first priority is to accommodate or to determine the population density. It must also consider the environmental load in terms of infrastructure. The urban form characteristic due to the high values of FAR, DU, and high population seems to be a good concept for the urban built environment. They can be very livable and comfortable environments if an efficient development pattern is used.

5. Acknowledgement

Authors would like to thanks to Fukuda Lab, The University of Kitakyushu for the simulation equipment.

Reference

[1] BPS-Indonesia, “Kepadatan Penduduk menurut Kota,” 2016.
[2] Kementerian-PU, “Persyaratan Teknis Pembangunan Rumah Susun Nomor 60/PRT/1992,” Jakarta, 1992.
[3] PU-Ciptakarya, “Pedoman Teknis Pembangunan Rumah Susun Sederhana Bertingkat Tinggi-05/PRT/M/2007,” Jakarta, 2007.
[4] BSN, “SNI 03-2846-1992 Tata Cara Perencanaan Kepadatan Bangunan Lingkungan Rumah Susun Hunian.” 1992.
[5] M. Bruse and H. Fleer, “Simulating surface–plant–air interactions inside urban environments with a three dimensional numerical model,” *Environ. Model. Softw.*, vol. 10, 1998.
[6] M. Bruse, “Simulating human thermal comfort and resulting usage patterns of urban open spaces with a Multi Agent System,” in *PLEA 24*, 2007.
[7] T.-P. Lin, A. Matzarakis, and R.-L. Hwang, “Shading effect on long-term outdoor thermal comfort,” *Build. Environ.*, vol. 45, pp. 213–221, 2010.

IOP Conf. Series: Earth and Environmental Science **152** (2018) 012002 doi:10.1088/1755-1315/152/1/012002

Potential of Rainwater System for Domestic Building in Jakarta

Susy Prajna Sari[1], Suhendri[2]

[1,2]Architecture Programme, Institut Teknologi Bandung, Ganesha Street No 10, Bandung. Indonesia

[2]suhendri.aritb@gmail.com

Abstract. Rainwater harvesting is the process of collecting, storing, and reusing rainwater. Different kinds of rainwater harvesting techniques have been used in developed and developing countries. Singapore, for instances, which has limited land sources and high demand of water, use rainwater harvesting technique on the high-rise buildings' roofs to get an alternative source of water. Also, in Japan, rainwater harvesting technique has been applied to residential areas and able to provide clean water for t oilet flushing, firefighter, and water supply in an emergency case. Nonetheless, rainwater management in Jakarta, compared to the countries mentioned above, is obviously left behind. With the high rainfall intensity, buildings in Jakarta are quietly possible to apply the rainwater harvesting technique. This paper is trying to analyse the potential of rainwater harvesting application in Jakarta. Starting from house buildings, the paper uses numerical calculation to study the environmental benefits of the rainwater harvesting implementation. The case study was taken in Gading Griya Residence, North Jakarta. The calculation results show that rainwater harvesting scheme can provide enough water supply for washing, flushing, fire extinguishing, and drinking water.

1. Introduction

Jakarta, with its rapid development, is facing the lack of fresh water supply. The Municipal Office of Water Supply (PAM Jaya) recorded Jakarta's annual water demand is 29.400 L/s. Meanwhile, the PAM Jaya could only supply 60,6% of the demand, and thus left 10.000 L/s deficit [1]. This deficit does not include the population growth which implies the demand growth every year.

On the other hand, Jakarta is hit by flood every year in the heavy rain season, between December to February. Flood that occurs in Jakarta is caused by several factors. One of the factors is the less number of available green open spaces in the city. Green open space acts not only as the source of fresh air supply but also as rainwater catchment areas. Another factor is caused by poor drainage systems [2].

One of the aforementioned problems can be a solution for the other. High precipitation rate in tropical city like Jakarta, should be potential to overcome the fresh water issue. But, the existing condition is quite contradictive. It is more a disaster instead of a blessing. Rainwater management and technologies that have been developed recently may be a solution for this contradictory fact. Rainwater harvesting technology, for example, if used widely, could reduce runoff water, and hence reduce flood risk. Moreover, water collected from it can supply water for household or domestic use. Furumai (2016) argues Although Indonesia is a tropical country with heavy rainfall, the utilization of rainwater itself is still not maximal. For example Jakarta which has high rainfall. Based on data from

Meteorology, Climatology, and Geophysics Council, Jakarta's average rainfall is around 2.500 - 3.000 mm each year [4]. The incapability in utilizing the high rainfall makes it turns into a threat for the city. Thus, this paper is aimed to analyse the potential of rainwater harvesting application in Jakarta, by focusing on the domestic sector.

Rainwater harvesting itself is the process of collecting, storing, and reusing rain water from various surfaces such as rooftops [5]. This technique basically has been implemented for years in many countries such as India, Singapore, China, Germany, Japan, and other countries including Indonesia. Singapore, a developed country which has limited earth surfaces has used rainwater harvesting technology to maintain clean water supply. In Japan, rainwater harvesting technique has been implemented in residential areas and it is able to provide clean water for toilet flushing, firefighting, and drinking water in emergency situation [2]. Benefits of using rainwater harvesting system are:

- economical alternative water supply;
- reduce flood risk;
- increase ground water supply;
- reduce soil erosion;
- overcome drought.

There are six basic components in rainwater harvesting system [6]. It explains as follows.

1.1. Catchment surfaces
Catchment surface represents the amount of rainwater that can be caught by a surface. It depends on the size and texture of the surface. Roof surfaces that made of porous and textured material like clay tiles will hold more rainwater than softer roof surface such as aluminum and metal roof. Roof materials also have a role in cleaning rain water from harmful things. As an example, rain water that is harvested from asphalt, wooden and shingle roof only suitable for irrigation needs.

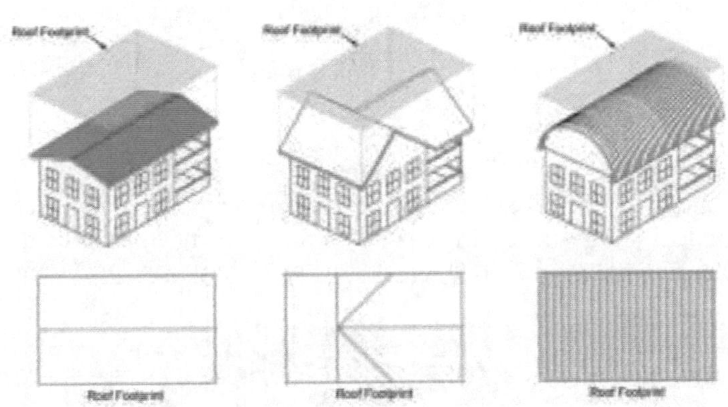

Figure 1. Catchment Surface. Source: rainwaterharvesting.tamu.edu

1.2. Gutters
Gutters serve as waterways for the rainwater. One of the examples is the roof's gutter.

IOP Conf. Series: Earth and Environmental Science **152** (2018) 012002 doi:10.1088/1755-1315/152/1/012002

Figure 2. Gutter for roof. Source: thremodeling.com

1.3. Filter

To separate the collected rainwater from its waste, a rainwater harvesting system needs filter. There are various filters which can be used in rainwater harvesting technique. The utilisation of the filter types depends on the amount of the dirt on the roof. Filters can be applied in the gutters to prevent dirt such as leaves, insects, bird's feces, and others into the storage tank.

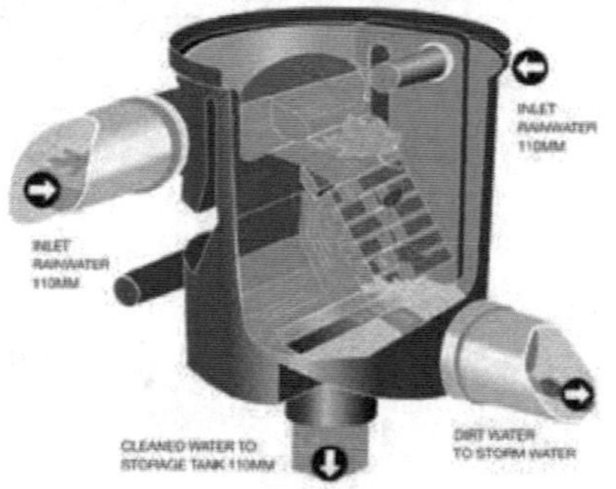

Figure 3. Rainwater harvesting's filter. Source: pinterest.com

1.4. Storage tank

Storage tank is used for storing the filtered rainwater. It can be made of various materials like fiberglass, polypropene, concrete, or steel. Storage tank better made of non-reactive materials and it has to be cleaned gradually to remove dirt deposition and prevent moss growth. Storage tank can be put above ground level and below ground level.

Figure 4. Storage tank for rainwater. Source: watertank.com

1.5. Distribution system

Stored rainwater later will be distributed by using a water pump.

1.6. Rainwater care system

Rainwater used for non-drinking water such as irrigation and toilet flushing can be cleaned just by a simple filter. Meanwhile, rainwater used for drinking water need to have further and complex filtration process.

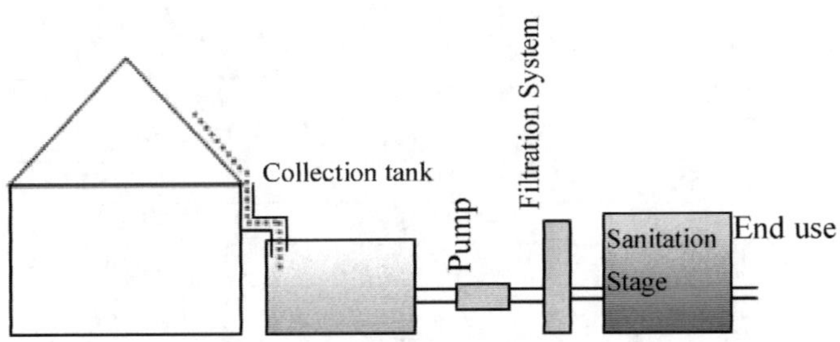

Figure 5. Rainwater harvesting scheme for house

2. Methodology

2.1. Assessment Method

Assessment of the potential of rainwater harvesting is done quantitatively by numerical method. The research identifies monthly average rainfall in Jakarta, as well as the water demand in domestic sector. For the fresh water demand, the data is taken from BPS Jakarta (2016) shows the data from 1999 to 2014. Table 1 shows the water demand data. Further, the average daily water consumption for every household in Jakarta is derived from the water demand data. The formula to calculate the average water consumption (Q_d) is shown by the following formula.

$$Q_d = \frac{\text{Average number of household}}{\text{Average water volume sold}} \quad (1)$$

The formula above gives a value of 1.08 m3 per household per day or 394.52 m3 per year.

Furthermore, numerical method is used to analyse the possibility in rainwater utilisation. In this case, the rainwater harvesting technology calculated is the one that collects rainwater from house roof. The formula used in this analysis is shown as follows.

Average Rainfall per rain days (mm/day):

$$Q_i = \frac{C}{H} \quad (2)$$

where:
C: Monthly Rainfall (mm)
H: Rain days (days)

Water collected from the roof (m^3/day)

$$V_i = \left(\frac{Q_i}{1000}\right) \times A \quad (3)$$

where:
A: roof area (m^2)

The Equation 3 is then plotted into a table and graph against the daily water consumption.

Table 1. Water demand for domestic sector in Jakarta [7]

Year	Number of Households	Water Sold (m^3)
1999	511,548.00	207,629,131.00
2000	562,255.00	228,350,012.00
2001	610,806.00	237,192,219.00
2002	649,429.00	255,161,069.00
2003	690,456.00	274,102,317.00
2004	705,890.00	270,908,257.00
2005	708,919.00	267,080,481.00
2006	725,441.00	266,221,436.00
2007	755,555.00	252,017,908.00
2008	778,044.00	258,939,302.00
2009	795,149.00	266,866,640.00
2010	805,153.00	283,965,498.00
2011	802,636.00	297,819,232.00
2012	800,093.00	310,004,991.00
2013	803,601.00	314,318,816.00
2014	813,356.00	320,883,521.00
Average	**719,895.69**	**269,466,301.88**

2.2. Case Study

The house is located at Gading Griya Residence block A3/65. The picture below shows the location map.

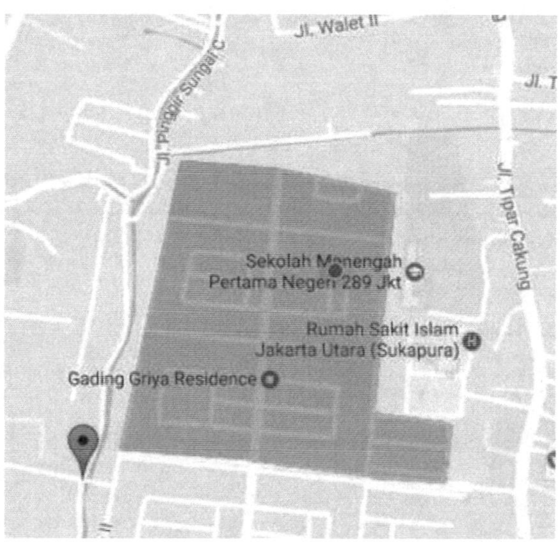

Figure 6. Case study location

Gading Griya Residence RW 12 is located at Kelurahan Sukapura, Kecamatan Cilincing, North Jakarta. Gading Griya Residence is famous for always flooding when heavy rainfall happens. It happens because of the less ground surfaces and the decreasing of ground level. Gading Griya Residence RW 12 is located at Kelurahan Sukapura, Kecamatan Cilincing, North Jakarta. Gading Griya Residence is famous for always flooding when heavy rainfall happens. It happens because of the less ground surfaces and the decreasing of ground level.

Figure 7. Flood happened in Gading Griya Residence

The average size of the houses in the residence is 9 x 15 meters equal to 135 m^2. The width of the road is around 3,5 meters and use paving block as the road cover. The residence is always flooding every heavy rainfall and the receding process is very slow although there are a jogging park, empty fields, and fishpond in the residence.

Figure 8. The case study house

3. Results & Discussion

Table 2. Rainfall data and calculation

Month	Rainfall (mm)	Rain days (days)	Rainfall per rain days
Jan	1075	26	41.35
Feb	689	22	31.32
Mar	174	20	8.7
Apr	168	16	10.5
May	47	10	4.7
Jun	174	12	14.5
Jul	214	16	13.38
Aug	39	4	9.75
Sep	0	1	0
Oct	52	4	13
Nov	65	11	5.91
Dec	211	15	14.07
TOTAL	**2908**	**157**	

Table 2 displays the monthly rainfall in Jakarta for the most recent year (2014). It contains also the calculation results of rainfall rate for every rain days. From the data, it can be seen that January-February period has the highest rainfall in a year. Total rainfall for a year in Jakarta is 2098 mm. It means, the possible water collected for a year according to Equation 3 is 392.58 m3. Compare to the annual water demand, which is 394.52 m3 the possible water collection could cover almost all of the need (Figure 5). It quite prospective, but a detail analysis is needed to see the real distribution of the rainfall. The more detail analysis is done by calculating the monthly distribution of the rainfall rate and the monthly water demand.

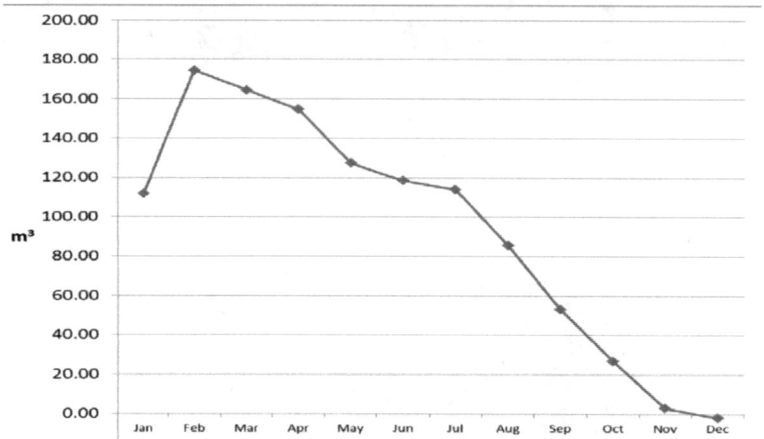

Figure 9. The potential reserved water from the rainwater harvesting after used for the daily needs

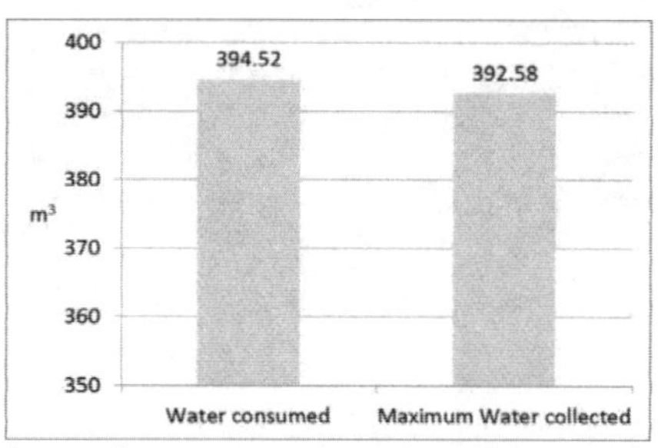

Figure 10. Comparison of water consumed and water collected fromthe roof

Table 3. Comparison of monthly water collection and consumption

Month	Maximum water collected (m3)	Water consumed (m3)	Collected minus Consumed (m3)
Jan	145.13	33.51	111.62
Feb	93.02	30.26	174.37
Mar	23.49	33.51	164.35
Apr	22.68	32.43	154.61
May	6.35	33.51	127.44
Jun	23.49	32.43	118.51
Jul	28.89	33.51	113.89
Aug	5.27	33.51	85.65
Sep	-	32.43	53.22
Oct	7.02	33.51	26.74
Nov	8.78	32.43	3.08
Dec	28.49	33.51	-1.94

In addition, Table 3 shows the monthly analysis. Along with Figure 6, they display a simulated distribution of water collection and water consumption. The problem is clear after Table 3 plots the distribution. Water demand is constant but the rainfall availability is not. Thus, in order to optimise the rainwater harvesting schemce, a reservoir is needed to conserve the collected water. Figure 6 shows the volume of the reservoir if used.

However, if such a reservoir is used, it needs a large space. The maximum volume to be reserved is 174.37 m3. It equals to a 5m x 5m x 7m box under the yard. Nonetheless, the big reservoir is not quite feasible in the most houses in Jakarta, since they yard is usually limited. If there is a reservoir used, it maybe not as big as the optimum size. And hence it needs to channel the excesive water to the city sewage. Illustration of the reservoir in the case study house is shown by Figure 7.

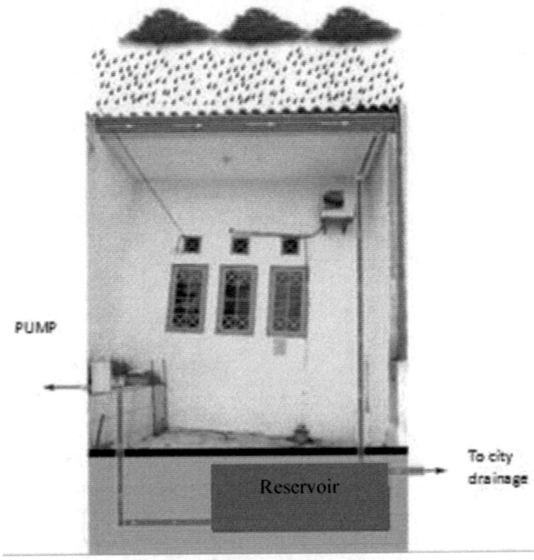

Figure 11. Illustration of the rainwater technology scheme in the case study house

4. Conclusions

The application of rainwater harvesting system aims to reduce run-off from the rainwater that falls to the ground and thus reduce the flood risk. Besides its function to reduce flood risk, rainwater harvesting is also usable for domestic needs. Rainwater harvesting scheme analysed in this study indicates that it is potential to supply almost 100% of an annual household's water demand in Jakarta. However, in order to get that optimum potential, a big volume of reservoir, relative to the site size, is required. Hence, it might be not economically feasible to make a big reservoir for a house. Nevertheless, a small reservoir still benefits the household. In a small volume compare to the daily water demand, the collected rainwater can be used for at least for activites such as car washing, plants watering, toilet flushing, or clean water supply in emergency situation.

Moreover, the large volume reservoir maybe feasible if located below the house building. Yet, it costs higher than under the yard reservoir. On the other hand, if there is a scheme that can manage the rainwater harvesting communally, it might be more feasible to get the optimum potential from the rainfall. A large scale reservoir is easier to get and maintain when it is a communal act.

In terms of sustainability, rainwater harvesting contributes in at least two of three sustainable aspects. It helps environement and economy of the households. For environment, it reduces run-off water and supply "free" fresh water for domestic use. Also, for the economy aspect, houses with rainwater harvesting technology can save money for fresh water, because they have alternative water supply.

IOP Conf. Series: Earth and Environmental Science **152** (2018) 012002 doi:10.1088/1755-1315/152/1/012002

References

[1] Kompas, “Menanti Air Bersih Murah dan Layak Konsumsi,” *Megapolitan Kompas*, Jakarta, 29-Apr-2015.
[2] UNEP, “Rainwater Harvesting and Utilization Around The World,” 2017. .
[3] H. Furumai, “Evaluation of rainwater harvesting and use potential considering climate change in Arakawa watershed Evaluation des potentialités de stockage et réutilisation des eaux pluviales face au changement climatique dans le bassin de la rivière Arakawa,” pp. 1–4, 2016.
[4] BMKG, “Data Curah Hujan dan Hari Hujan di DKI Jakarta Menurut Bulan,” 2016. .
[5] A. Pacey and A. Cullis, *Rainwater Harvesting: The Collection of Rainfall and Runoff in Rural Areas*. Intermediate Technology Publications, 1986.
[6] F. A. Abdulla and A. W. Al Shareef, “Roof rainwater harvesting systems for household water supply in Jordan,” *Desalination*, vol. 243, no. 1–3, pp. 195–207, 2009.
[7] BPS, “Data Pelanggan PAM Jaya,” 2016. .

Recurring Design Concepts for Resiliency in Asia

Heru Wibowo Poerbo

School of Architecture, Planning and Policy Development,
Institut Teknologi Bandung, Ganesha Street No 10, Bandung. Indonesia

Email: heru@ar.itb.ac.id

Abstract. Community resiliency has not been a topic that consciously applied in architectural design, although it may have been addressed as other design aspects. Ten leading universities from around the world undertook design competition and conference focusing on designing resilience. Cases for the competition were village or district in Asia that are prone to natural disaster. This paper investigates most common design concepts proposed by the participating universities. The investigation is performed on the design proposals in the 2015, 2016 and 2017 Designing Resilience in Asia competition using theme analysis method. Concepts like protecting the main production facility and providing shelter for the residence during the peak of the natural disaster might be common to several of the participants. The identification of recurring design concepts in the competition entries can help us understand the important issues that have to be dealt with in designing resilience in settlements in our region. New design concepts for resilient community can be learnt from the settlement design by participants from ten countries around the world.

1. Introduction

Architects have not been designing for resilience until now, because architectural design principles mainly concern the conception of building shape and space for its aesthetic and functional use. Newer design principle of green building for environmental conservation has been adopted and generally accepted by architects. In designing larger area, such as site plan of a residential compound or a village, the ecological consideration in design is translated into sustainable design. However, resilient design calls for more than just sustainability because by definition a resilient place must be able to recover and resume its life after being hit by catastrophe.

Architects still need to learn more about designing resilience due to the scarcity of design theory about this subject. Moreover, designing resilience that entails community involvement in strengthening their living place might be different from one region to the other. Therefore it is imperative to study the issues that have to be dealt with in designing resilience in settlements in our region. The annual design competition and conference focusing on designing resilience in Asia provides an excellent opportunity to learn more about this subject.

The Designing Resilience in Asia event is organized by the National University of Singapore. Ten leading universities from Asia, Europe, America and Australia participated in the closed design competition since 2015. The participating universities are: Bangladesh University of Engineering and Technology (BUET), Center for Environmental Planning and Technology University, India (CEPT), Institut Teknologi Bandung, Indonesia (ITB), Louisiana State University, USA (LSU), National Cheng Kung University, Taiwan (NCKU), National University of Singapore (NUS), Royal Melbourne

Institute of Technology, Australia (RMIT), South China University of Technology, China (SCUT), Technische Universität Darmstadt, Germany (TUD), University of Pennsylvania, USA (UPENN), Université de Montréal, Canada (UDM). Cases for the competition were village or district in Asia that are prone to natural disaster.

This paper investigates most common design concepts as well as the winning team's design concept proposed by the participating universities. The investigation is performed on the design proposals in the year 2015, 2016 and 2017 Designing Resilience in Asia competition using thematic analysis method.

2. Theory on Resilience

In the DRIA International Research Programme, 'resilience' is viewed as a model for anticipating, preventing and preparing for the effects of climate change. Due to the global climate change such as sea level rise, flooding, typhoons, coastal erosion, storm surges, subsidence or droughts, resilience is not only timely but critical under the exigencies of those environmental effects.

The objective of the design is to return the original environmental qualities to the urban milieu. This is done by designing practical actions to adapt to the impacts of climate change, protecting communities and reducing vulnerability, while mitigating the causes of global warming and climate change through specific solutions that actively contribute to the environmental recovery. DRIA uses the definition by Cutter [1]: "Resilience is the ability of a social system to respond and recover from disasters and includes those inherent conditions that allow the system to absorb impacts and cope with an event (...) adaptive processes that facilitate the ability of the social system to re-organize, change, and learn in response to a threat."

The 100RC (Resilient Cities Network) defines urban resilience as "the capacity of individuals, communities, institutions, businesses, and systems within a city to survive, adapt, and grow no matter what kinds of chronic stresses and acute shocks they experience." Building urban resilience requires looking at a city holistically: understanding the systems that make up the city and the interdependencies and risks they may face. By strengthening the underlying fabric of a city and better understanding the potential shocks and stresses it may face, a city can improve its development trajectory and the well-being of its citizens. [2]

3. Research Method

Thematic analysis is used to examine the theme within data (Boyatzis 1998) [3]. Data is collected from the design proposals in the 2015, 2016 and 2017 Designing Resilience in Asia competition. Each year there were 17 to 19 entries in the competition, as each participating university may submit one or two proposals. Altogether there were 54 design proposals in the three years of competition. Coding of the data is done with regard to the resilience theory and particularly to the brief of each year's competition. The next phase is combining the codes into overarching themes that accurately depict the data. Analysis looks at how the themes support the data and the main theoretical perspective. Finally, the conclusion of this research is presented as design issues that should be taken into consideration when we design for resilience in Asian countries.

The objective of the design is to return the original environmental qualities to the urban milieu. This is done by designing practical actions to adapt to the impacts of climate change, protecting communities

3.1. Method for Data Collection

The data collection is done by archive study available in the repository of DRIA international contest website at http://www.designingresilience.com. The contest entries of each participating team are documented in the form of high resolution panel images, short narratives on proposed draft concepts and presentation files. Archive studies can be performed on DRIA competitions in 2015 and 2016, whereas in 2017 it can only be done by taking note directly during the presentations because the archives for the 2017 competition were not available at the time of writing of this article. In 2015 and 2016 the author followed the presentation of all participants as well as in the year 2017.

3.2. Analysis Method
Analysis is done by reading the concepts of planning and design of the proposed area of each participant. Coding and classification of data are made into tables for each participant in each competition year. The frequency of concepts put forward by each participant is an indicator that the concept is important for the design's resilience of the location.

4. The Designing Resilience in Asia Competition Case
The 2015 DRIA case was Xinxing Harbour Village. The village is located in the Hainan Province, China. There are 350-370 household in the village, with the main local culture and livelihood is premised on small scale farming and shallow ocean fishing. The village is located in a typhoon-prone region.

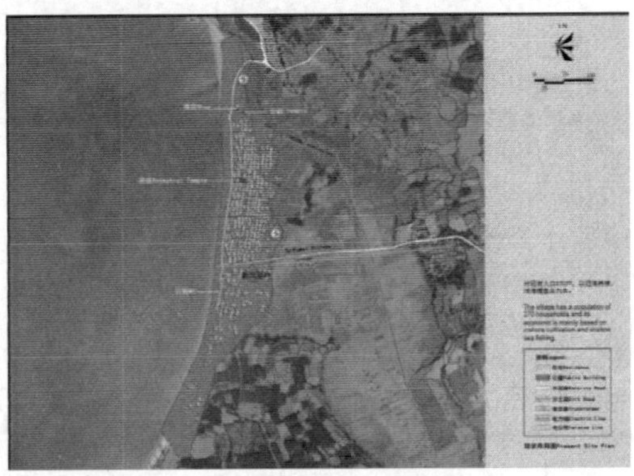

Figure 1. Map of Xinxing Harbour Village.

Source: local government of Chengmai Prefecture, Hainan Province, China

In 2016 the case was Valenzuela City, located north of Manila, the Philippines. It has access to Manila Bay via some wetlands. The site is 3.2 Km long area along the Polo River. It is still affected by flooding episodes during light and heavy rain, and has been touched by several typhoons in the last years. Livelihood of the population here is from the fish ponds and working in nearby city.

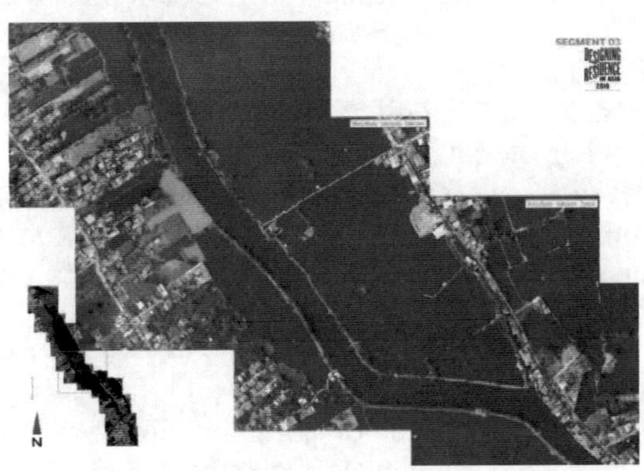

Figure 2. Map of Valenzuela City [4]

The competition site for 2017 is Kampung Kali Bahru located in the Kelurahan Bandarharjo on the northern part of Semarang, Indonesia. About a third of the population living in the area is dependent on some kind of manufacturing or industrial activity for their livelihood. There are fish smoking (*bandeng asap*) industry in this village, as well as some heritage buildings. The village faces

IOP Conf. Series: Earth and Environmental Science **152** (2018) 012003 doi:10.1088/1755-1315/152/1/012003

land subsidence problem due to excessive ground water extraction. It also suffers from tidal flooding that regularly rises into this site.

Figure 3. Map of Kampung Kali Bahru. [5]

The competition brief calls for preventive design paradigm for an Asian community's resiliency that integrates urban design/planning, architecture and building technology that particularly addresses the following issues:

1. URBAN DESIGN / URBAN PLANNING
- Strategy - The relevance and significance of the preventive urban design/planning strategy to promoting and fostering resiliency locally, regionally and globally
- Plan - The relevance and significance of the preventive urban design / urban planning plan in assuring resiliency through an engagement with topography, resources management, site planning, urban morphology, building typology and infrastructural management
- Policy - The relevance and significance of the preventive urban design / urban planning policy in promoting and fostering resiliency through novel forms of community engagement and innovative decision and policy making strategies

2. ARCHITECTURE
- Program - The relevance and significance of the architectural program in promoting everyday community engagement while also fostering the community's resiliency
- Morphology - The relevance and significance of the architectural form and structure as exemplars of artistic innovation while equally assuring technical resiliency
- Tectonics - The relevance and significance of the architectural constructability in light of the local community's cultural, social and technical assets

3. BUILDING TECHNOLOGY
- Materiality - The relevance and significance of the materials for construction in light of the community's access to material resources
- Technology - The relevance and significance of the construction methods as exemplars of creativity and innovation, as well as, of the community's traditions and conditions
- Performance - The relevance and significance of the construction methodology to assuring a building's performative resiliency.

5. Analysis of Design Proposals

Analysis of the design entries is performed for each year to allow differences in the coding of the proposal due to variations of the environmental challenge that the location faces each year. Table 1 shows the summary of the main concepts in the 2015 DRIA competition. Most of the teams focus on reforestation of mangrove area as protection from the storm surge. Architectural design concept includes redesign of the existing houses and new community hall that also serves as shelter during the storm. Peculiar of the 2015 competition is the proposal for power or electricity generation to make the Xinxing village more resilient after a storm attack.

Table 1. Design concept of DRIA 2015

	BUET		CEPT		ITB	LSU	NCKU		NUS		RMIT	SCUT		TUD		UDM		sum
Reforestation (upstream)																		0
Reforestation (downstream/mangrove)		1	1	1		1	1			1		1	1		1		1	10
Wetland (natural, constructed)	1		1				1	1		1			1					6
Pond, reservoir, bioswales			1			1	1		1	1								5
Water treatment & management	1								1						1	1		4
Rainwater harvesting, groundwater recharging																		0
Sea wall (natural), coastal protection	1				1	1		1			1		1					6
Drainage, canal, waterway	1		1												1	1	1	5
House: floating, cantilever, stilt, vertical, modular	1	1	1	1	1	1		1	1	1	1	1	1		1	1	1	15
Architectural, cultural heritage																		0
Community shelter		1			1	1	1		1	1			1	1		1	1	10
Natural bldg material (bamboo, coconut trunk)												1					1	2
Raised platform for public place		1			1	1		1		1	1		1	1		1	1	10
Open space, community space, marketplace	1						1	1	1		1						1	6
Mixed use																		0
waste management/recycling	1						1	1						1	1			5
Tourism	1													1		1		3
Emergency evacuation, medic			1		1		1						1					4
Skill, craft for economy			1								1			1				3
Fishery economy						1	1			1					1			4
Agriculture, livestock, fishery/oyster		1				1		1	1									4
Community participation		1	1		1		1					1		1		1	1	8
Education							1	1										2
Power/electricity			1				1	1	1									4
				Special acknowledgement			Winner Planning	Honorable			Honorable	Winner Architecture		Winner Planning+Arch	Honorable			

Main concepts of the 2016 DRIA competition are presented in Table 2. Most of the proposals focus on reforestation of the mangrove, canal and drainage improvement, as well as transportation network (to connect the Valenzuela site with nearby city as place of work). Community center as shelter is often proposed for this location, along with improvement of housing design.

Table 2. Design concept of DRIA 2016

	BUET	DLS	ITB		LSU		NCKU		NUS		RMIT		SCUT		TUD		UDM		sum
Reforestation (upstream)																			
Reforestation (downstream/mangrove)	1			1		1				1				1	1	1	1	1	9
Wetland (natural, constructed)															1				1
Pond, reservoir, water retention basin					1				1	1				1	1		1		6
Water treatment & management					1	1						1			1				4
Rainwater harvesting, groundwater recharging			1	1											1				3
Dike, coastal protection	1								1		1					1	1		5
Drainage, canal, waterway	1			1		1					1		1	1	1	1	1		9
House: float,stilt, vertical, modular, flexible	1		1	1	1		1	1	1	1	1	1		1	1	1	1	1	15
Architectural, cultural heritage																			0
Composite material (concrete, bamboo), recycled	1		1		1			1				1		1				1	7
Natural bldg material (bamboo, coconut trunk)										1							1	1	3
Raised platform for public place			1			1	1		1				1				1	1	7
Open space, community space, marketplace					1		1			1		1	1		1	1			7
Community shelter (Multi function)			1	1					1	1			1				1	1	7
Manual power generation + electrical network								1											1
waste management/recycling		1				1									1				3
Transportation network	1	1		1	1		1		1		1		1		1			1	10
Emergency evacuation, medic										1						1		1	3
Tourism for economy				1											1				2
Enterpreneruship, trade					1							1		1					3
Paddy field, roots vegetation,agriculture, fishery				1						1		1		1				1	5
Information, public participation/awareness	1						1		1	1						1			5
Coop, community mortgage														1				1	2
Education, training				1			1					1	1				1		5
Application (GRAB, sproud)																			0
				Honorable	Honorable			Honorable		Honorable		Honorable	Winner Architecture	Honorable		Winner Planning	Honorable	Winner Planning+Arch	

In 2017, DRIA competition proposals are focused on housing design as a response to the sinking houses in the Semarang site. Environmental concepts include natural or constructed wetland, reforestation of mangrove and water retention basin. Rainwater harvesting is proposed mainly as a way to reduce groundwater extraction that has accelerated the land subsidence process. Revitalization of drainage and canal is proposed as part of waterway transportation. Adaptive reuse of historic warehouse and the creation of public open space as market are proposed in relation to community's economic improvement. Some proposal includes tourism to complement the traditional fish smoking industry in the community. Different from the DRIA 2015 and 2016, in 2017 some participants propose education and online application (such as GRAB) as part of the way to enhance the population's socio economic resilience. There is a change in the participating universities in DRIA

2017, with the Louisiana State University, USA (LSU) replaced by the University of Pennsylvania, USA (UPENN).

Table 3. Design concept of DRIA 2017

	BUET		CEPT		ITB		NCKU		NUS		RMIT		SCUT		TUD		Upenn		UDM		sum
Reforestation (upstream)															1		1				2
Reforestation (downstream/mangrove)			1						1	1	1			1	1	1	1		1		9
Wetland (natural, constructed)	1		1							1		1		1							5
Pond, reservoir, water retention basin	1		1							1	1			1	1		1			1	8
Water treatment & management				1			1								1	1	1				5
Rainwater harvesting, groundwater recharging		1		1			1			1										1	5
Sea wall, coastal protection	1								1								1				3
Drainage, canal, waterway	1							1	1				1	1			1		1		7
House: floating, cantilever, stilt, vertical, modular			1	1	1		1	1	1	1	1	1	1	1		1	1				13
Architectural, cultural heritage									1						1						2
Warehouse (heritage) activity				1								1		1	1						4
Natural bldg material (bamboo, coconut trunk)																			1		1
Raised platform for public place	1																			1	2
Open space, community space, marketplace		1		1			1	1		1		1					1				7
Mixed use																	1				1
waste recycling (Plastic --> brick)	1															1			1	1	4
Transportation network			1		1	1	1	1													5
Emergency evacuation, medic																1					1
Tourism for economy	1				1	1					1		1								5
Fish smoking economy		1								1	1	1								1	5
Paddy field, duck, fishery				1									1								2
Information, public participation (all levels)	1				1						1		1		1						5
Education									1		1	1									3
Application (GRAB, sproud)							1									1					2
	Honorable		Honorable		Honorable				Winner Architecture		Winner				Winner Planning						

The three year design competition shows that designing resilience encompasses broad perspective such as regional ecology, economy and transportation down to technical detail of house unit design, construction and its material. The broad viewpoint is also dictated by the competition brief in urban design/planning, architecture and building technology. However, the recurring concepts can be grouped into three or four categories. First is the environmental concept that addresses nature and infrastructure, the second category is building or architectural design and the third category pertains to the community's social and economic resilience. The most common design concept in the environmental category is reforestation in mangrove area and other concepts such as sea wall (in Xinxing), drainage/canal (in Valenzuela) and water retention pond (in Semarang). In building or architectural design category the most common design concept is the house design, whether floating, on stilts, using cantilever, modular, vertical etc. Another building that is often proposed by the participants is the community hall, either as mixed-use building for education and economy or as shelter during the storm (in Xinxing and Valenzuela). The third category of social and economic resilience is improved by providing public space for the community, education and training, community participation and information. Economic resilience is strengthened by focusing more in the community traditional activity in fishery and agriculture, or introducing new activity such as tourism or improved skill by training.

Figure 4. DRIA 2015 Winner in Overall Design Excellence [6].

6. Conclusions

There are three categories need to be considered in designing resilience. The aspects in each category are selected to suit the problems faced, e.g. typhoon/storm with community shelter that serves as a place for education and training to improve skills, provision of clean water from rain water storage to reduce land subsidence in Semarang. In general water management is crucial in all cases as water is the source of life. The three DRIA cases are in the coastal region, so that shore protection is necessary with mangroves and wetlands. If necessary, protection of water supply is conducted in upstream region. Community resilience requires cohesiveness of the people to utilize their natural resources and skill wisely.

A combination of physical, natural and socio-economic considerations can maintain community resiliency. The winning proposal in the overall category (planning + architecture) shows the integration of the concept, for example, building with natural ingredients of bamboo which is easy to grow.

References

[1] Cutter SL, Barnes L, Berry M, Burton C, Evans E, Tate E, Webb J. A place-based model for understanding community resilience to natural disasters *Global Environmental Change* Volume 18, Issue 4, October 2008

[2] http://www.100resilientcities.org

[3] Boyatzis, R E 1998. *Qualitative Information: Thematic Analysis and Code Development.* (Thousand Oaks, CA: Sage Publications)

[4] School of Design and Art. De La Salle – College Saint Benilde, the Philippines 2016. DRIA data prepared by Ar. Harry Joseph R. Serrano

[5] Department of Architecture, the National University of Singapore 2017 *Designing Resilience in Asia*. Brief unpublished.

[6] http://www.designingresilience.com

IOP Conf. Series: Earth and Environmental Science **152** (2018) 012004 doi:10.1088/1755-1315/152/1/012004

The Study of Attachment between Distribution Pattern of Land Price and Apartments Price (Case Study: Vertical Housing Development in Administration City of South Jakarta)

Sugiyantoro [1], B. Christian [1], R. Arianto [1]

[1]Department of Urban and Regional Planning, Institut Teknologi Bandung, Indonesia

torolondon@yahoo.co.uk; brianchofficial@gmail.com & rama.arianto@hotmail.com

Abstract. Land and it's utilization on housing development undoubtedly have become an essential issue in various studies. The comprehension on each locations of the case has allowed several important studies of particular patterns related to the capacity of land and housing market. Especially Jakarta as the most promising area for property business, also has the highest lands and apartments price in Indonesia. Land and apartment are considered as the most unique commodities, highly profitable. But since 2014, there has been lots of warnings about possibility of apartments oversupply; stagnation of apartment sales also has been indicated in locations at the edge of this city, which will enable the demand's transition from landed house to apartment, as form of vertical housing. Based on fundamental theories of land and apartment pricing, the research presumes that apartment development is considered as an effort of land intensifications, to trigger the enhancement of land prices in particular locations. Therefore, this research means to comprehend the correlation of distribution pattern of lands price with apartments price. The location's selection is based on expansion of the urban development of Jakarta Metropolitan Area to southern region which has become a general tendency.

1. Introduction

The essential issue of attachment between land and apartment pricings is derived from the rapid rate of development on various aspects in urban area; in this case, Jakarta as the capital city and still-developing-megapolitan of Indonesia [1]. From the physical aspect of urban development in the related location, scarcity of land supply because of the increasing rate of property development has always become a common condition in areas with extremely high density population. While from the non-physical aspect of urban development in related location, people of Jakarta tend to consider the property as most unique and highly profitable commodities especially with the dominant market mechanism.

1.1. Issue

In these recent years, there has been many warnings about the possibility of apartments oversupply in Jakarta. Referred to the market speculation by various property consultant in Indonesia, stagnation of apartment sales also has been indicated in settlement locations at the edge of Jakarta City, which will enable the demand's transition from apartment, as form of vertical housing, to landed houses; there is a possibility that this stagnation will influence the sales of related apartments.

1.2. Term
As the research presumes that apartment development is considered as an effort of land intensification, also to trigger the enhancement of land prices in particular locations, there are still needs for further explanation of terms that will be used here. Especially the term of "apartments" on this research, could be defined in three general meanings: (1) Apartment as a residential which are dominantly owned by private sector for market orientation; (2) Rusunami as a public housing which are owned by government for public needs with permanent ownership; and (3) Rusunawa as a public housing which are owned by government for public needs with temporary ownership [2].

1.3. Goal
While this research means to comprehend the correlation of distribution pattern of land prices with distribution pattern of apartment prices, the issue lies behind all of them, mainly case study in the city of Jakarta which is based on expansion of the urban development of Jakarta Metropolitan Area to southern region that has become the general tendency, is about relatively the misfunction between Apartments, Rusunami, and Rusunawa.

2. Theories
Based on theories that has become the fundamentals of it, the research presumes that the apartment developments are considered as an effort of land intensification, also to trigger the enhancement of land price at particular locations.

The flowchart on the figure 1 shows the relations of ideas and theories that are going to be included, associated, and referred in this research. Mainly, the ideas are regarding property value by private sectors, asset value by public sectors, and housing development (private and public); while the theories that are referred in this research dominantly based on Ricardian Law of Rent, Von Thunen's Land Rent, and Alonso's Bid Rent Theory, with further information and explanation on the public housing such as Rusunami and Rusunawa.

3. Methods
The data collecting method that will be conducted in this study is by interviews with reference, primary observations, and studies of related literature [3]. Considering the topic or issue about development of land prices and apartment prices in part of Jakarta, it is not only related to the market condition, which is commonly being researched by the private sector such as land and property developers, but also Rusunami and Rusunawa that have been mentioned before as part of them. It makes the references of this research, both stakeholders interview as primary survey and archived data as secondary survey, are primarily from the government executive institutions: (1) Kementerian Agraria dan Tata Ruang Republik Indonesia (Agrarian and Spatial Planning Ministry of Indonesia Republic) with Badan Pertanahan Nasional (National Land Council), (2) Kementerian Pekerjaan Umum dan Perumahan Rakyat Republik Indonesia (Public Works and Housings Ministry of Indonesia Republic), (3) Dinas Perumahan dan Gedung Provinsi DKI Jakarta (Housings and Buildings Agency of DKI Jakarta), and (4) Dinas Cipta Karya, Tata Ruang, dan Pertanahan DKI Jakarta (Development, Spatial, and Lands Agency of DKI Jakarta).

IOP Conf. Series: Earth and Environmental Science **152** (2018) 012004 doi:10.1088/1755-1315/152/1/012004

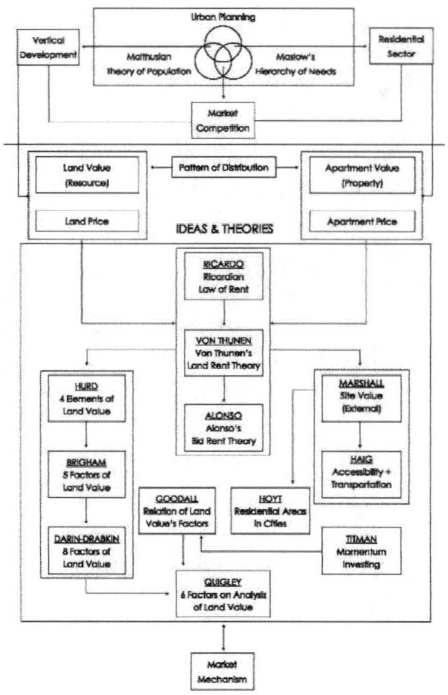

Figure 1. Research Theoritical Scheme.

3.1. Scope

This research will be done strictly within Jakarta, it will be done periodically from year to year and viewing the progress each year. The time scope of research has to be precise in as far as the data collecting method may able to retrieve. Considering the present time of this research is being conducted is the mid of 2017 (before the end of third quarter), the oldest data that has been collected through networks, mostly by secondary survey, is from 2009; then we have approximately 8 years of time range of data.

3.2. Phase

With the new condition of dividing apartment term into three categories: Apartment, *Rusunami*, and *Rusunawa*, the methods of analysis that will be conducted in the study first started with categorization between those three. The categorization is simply conducted by the basic function or market target between: (1) Apartment: general market by profit orientation, (2) *Rusunami*: specific market of public needs for permanent ownership, (3) *Rusunawa*: specific market of public needs for temporary ownership. This phase will decide the more advanced or further phase to analyze the sample in order to recognize their true function in the context of vertical housing development, part of the residential sector. After the first phase of categorization, then it is named characterization as the samples are going to be analyzed by their characters based on various elements or variables related to the urban development such as: location, accessibility, urban design, regulation, supply and demand, etc. [4, 5, 6]. With the result of characterization per categories of samples, the data is prepared to be included in the final phase of comparison by interpolation using Geographic Information System which is capable for the formulation of spatial data [7].

The result of those three phases: categorization, characterization, and comparison are capable to recognize the distribution pattern of housing-related-elements in rapid-developing-urban-area and comprehend how it is going to influence the sustainability of one region through settlement technology.

4. Situation

Around 8,500 units of apartments have been released or published to the market with price range of more than Rp 50,000,000/m^2, in this first quarter of 2017 have been taken up to 72% [8]. There is a lot of apartments in Jakarta with price range of approximately Rp 60,000,000/m^2 which could be considered to be in the category of premium-level property in Singapore, the neighboring country of Indonesia which has a more advanced and higher rate of economic level. It is clearly stated that those overpriced apartments in Jakarta are not only influenced by the variable of location and accessibility, but also the internal facility, unit capacity, and building quality.

Table 1 Apartments in Jakarta, 2017

	2Q 2017	Prediction 2017
Supply	3,130	21,167
Take-Up Rate	84.86%	87.00%
Average Price*	Rp 32/m^2	Rp 33/m^2

Until July 2017, there has been 3,130 units of apartments from the total prediction of 21,167 units. The take-up rate or units of apartment that have been taken by market is 85%. It is projected to be increased to 87% along with the stable macroeconomics conditions and supporting by "upgrade rating" from S & P that is able to take FDI (Foreign Direct Investment) to Indonesia which is going to contribute directly in improvements of property development in Jakarta [9].

4.1. Problem

The supply trend of vertical housing development from mid-class to high-class economy population is considered to be contrast compared to the low-class economy population in Jakarta, which affordable housing facility is still their most important needs. From the new policy occurred in city of Jakarta related to residential sector, government has issued the higher rate of development of *Rumah Susun* (*Rusunami* and *Rusunawa*) as the low-class economy population, family with monthly income of less than Rp 7,000,000 should be the main priority and primary target. However, the program is proven unnable to define the problem for mid-low-class economy population with monthly income in range of more than Rp 7,000,000 to less than Rp 15,000,000. They are not suitable for the criteria of ownership on *Rumah Susun*, a form of housing for low income family in population who are not able to afford apartments which price range have been increased rapidly to unaffordable level or so called overpriced.

That kind of rate of imbalance development or contrast between two neighbouring sides made Jakarta faces its problem in distribution pattern of the lands price and apartment price, in this case also including *Rusunami* and *Rusunawa*.

Table 2 Rusunawa Total in Jakarta by Administrational Cities, 2017

Administrational City	Total Amount
Kota Administrasi Jakarta Barat	3
Kota Administrasi Jakarta Pusat	2
Kota Administrasi Jakarta Selatan	0
Kota Administrasi Jakarta Timur	13
Kota Administrasi Jakarta Utara	5

The total amount of *Rusunawa (Rumah Susun Sewa)* in the Jakarta by 2017 has become one illustration of how low the supply of vertical housing for populations which are considered the low-class economy population, family with monthly income of less than Rp 7,000,000. It is suitable with the master plan of Jakarta or RTRW (*Rencana Tata Ruang and Wilayah*) which has planned and aimed East Jakarta (*Kota Administrasi Jakarta Timur*) as a region for mostly residential sector. However, the high amount of population that was distributed all over the city, both in inner core and perimeter areas, is still in contrast with the affordable supply.

Moreover by considering the Master Plan of DKI Jakarta [10], there is still a missing target for residential sector development in areas such as South Jakarta (*Kota Administrasi Jakarta Selatan*) which also has a high amount of population but not even one Rusunawa on it. The residential sector development in South Jakarta is highly concentrated for apartment development; which is almost fully-controlled by the private sector with market orientation for the high-class economy population. It is somehow a contradiction to the general situation or demographic condition in several areas, except inner core or Jakarta Central Business District, in South Jakarta which still have lot of mid-low-class economy population and also commuters from Depok, Tangerang Selatan, and Bogor.

4.2. Quantitative Data

The existing amount of *Rusunawa* as form of vertical housing in residential sector, which is considered as an example for public housing in Indonesia, with the data from government executive institute, is still facing a huge gap compared to the demand from low-class and mid-low-class economy population which has been distributed all over the city of Jakarta.

Table 3 Total Unit of Rusunawa in Jakarta by Administrational Cities, 2017

Jakarta	Tower	Type	Unit
Jakarta Barat	15	18 - 36	1,969
Jakarta Pusat	6	18 - 30	540
Jakarta Selatan	0	-	0
Jakarta Timur	78	14 - 30	7,568
Jakarta Utara	62	21 - 30	6,274
Total	161	18 - 36	16,351

Total of 16,351 units, or it could be calculated with assumption of one average family with 5 persons, of Rusunawa obviously is inequal to the amount of population's demand. While there is estimated to be more than 100,000 families with primary and urgent needs in housing facility, in existing condition, there is no more than 20,000 units of public housing offered by the government in order to fulfill their population's needs. It is worsen by the average size of the Rusunawa's unit with just around 18 m^2 to 30 m^2; not really enough for families of four and more, which is still considered as the demographic issue in Indonesia, especially Jakarta, based on the standard of general quality living (SNI 03-1733-2004) [11]. It is also influenced by the fact that Jakarta has been successfully importing approximately 4 million people as commuters every day from the satellite cities as the impact of urban sprawl: Bekasi, Tangerang, Tangerang Selatan, and Depok; the result of this mismatch between job and housing has been a perfect example of problem in high-density urban area with no development control.

4.3. Qualitative Data

The qualitative data in this research was mostly retrieved from interviews with stakeholders from The Housings and Buildings Governmental Agency of DKI Jakarta. Discussed dominantly about *Rusunami* and *Rusunawa*, clearly it has not only differ in terms but also their function in society, different stewardship, physical requirement such the maximum of 5 floors per tower (defined as KLB or *Koefisien Lantai Bangunan*), etc. Many issues which are considered new has been found during this qualitative survey occurred in the context of vertical public housing in Jakarta. Such as the optimalization of *Rumah Susun*'s function as part of instruction from the governor until several cases of where there are more occupied units than the supply or offered units in one apartment.

The answer to that case or issue is somehow about trans-function between commercial units to residential units in some *Rusunami* and *Rusunawa*. It is obviously a form of coercion and proof of backlog in the supply of those public housing facilities compared to the demand from populations that urgently need housing facility but cannot afford more than the offered units which are categorized as *Rumah Susun*.

Establishment of *Rumah Susun*'s project also could be planned and executed by the private sector such as land and property developer as form of Corporate Social Responsibility with the calculation from proportion of constructed apartments (form of residential sector that has been categorized as mostly for the general market with profit orientation, unlike *Rusunami* and *Rusunawa*). Those kind of acts from government's policy for sure is aimed to balance the housing development for the high-class and mid-high-class economy population, also mid-low-class and low-class economy population. However, it needs to be restated again, which has been an issue in this research along with the actual problem in the system especially in Jakarta, that there are perpetrators who are using the liberal system and market mechanism for profit orientation. The example of such things is identification of subject or cases report about people who is marketing the units of Rusunami and Rusunawa. Either in small amount or huge scale, it has become a problem that ruins the system of public housing which is basically targeted for people who urgently need housings to fulfill their primary needs but considered with low economy capacity; now as the problem were occurring in various apartments which legally has been categorized as *Rumah Susun Milik* or *Rumah Susun Sewa*, the vertical public housing have become a form of long-term investment; not directly functioning as a place for living, but tools for mid-high-class and high-class economy population.

5. Analysis

5.1. Categorization

From the phase of categorization, the term of apartment, especially in Jakarta, has been defined based on its function in the city, price range that is referred to the population's affordability capacity, and also the specification of mainly physical elements on it.

Table 4. Apartments Categorization

Housing	Definition
Apartemen	Form of vertical housing, part of residential sector development, dominantly planned and executed by the private sector such as property developers, for general market (mid-class, mid-high-class, high-class) with profit orientation.
Rusunami	Form of vertical housing, part of residential sector development, dominantly planned and executed by the public sector such as government agencies, for specific market (mid-low-class, low-class) for public needs and permanent ownership.
Rusunawa	Form of vertical housing, part of residential sector development, dominantly planned and executed by the public sector such as government agencies, for specific market (mid-low-class, low-class) for public needs and temporary ownership.

Still with a lot of functional violations by individuals, the *Rusunami* and *Rusunawa* have failures that are still occurring in the system with misfunctionality of public vertical housing for the specific market in population with huge diversity of economics capacity.

5.2. Characterization

Referred to the historical data of apartment developments in Jakarta, by Housings and Buildings Agency of DKI Jakarta, it was identified that there is a major agglomeration of Apartments development instead of Rusunami or Rusunawa, which are more substantial for a lot of specific market in the population. From Rumah Susun Klender of Jatinegara, which has been established in 1990, to the newest one, Imperium Tower Apartment at this June 2017, the trend is showing how private sector dominates the development of vertical housing in this urban area.

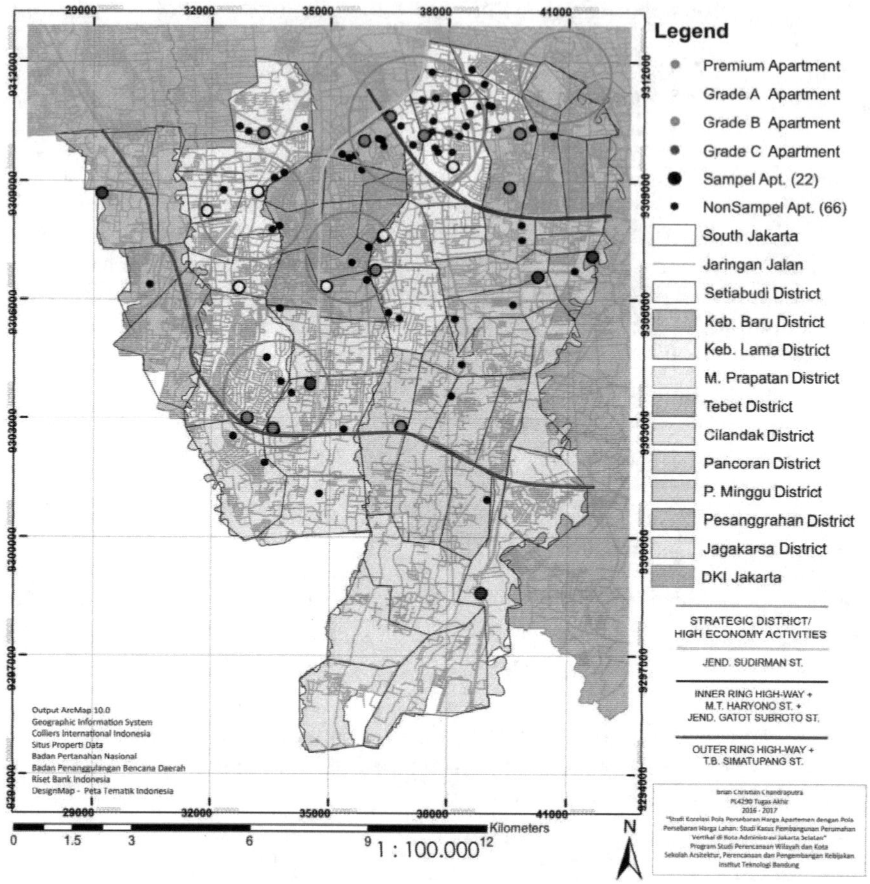

Figure 2. Character of Distribution

As the property developers become more dominant than government institutions in the matter of vertical housing development, it tends to create a plan and execution based on market preference; the agglomeration of it, which is considered as an inequal distribution pattern, could be easily identified from the map above. Not because of the basic needs, but location, accessibility, urban design, regulations, supply and demand, and others from the perspective of general market which is obviously influenced by the market mechanism. It creates the trend of development in some characters: (1) Location with potential to offer properties in high range of price based on its lands value, (2) Accessibility to working place or shopping center with potential to increase the market value, (3) Regulations that could be influenced in order to maintain stable increasing rate of properties value, and more economical reasons that in long term would be negatively impacted on the sustainability of one urban area.

IOP Conf. Series: Earth and Environmental Science **152** (2018) 012004 doi:10.1088/1755-1315/152/1/012004

5.3. Comparison

In order to understand the correlation between the distributional patterns of land prices with distributional patterns of apartment prices, it needs to be compared not just between the land and property but also the division of property itself. Using GIS, it is able to be conducted through spatial data with variables of price range and specific location [12]. The result of it is the statistically and scientifically processed distributional pattern between the elements such vertical housing: Apartment, *Rusunami* and *Rusunawa*.

This interpolation map below shows South Jakarta as one sample in the city of Jakarta, part of this primary case study, which has become the agglomeration of extremely high property price or overpriced apartments based on the affordability of the major population [13].

Those are only for the apartments which has been categorized in the previous phase as form of vertical housing for general market with profit orientation. A contrast condition both in quantitative and qualitative aspect with Rusunami and Rusunawa by estimating the existing also future situation.

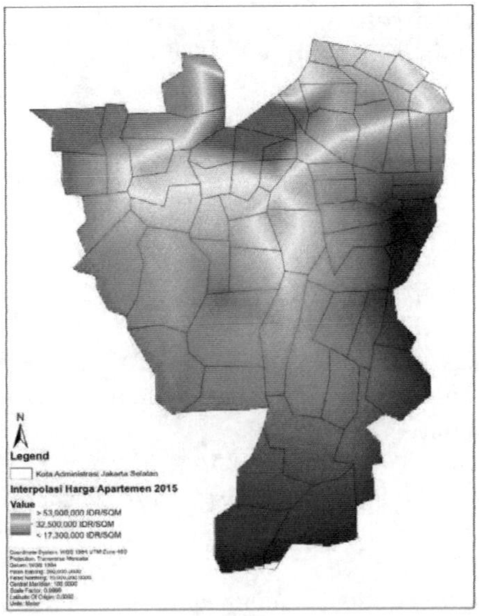

Figure 3. Interpolation of Apartments

Table 5. Recently Released *Rumah Susun*

Rumah Susun	Unit Released	Unit Occupied
Tubun	524	300
Cakung Barat	300	300
Rawa Bebek	400	400
Jalan Bekasi	200	300
Lokbin Semper	235	235
Marunda	300	300
Rawa Buaya	358	0
Total	2,317	1,835

6. Conclusions

There are difference both in terms, functions and specifications, between Apartment with *Rusunami* and *Rusunawa*, form of vertical housing development in Indonesia, especially the city of Jakarta. While apartments is considered to be marketed through the market mechanism with their general market target, *Rusunami* and *Rusunawa* basically should be aimed through government regulations for specific target within the population.

In reality, the amount of planned form of residential for general market such as Apartment is several times higher than the development or planned supply of *Rusunami* and *Rusunawa*. The similar situation also occurred to future condition as the increasing rate of demand for affordable *Rumah Susun* is still rising.

However, it gets worse because there are proven violations of *Rusunami* and *Rusunawa*'s marketing through market mechanism with increasing price by some perpetrators that ruins the system. The Government failed to fulfill the urgent and primary needs of those specific population because there are lack of control and monitoring for the related issues.

There is an attachment in positive orientation and similar direction between distribution patterns of land prices with the distribution patterns of apartment prices in urban area which has been the study case.

These kinds of conditions in rapid-developing urban area with high density and various aspects of discrepancy tend to fail the aim and plan to create a sustainable city, in this case from perspective of public housing as part of residential sector development. A monitoring act and development control through this research of distribution pattern and mapping technology are essential as the process to achieve the sustainable and maintain ideal condition.

References

[1]. Dowall D, Leaf M. *The Price of Land for Housing in Jakarta*. Berkeley: University of California Press; 1991.

[2]. Undang-Undang Republik Indonesia Nomor 20 Tahun 2011 *tentang Rumah Susun*. Jakarta: Dewan Perwakilan Rakyat; 2011.

[3]. Vanderstoep, SW, Johnston, DD. *Research Methods for Everyday Life Blending Qualitative and Quantitative Approaches*. San Francisco: Jossey-Bass; 2009.

[4]. Ottensman, J. R. *Using Geographic Information Systems to Analyze Library Utilization*; Columbia: Library Quarterly; 1997.

[5]. Adisasmita, Rahardjo. *Ekonomi Tata Ruang Wilayah*. Yogyakarta: Graha Ilmu; 2014.

[6]. Winarso, H. *Access to Main Roads or Low Cost Land? Residential Land Developers' Behaviour in Indonesia*. Journal of the Humanities and Social Sciences of Southeast Asia, Volume 158, Issue 4, Page 653-676; 2002.

[7]. Tomlinson R. *Introductory Readings in Geographic Information Systems*. Columbus: Ohio State University; 2003.

[8]. Research and Consultancy Division. *Property Market Analysis*. Jakarta: Savills Indonesia; 2017.

[9]. Research Department. *Jakarta Property Market Report*. Jakarta: Colliers International Indonesia; 2017.

[10]. Peraturan Daerah Nomor 1 Tahun 2012 tentang *Rencana Tata Ruang Wilayah 2030*. Jakarta: Pemerintah Daerah Provinsi DKI Jakarta; 2012.

[11]. Standar Nasional Indonesia 03-1733-2004 tentang *Tata Cara Perencanaan Lingkungan Perumahan di Perkotaan*. Jakarta: Badan Standardisasi Nasional; 2004.

[12]. Favian, Fauzan. *Studi Pengaruh Karakteristik Spasial Faktor Harga Lahan terhadap Tingkatan Harga Lahan Kota Jakarta Selatan*. Bandung: Institut Teknologi Bandung; 2016.

[13]. Christian, Brian. *Studi Korelasi Pola Persebaran Harga Apartemen dengan Pola Persebaran Harga Lahan: Studi Kasus Pembangunan Perumahan Vertikal di Kota Administrasi Jakarta Selatan*. Bandung: Institut Teknologi Bandung; 2017.

Technological Contexts in Developing Bamboo Wall System for Housing in Indonesia and Philippines

N B C Auman, A Widyowijatnoko, S Wonorahardjo

School of Architecture, Planning and Policy Development, Institut Teknologi Bandung, Indonesia

iambryanauman@gmail.com, andry@ar.itb.ac.id, titus@ar.itb.ac.id

Abstract. The development of bamboo wall technology for housing depends on its technological context. Reviews on a number of literatures indicated that there are some major issues about bamboo wall technology such as: construction technology, performance of construction materials, strength and durability of components and building systems. The aims of developing the technologies could be defined such as: post-disaster conditions, affordable housing, wind and earthquake resistance, weather and climate resistance. This paper will discuss about the comparison of several approaches in building bamboo wall technology in Indonesia and Philippines. In Philippines, the building technology must have optimum resistance to both typhoons and earthquakes, while in Indonesia the building must have good resistance against earthquakes. Both have a big challenge in promoting social status and long durability.

1. Introduction

1.1 Why Bamboo?

In South East Asia, bamboo has always been a readily used material. In Bangladesh and Burma, there are enormous numbers of bamboo houses with about 70% [1]. In 2003, Indonesia Agricultural Census indicates that in Indonesia there are about 4.73 million households that harvest bamboo crops for about of 37.93 million families or an average of household control of 8.03 clumps [2]. This is a potential resource to utilize bamboo as a building material and construction.

Commonly regarded as a poor man's timber, it is a no surprise that this green grass has been used since ages. Bamboo has been able to provide shelter for families, livelihood for craftsmen and food for rural regions [3]. Bamboo is a fast-growing renewable resource [3]. With only 3-5 years to fully mature, this sustainable material has no supply problems and can be harvested readily for building material use, not like timber that needs 10-20 years for maturity. The biggest advantage of this material is that it is very cheap and in some cases free. In rural areas, even wild bamboo that is available in rural banks can be used as a building material. This allows bamboo to be harvested, carried, transferred and stored easily anywhere which saves more time and labour costs. Bamboo is easy to cut, handle and repair without the use of any sophisticated tools and it is suitable for all types of building structures and construction. Its hollow sections make bamboo a lightweight material more than any modern material today [4].

1.2 Why Plastered Bamboo Housing?

A document published by the International Strategy for Disaster Reduction (ISDR) identified four phases of shelter recovery transitions [5]. These are:

1. Spontaneous Shelter - This provides an interim, safe haven while the situation stabilizes. This applies in the first 72 hours.
2. Emergency Shelter - This provides emergency shelter and feeding to displaced population requiring shelter. This happens in the first 60 days.
3. Interim Housing - This provides temporary or transitional housing with safe and secure shelter, water and power to displaced disaster victims while efforts are underway to make permanent repairs to dwelling or find other suitable permanent housing. This phase happens one year after the disaster and goes beyond.
4. Permanent Housing - This provides long-term, permanent housing solutions for disaster victims.

The aim is to create an upgradable housing solution from interim or transitional to permanent and durable. This can be achieved by using plastered bamboo housing construction that was originally from Colombia known as *bahareque encementado*. In Colombian construction code, it is defined as system composed of Guadua bamboo or Guadua bamboo & timber skeleton, and a sheathing of riven Guadua estrilla boards nailed to the skeleton and covered with a cement render applied over steel mesh. According to Andry Widyowijatnoko, an Indonesian architect who has been promoting plastered bamboo housing construction said [6]:

"The development of plastered bamboo construction was also to provide a better concept of housing in emergency cases. With the prefabricated bamboo panel that could be made and stored long before disaster occurred, the quick response building can be built in a day after disaster to provide emergency shelter to the victims and can be plastered gradually on their own to become permanent."

2. Bahareque Wall System

2.1 What Is Bahareque Wall System?

Baharaque is a construction technique refers to a mixed timber, bamboo, mud and or cement-mortar wall construction technique that originated in Colombia. Houses are built with a concrete base; bamboo frames with woven walls are plastered with mud or cement-mortar to create a permanent shelter. This traditional bamboo architecture not only protects the bamboo within the wall, it also improves the appearance and quality of houses specially when they are plastered and painted. The term bahareque has no precise equivalent in English. In Peru and Chile, this type of construction is called *quinca*.

2.2 Classifications of Bahareque

There two types of *bahareque* wall construction system, hollow and solid type. Although these two have different building methodologies, they have the same skeletal framing. The framework can be made of wooden studs, wooden braces or bamboo wooden grid with bamboo strips.

2.2.1 Hollow Type

Hollow type *bahareque* is the oldest type and was used before the Spaniards conquered Colombia. These walls have a timber or bamboo frame as wall base with flattened bamboo on both sides while hollow in the middle then covered with plaster on both sides.

Using flattened bamboo or locally known as *estrilla* in Colombia, there are two types of sub classification in hollow type *bahareque* namely; single skin which is plastered only on the exterior for moisture control and extreme weather conditions while the interior is barely exposed and double skin wherein both side are plastered. Due to its hollowness of the wall it creates an air gap in between stabilizing the air temperature in the interior. This lowers the temperature inside the house which creates a more desirable thermal comfort for human environment.

2.2.2 Solid Type

Solid type *bahareque* is the second type with a lot of variant woven techniques depending on what country it is made. These walls have a timber or bamboo frame with woven split bamboo covered with plaster on both sides.

Solid type *bahareque* has two sub classifications these are: bamboo split both horizontal strip and vertical strip which are interlaced alternatively to create a rigid reinforcement for the plaster on both sides and bamboo woven mat which are divided into bi-axial weave, tri-axial weave and multi-axial weave. Bamboo woven mat uses flattened bamboo interwoven in two, three or more directions. Unlike hollow type, this variation creates a more rigid wall system that can be used as structural load bearing member.

Table 1. Classifications of *Bahareque*

Type	Sub type
Hollow Type	Flattened Bamboo • Single Skin • Double Skin
Solid Type	Bamboo Split - Horizontal Bamboo Woven Mat - Bi-axial - Tri-Axial - Multi-Axial

2.3 Components of Bahareque

Bahareque wall construction system is composed with three components, namely; weaving patterns, wall framings and plaster composition. These three components are dependent from each other and can directly affect the structural durability of the shelter. *Bahareque* buildings are characterized by high flexibility when carefully constructed and well-maintained.

2.3.1 Weaving Pattern

Weaving Patterns are only applied on the second kind of *bahareque* wall, solid-type. Different weaving patterns have developed from vernacular architecture and traditional cultures have been documented from many studies. All these variety can be classified into three kinds; Bi-axial Weave, Tri-axial Weave and Multi-directional Weave. Bi-Axial weave consist of two directional axes arranged in horizontal and vertical patterns. Tri-Axial weave adds one more axis diagonally creating three directional axis which can be horizontal-based or vertical-based diagonal. Multi-directional weave consists of four or more axis which gives the pattern a complex design.

2.3.2 Plaster Composition

Plaster Compositions may vary from different countries but most common types mainly consists of cement. A cement plaster is a mixture of sand, Portland cement and water which is applied as masonry for interiors and exteriors to create a smooth surface finish. Some plasters use gypsum also known as plaster of Paris and some use a more natural approach made of lime. Lime plaster consists of sand, water and lime. Compared to cement plaster, lime plaster is more elastic and cracks much less, requiring no expansion joints [8]. Compared to gypsum plaster also, lime plaster is more durable and more resistant to natural elements [8].

2.3.3 Wall Framing

Wall Framings are divided into four types; no bracing, vertical bracing, horizontal bracing and diagonal bracing. These are the basic configurations in wall system and the most practical shape in terms of building construction. Wall Framings are the framing system that holds the walls enclosure by using vertical or horizontal wall studs and or a combination of both. One important factor in wall framings is the bracing system because it holds and strengthens the stud to counter-act any structural load applied to it. These can be structural (load bearing wall) or non-structural (non-load bearing walls).

Table 2. Components of *Bahareque*

Components	Sub-components
Weaving Pattern	Bi-Axial Tri-Axial Multi-Axial
Plaster Composition	Binding Agent • Cement • Lime • Clay Structural Filler • Sand Additive Fiber • Animal hair • Rice husk/coir • Coco huck/coir • Hemp • Water
Wall Framing	No Bracing Vertical Bracing Horizontal Bracing Diagonal Bracing

3. Comparing Indonesia and Philippines

3.1 Bamboo Species

Most common species used by local carpenters and craftsmen in Indonesia are *Dendrocalamus latiflorus* (*bambu wulung*), *Dendrocalamus asper* (*bamboo petung*), *Gigantocloa pseudoarundinacea* (*bamboo gombong*) and *Gigantocloa atroviolacea* (*bambu hitam*). With the wide variety of bamboo species to choose from, Indonesian artisans have explored various combinations of these species in building construction. Meanwhile in Philippines, the most common and readily available bamboo specie nationwide is *Bambusa blumeana* (*kawayang tinik*). In some regions like Mindanao, located in the southern part of the country, *Dendrocalamus asper* (*kawayang apos*) and *Dendrocalamus latiflorus* (*kawayang buho*) are abundant and used.

Comparing both countries, Indonesia has better and more bamboo specie selection that can be used in building construction. Each bamboo species have different set of physical and mechanical properties. And since Indonesian has a wide variety of bamboos to choose from, they can explore more possible combinations in bamboo construction.

3.2 Plastered Bamboo Housing
Colonial Dutch era in Indonesia introduced this kind of method in the past and is still continued to be used up until today. Many new architects in Indonesia have constructed several plastered bamboo housing in Indonesia for the past few years. Dr.-Ing. Andry Widjowijatnoko has built housing projects using *bahareque* wall system. To name a few these are [7]:

a. The Prototype House in di Pasir Impun, Bandung, West Java;
b. The Earthquake Refugee's House in Sukabumi West Java;
c. The Plastered-Bamboo Wall Prototype in Environmental Bamboo Foundation, Bali;
d. The Community Learning Centre in Jatinangor, Sumedang, West Java;
e. And the Community Centre in Nagalawan, North Sumatera.

Most plastered bamboo wall systems are found in old churches that were built by the Spaniards during colonial era using a modified version of the *bahareque* wall. *Tabique pampango* as it is called in Philippines; a walling technique made of bamboo woven mat plastered with lime plaster said to be of crushed seashells mixed with molasses and egg white. Unfortunately, today fewer architects use this construction method in housing projects due to preferable conventional construction.

3.3 Disaster
The entire archipelago of Indonesia sits on the Pacific Ring of Fire and is considered as the most active seismic region in the world with approximately hundreds of active volcanoes. Due to this fact, Indonesians have high seismic activity that causes earthquakes, volcanic eruptions and tsunamis. Indonesian architecture, specifically traditional houses, adapted to these catastrophes and has been improving their structural quality for earthquake resistance over the past decades.

Philippines, on the other hand, are more exposed to tropical storms with approximately twenty tropical cyclones hit the archipelago annually. In rainy season months, June to September, are the most active months for typhoons. With this, Philippine architecture, have relied on heavy roof structures with low eaves to survive strong wind currents. Due to this fact, Philippines have huge difficulty in building wide roof structures like the bamboo architecture in Bali, Indonesia.

4. Conclusion

Indonesia and Philippines are two archipelagic countries located in tropical areas of South East Asia. Seemingly, these two nations sit on the bamboo belt of the world where Asia is considered as the richest continent of bamboos. It is no surprised that traditional and vernacular architecture used bamboo to build their houses. Bamboo construction is easy to harvest, transport, use and build. It is strong enough to resist wind and flexible enough to conform to seismic forces. Both nations are also exposed to active natural calamities that wreak havoc and destruction for thousands of inhabitants e.g. Indonesia with earthquakes and Philippines with typhoons.

Bamboo is still perceived as a "poor man's timber" by many people. However, this perception can be changed by continually using bamboo in modern architectural design context and by conducting conference and seminars in academic sectors and as well as hands-on trainings and workshops in the communities.

Although, Philippines is behind Indonesia in terms of bamboo architecture, both countries can learn from each other through comparison and contrast of bamboo wall system of *bahareque*. This can be achieved by international collaboration between two countries through applied research and joint projects not just by professional architects but also by local artisans and craftsmen in the community. The author believes that this humble plant can unite not just Indonesia and Philippines but as well as the South East Asia and the world.

5. References

[1] Vries S K *D* 2002 *Bamboo Construction Technology for Housing in Bangladesh* (The Netherlands: Eindhoven University of Technology)
[2] Soemarno 2010 *Model Perencanaan Kawasan Agroforestry Bambu* (https://bamboeindonesia.wordpress.com/bambu-untuk-konservasi/makalah/soemarno/)
[3] López O H 2003 *Bamboo: The Gift of the Gods* (Bogota: D'vinni Ltda)
[4] Paudel S 2008 Engineered Bamboo as a Building Material *First International Conference on Modern Bamboo Structures* **01** p 33-40
[5] International Federation of Red Cross and Red Crescent Societies. 2011. Post-Disaster Shelter: Ten Designs. Geneva, Switzerland.
[6] Tanuwidjaja G, Mustakim, Widyowijatnoko A, Faisal B, Tallar R and Hidayat M 2010 Eco Bamboo Plastered House for Sustainable and Affordable Homes in Indonesia *Tropical Eco-settlements*
[7] Widyowijatnoko A 2005 *Plastered Bamboo Project (*Bandung: Institut Teknologi Bandung)
[8] Widyowijatnoko A 2012 *Traditional and Innovative Joints in Bamboo Construction* (Aachen: RWTH Aachen University)

6. Acknowledgements

This paper would not be possible without the support and guidance of Dr.-Ing. Andry Widyowijatnoko ST, MT and Dr. Ir. Surjamanto W. M.T. of Institut Teknologi Bandung, Indonesia.

IOP Conf. Series: Earth and Environmental Science **152** (2018) 012006 doi:10.1088/1755-1315/152/1/012006

Incremental Housing Development; An Approach In Meeting the Needs Of Low Cost Housing In Indonesia

A H Wibowo[1] and D Larasati[2]

School of Architecture, Planning and Policy Development,
Institut Teknologi Bandung, Ganesha Street No 10, Bandung. Indonesia

[1]E-mail : ardian.hario@gmail.com

Abstract. As a country with a rapid population growth, there is a very high shortage of homes and need a quick solution to build houses for the community. The emerging solution is mass housing with industrialization system. As time goes by, this mass housing solution raises a new problem, the mass housing users are not satisfied with the existing home. Incremental development approach is considered as one of the solutions for solving the mass housing problem. Incremental development is a constructive approach that allows the separation of dwellings to be built, altered and dismantled without disturbing others. With this approach, dwelling is not seen as a finished product, but it's a process where residents can participate in designing their own house according to the needs and economy capabilities. Furthermore, Housing provision is built according to minimal needs and it's designed as a 'permanent longlife' and adaptable base. This paper will discuss the criteria of incremental house for low-income communities provided by the government. Literature studies and case studies are used to find the criteria for incremental house. Some criteria can be used as a reference for incremental house construction as a housing solution in Indonesia.

1. Introduction

The lack of occupancy level (backlog) in Indonesia is still very high. The data from BPS demonstrates that the estimated number of houses built per year is only about 400 to 500 thousand of housing units. On the other hand, the demand for community housing per year reached over 800 thousand housing units.

John F. C. Turner claims that the user involvement from procurement process until the finished product is interrelated with the level of user satisfaction. In the process of building a house when the user is not involved in, this will probably lead to a massive problem for the finished product does not match the expectations of its users. Dwelling/housing as a design product will achieve a great success when the potential users are involved in the design process [1].

There is a simple concept that can perform as a solution for the mutual benefit of future society and the government. This particular solution is called the incremental housing. Incremental housing concept is a housing system that is designed for the occupants to be able to develop their own house in the future. The earliest stage of the house will start with the most crucial rooms such as bathroom, bedroom, and kitchen which are provided by the government beforehand. Later on, the house will be developed in accordance with the needs and economic conditions of the residents so that the incremental house has a very participatory design approach. This particular concept is actually not a new concept, some countries including Indonesia have applied this concept. Moreover, this concept

IOP Conf. Series: Earth and Environmental Science **152** (2018) 012006 doi:10.1088/1755-1315/152/1/012006

has also been written in the agenda of the RPJM (Rencana Pembangunan Jangka Menengah) Technocratic Plan sub-Division Housing for the year 2015-2019 [2].

This paper aims to identify which aspects that need to be considered in designing an incremental housing in order to meet the demand for community housing in Indonesia.

2. Methodology

The methodology used in this paper is comparative study. It will comparing a several projects that are considered to have successfully implemented the concept of incremental housing. Design criteria from literature are used to compare the projects.

.

3. Design Criteria

According to Wainer (2016), incremental house is not the same as build a complete small house. Incremental House provide some homes that can be developed by the residents themselves. The main purpose of this approach is to ease the burden of development costs at the earliest. So that the dwelling can be built in a location that is really desirable [3].

Aravena (2016) on an Elemental project, said that in planning a dwelling with the concept of an incremental house must pay attention to some ideal conditions [4], including:

- Good location: dense enough projects able to pay for expensive well located sites.The text should be set to single line spacing.
- Harmonious growth in time: build strategilly the first half (partition structure and firewalls, bathroom, kitchen, stairs, roof) so that expansion happens thanks to design and not despite it.
- Urban layout : introduce in between private space (lot) and public space (street), the collective space, not bigger than 25 families, so that social agreements can be maintained.
- Provide structure for a final plan growth.
- Middle-class DNA: plan for a final scenario at least $72m^2$ or 4 bedrooms.

According to Greene & Rojas (2008), incremental house process is divided into 3 stages. Each stage has an opportunity for the residents to create an efficient and equitable adjusting to their character [5]. These three stages are :

- Search for location; Residents are given the opportunity to choose a site that suits their needs. So they can have an opportunity to work in the city.
- Basic building construction; From the residents' point of view, the primary function of the house is to provide protection against cold, rain, sun, wind. This is a priority after getting a land or location.
- Residential development; After having a house, the occupants begin to develop their dwellings with potluck material, especially if the need is very urgent such as the need for space for family members.

Based on literature study, it can be concluded that design criteria for incremental housing is location, core house, expansion posibilities, building type and allotment of residents.

4. Case Study Data

Case studies to be taken are incremental house projects that are considered successful implemented. Here are 3 incremental house projects that have been built and 1 project concept for vertical dwelling.

4.1. Quinta Monroy Housing

These modules will be able to hold up to five families and come with two different types. Here is the explanation of both units;

The housing type; which is shown blue on above picture, has got direct access from the front door. The space growth only happens to sideways and backward. This kind of type provides the possibility for the occupants to open a business in order to help their economy. In one designed module, follows two types of the housing unit.

The Duplex type; which is shown orange, blue, and red on above pictures. This type is located right above the housing type, this causes each of the units has got their own stairs. The space growth is not limited to sideways and backward, but also upward. Therefore, each of the units is provided with stairs. In one designed module, follows three types of the housing unit.

Figure 1. Before & After Post-Habit QM housing

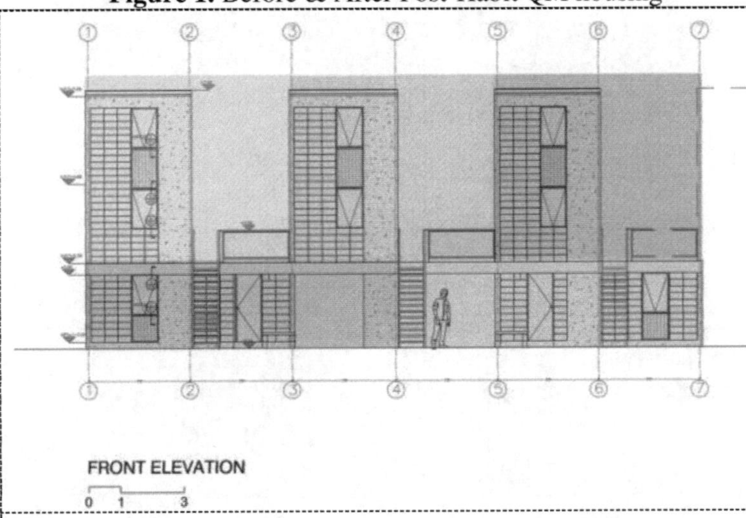

Figure 2. Dwelling Types In Design Module

4.2. Villa Verde Housing

Villa Verde is a housing project owns by Elemental that collaborates with a forestry company named Arauco. This company desires housing for their labours. The location itself is not in downtown. This is because they want the labours to stay close to them, so they choose a site that is just nearby the workplace.

With landed housing type, the ownership of land in each unit looks more clear and also create a larger built area. The space for expansion can hold up to 2 bedrooms and 1 family room. In total, Villa Verde can hold up to 4 bedrooms including the dining room, family room, and also a terrace to host some guests. Identical to the previous design, the service area is provided on one side to ease the piping works and one staircase for the vertical access. Therefore, when it comes the time for the occupants to develop their house, they will not need to think about the circulation of stairs.

Figure 3. Before & After Post-Habit VV housing

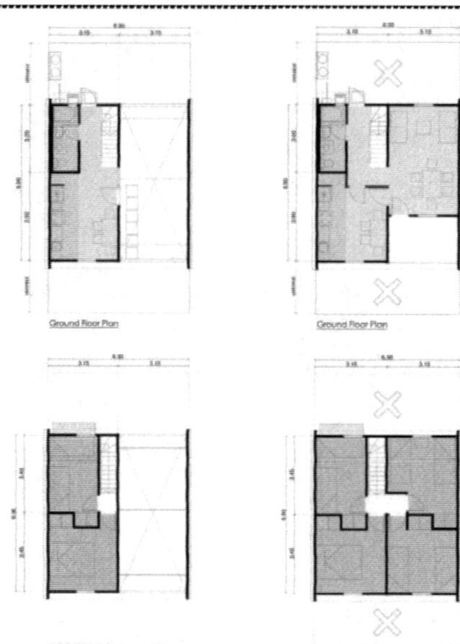

Figure 4. Before & After Post-Habit VV housing

4.3. Tila Housing block

The Tila housing block possesses 39 apartment units with all of them facing south. The circulation that is used is a single loaded corridor located on the north side. With a 5 meters height of each story and beams as the support system provided beforehand, the occupants can surely add another mezzanine level. The occupants are free to fill the interior with whatever they like with a help of partition wall as the infill system. This kind of design sets the boundary so that the expansion will not happen outwards, but inwards instead. As a supporting system besides structural, the architect provided one bathroom for 50m2 unit type and 2 bedrooms for 102m2 unit type. The supporting system and infill will not give any chance to the occupants to identify their dwellings from outside. In contrast to the outgrowth design where the residents are able to design their facade as they please, the entire facade is already designed beforehand. Therefore, this design is intended for the upper middle class.

Architects as providers of support system only provide a free space which can be designed by the residents as an infill systems. The advatages of this concept are not expensive for build and residents will be more satisfied with their residence because they can desaing their own interior. Infill systems not only put the furniture in space, but also set the layout of space on the unit. Such as putting the bedroom, work space, stairs, mezzanine, family room, and others.

Figure 5. Before & After Post-Habit Tila housing

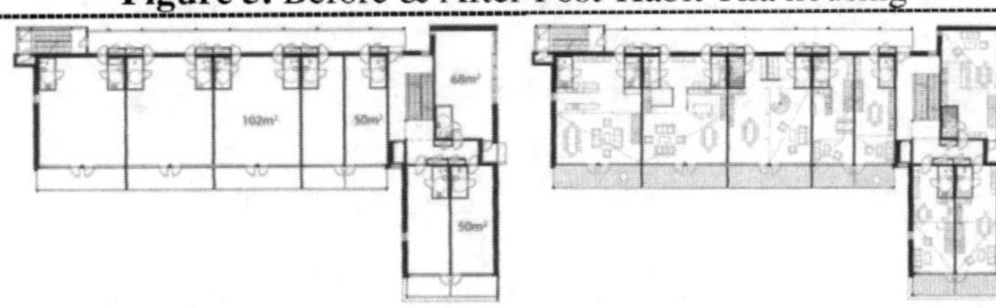

Figure 6. Tila housing Before & After Plan

4.4. Weston Williamson Concept

British Weston Williamson Studio brings the concept of incremental house in vertical dwellings to solve the problem of shortage of homes in Palestine. The idea of a n incremental house is inspired by Chilean architect Alejandro Aravena, where people can add some space according to their economic capabilities. According to a survey conducted by the Israeli-Palestinian mediation organization Quartet and World Bank offices, the need for homes in Palestine is not met by the market. It is estimated that 70 percent of households can not afford the average house price and many of them do not require the size of the house that is generally offered. The apartments will be built with cores in the middle and each unit will have a large balcony. The residences are designed to provide long term homes that "grow with its residents". Residents are allowed to extend out onto balconies using prefabricated panels. Pre-fabricated panels allow a variety of different configurations to create attractive shapes. This apartment has a variety of sizes and layouts, ranging from one to three bedrooms. While the communal space is designed on the ground floor. The expansion can only add 3 rooms and it can be done on the balcony that deliberately designed large enough.

Figure 7. Weston Wiliamson's Design

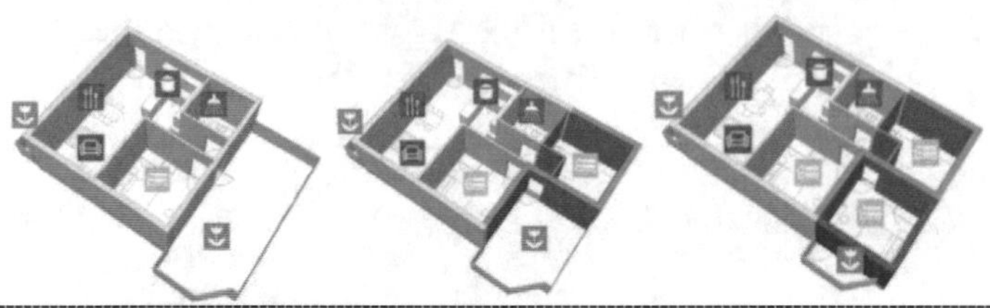

Figure 8. Incremental Process

5. Analysis

The analysis will be done through the case study comparison table adjusting to the design criteria.

Table 1. Precedents Comparison.

Project's Name	Quinta Monroy	Villa Verde	Tilla Housing	Weston Williamson
Location	Chile, Urban	Chile, Urban	Finland, Urban	Palestine
Type	House & Duplex	House	Loft	Apatment
Core Area	36 m^2	43.5 m^2	50 m^2	25 m^2
Expansion Area	72 m^2	76.6 m^2	50 m^2	37.5 m^2
Expansion Types	Sideways	Up & Sideways	Ingrown	Sideways
Main Function	Toilet, Kitchen, Multifunctional Room	Dinning Room, Kitchen, Toilet, Bedroom	Toilet & Multifunctional Room	Toilet, Kitchen, Bedroom, Living Room
Additional Room	Bedroom, Dinning Room, Workspace	Terrace, Bedroom, Living Room, Workspace	Dinning Room, Bedroom, Kitchen, workspace, Living Room	Bedroom
Floor to Floor	2.34 m	2.48 m	5 m	3 m

Location: Three of the four projects chose locations in urban areas. It is because the incremental house is a long term occupancy, so when the residents decide to settle on the dwelling then they will not move to another place again. In addition, the presence of supporting facilities on site also be considered, such as near the workplace, close to educational facilities, health, shopping, and others.

Building Type : two of the four projects are low-rise resident. it considers the ease of community participation when developing their dwellings.
Core & Expansion Area : three of the four projects have an expansion area nearly 2 times from the core house. while in high-rise resident, the expansion area is ingrowth the unit or just give a little space behind the unit.

Expansion Type : two of the four projetcs have expansion areas aside from the unit because it is considered to be the most balanced in development. This provides an opportunity for residents to be able to open a business in fornt of the expansion area but still has a privacy space behind it. Other expansion options include upward development or inward development. For inward development, the provider has to prepare a large empty space to be filled by residents.

Main Function : from the entire project, the bathroom becomes the most important space that has to be provided at the core of the house. Included with wet area for kitchen. Two of the four projects provide bedroom in the core space, two other projects provide multifunctional space which is more flexible for occupan.

Additional Room : from the entire project, the bedroom becomes additional space in the expansion area. This is related to the increasing number of family members. The other rooms adjust the needs of residents, such as, terraces, workspaces, and additional kitchens.

6. Conclusion
Based on the analysis of case studies that have been described, it can be concluded that the concept of incremental house should be concerned with several main things :

Location, urban area are more appropriate to apply the concept of incremental house.

Core house, the area of the core house depends on the standards that have been applied to each country. But at least, providers should prepare the initial shelter in the form of building structures, utilities and vital room (bathrooms and kitchens). These rooms is important considering the water plumbing is a part of building utility. The bedroom or living room can be an alternative choice of space in the core house.

Expansion, side development has more advantages than other development directions. It can balance the benefit of forward and backward development.

Building Type, incremental house is more appropriate to be applied on low rise residential. while in high rise residential, The designer can plan the expansion of the dwelling into the unit as in the example precedent.

Participation, to achieve success on the concept of incremental house, communities should be a participator in the construction of their own homes. Residential provider can provide socialization in advance and control every home development.

7. References
[1] Turner John F C 1976 *Housing By People : Towards Autonomy In Building Environments* (New York: Phanteon Books)
[2] Anonymous 2015 *Penyediaan Hunian Layak bagi Masyarakat Berpenghasilan Rendah (MBR) dalam Rencana Pembangunan Jangka Menengah Nasional 2015-2019.*
[3] Wainer L S, Ndengeioma B, Murray S 2016 *Incremental Housing and Other Design Principles For Low-Cost Hosuing* (Londong: International grwoth Centre)
[4] Aravena A 2011 *The Forces In Architecture* (Tokyo: Toto Publishing)
[5] Grene M and Rojas E 2008 *Incremental Construction: A Strategy To Facilitate Acces To Housing.* (London: International Institute for Environment and Development)

IOP Conf. Series: Earth and Environmental Science **152** (2018) 012007 doi:10.1088/1755-1315/152/1/012007

Water and Sanitation Technology Citizen Needs Assestment in Kolorai Island

Yudha Pracastino Heston[1] and Yonanda Rayi Ayuningtyas

Balai Penelitian dan Pengembangan Sosial Ekonomi Bidang Permukiman.
Jl. Laksda Adisucipto No.165 Yogyakarta. Telp/fax (0274) 555205/546978

[1]pracastino@yahoo.com

Abstract. Kolorai is an island located in Pulau Morotai Regency. Its existence as an island in the National Tourism Strategic Area (KSPN) of Morotai. It has a land area of 4.33 km^2, with a population of ± 550 people. The citizen relying on wells as a source of clean water, and has been developing a piping system to distribute clean water, using a water reservoir which is located in each resident's house. There are 11 public toilets scattered in Kolorai Island. The emerging problems are related to the availability of energy for the distribution and adequacy of clean water source quality. The research was conducted to answer the problem solution needs in a participatory method, with technology product support from Ministry of Public Work and Housing, Research and Development Agency. The research was conducted with qualitative approach, using mix method. The research result is a comprehensive solution to fulfill the needs of clean water and sanitation.

1. The first section in your paper

Kolorai Island is one of the villages in Pulau Morotai Regency. It has a land area of 4.33 km2, with a population of ± 550 people from 113 families [1]. It is one of the supporting tourist attractions, which supports Dodola Island as one of the main attractions in Morotai Island Tourism Area. It developed into a tourist village, while most of its citizens have tourism awareness. The tourist village function is supported the existence of homestays, which managed by the citizens. It have been developed as a part of Sail Morotai 2012 event [2].

The water distribution and sanitation system facilities condition in Kolorai Island has not been able to fulfill domestic and non-domestic needs. A common problem with the clean water provision is the lack of energy for the distribution. Currently available water sources do not have sufficient capacity and quality [2]. There are several households in Kolorai Island do not have a private toilet, and few residents defecate on the beach. There is no adequate waste management system, either for black water or gray water yet.

IOP Conf. Series: Earth and Environmental Science **152** (2018) 012007 doi:10.1088/1755-1315/152/1/012007

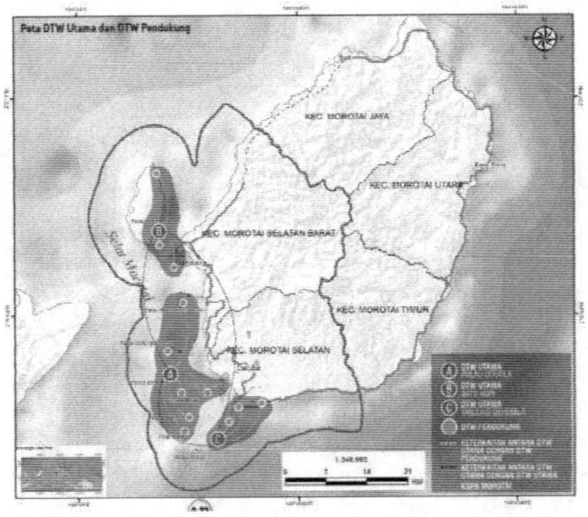

Figure 1. Map of Morotai Island Tourism Attractions [2]

2. Another section of your paper

Brikke dan Bredero describe the factors that influence community-based water supply technologies selection are [3]:

Table 1. Factors affecting community-based clean water technology [3]

Factors	Substances	
Technical factors	1. Demand 2. Capital costs 3. Extension capacity 4. Compability with norms and legal frameworks	5. Compability with existing water-supply systems 6. Comparative advantages 7. Technical skills
Environmental factors	1. Availability, accessibility, and reliability of water 2. Seasonal variations 3. Water quality and treatment	4. Water source protection 5. Risk of a negative environmental impact
Institutional factors	1. Legal framework 2. Regulatory framework 3. National strategy 4. Existing institutional set-up	5. Support from government, NGOs, external support agencies (ESAs) 6. Stimulation of private sector 7. Transferring know-how
Community and managerial factors	1. Local economy 2. Living patterns and population growth 3. Living standards and gender balance 4. Household income and seasonal variations 5. Users preferences	6. Historical experience in collaborating with different partners 7. Village organization and social cohesion
Financial factors	1. Capital costs 2. Budget allocations and subsidy policy 3. Financial participation of users	

While the factors that influence community-based sanitation technology selection are:

Table 2. Factors affecting community-based sanitation technology [3]

Factors	**Substances**
Technical factors	Design preference Technical standars and expeted lifetime of the the technology Availability of construction materials Cost of construction
Environmental factors	Soiltexture, stability, permeability Groundwater level Control of environmental pollution Availability of water Possibility of flooding

Institutional factors	Existing national/local strategies Roles and responsibilities of actors implied Training capacity Availability of subsidies and loans Avaibiltty of workers,
Community factors	Sociocultural aspects Motivational aspects Discouraging factors Social organization factors Other factors

Tim Teknis Pembangunan Sanitasi explains that sanitation mapping process is influenced by other factors, to provide system and technology selection [4]. The system selection are influenced by (1) physical, ie population density, land use, and land suitability (2) arrangements, ie sanitary, management, institutional, O&M and asset ownership arrangements, (3) and finance - funding, this is related to the city's fiscal capacity, support and funding mechanism. The technology selection is influenced by (1) environment, ie health risks, utilization of ground water and surface water, (2) culture - behavior, by looking at the awareness level and community management skills, (3) Investment costs and recurring, related to affordability and technological precision.

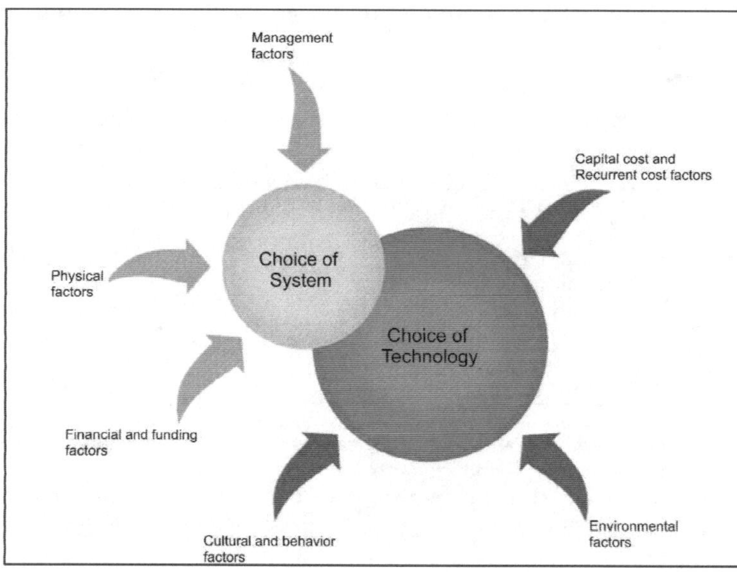

Figure 2. Variable Determination of Technology Selection [4]

3. Data and Methods

This research used qualitative data collection method, direct observation, questionnaire, and direct interview. Qualitative data collection methods may take several considerations [5] such as the situation, time, researcher involvement, participant involvement, participant anonymity, software requirements, and data forms. The application of clean water and sanitation technology is based on public perception needs, with participatory method in Kolorai Island.

The need of technologies synchronize with Ministry of Public Work and Housing, Research and Development Agency (Balitbang PUPR) water and sanitation technology products suitability and availability.

The analytical method used qualitative analysis with mix method. The analysis has done with the scope of the discussion factors, that is:

1. Environmental, Kolorai Island geographical condition, clean water condition by identifying potential sources, reservoir, distribution, and utilization of clean water, as well as waste water disposal, collection and treatment systems, and existing waste management systems.
2. Community, Kolorai Island socio-economic conditions, settlement patterns, and technology used based on public preference.
3. Financial, public economic conditions in order to see ability and willingness to pay more money to better facilities.
4. Institutional, regional government and realted institusional clean water and sanitation program and activities.
5. Technical, Balitbang PUPR technology availability and readiness level.

4. Results

4.1. Environmental Factors

4.1.1. The geographical condition of Kolorai Island

Kolorai Island is administratively part of Morotai Selatan District, Pulau Morotai Regency. It has beautiful coastline as a tourist attractions. Tourists can experience daily life activities of local communities. This island located approximately 1 hour drive by using a motor boat or speedboat from Daruba Port. It has plain topography with white colour sand beach [6].

4.1.2. Condition of clean water

Kolorai Island communities rely on well as a source of clean water. The main water reservoir source is obtained from the wells with solar powered pump, then distributed to the resident's houses with a piping system.

Figure 3. Solar Powered Water Reservoir

The emerging problems are related to the availability of energy for the distribution and adequacy of clean water source quality. There are wells in each 3-4 houses, that taken with electric water pump which powered by a generator set, or manually using a bucket. Water that distributed from the main well is in good quality, tasteless, colorless, and odorless, and there is no significant difference between the dry season and the rainy season for its quality. Some respondent (16%) taste salty or brackish.

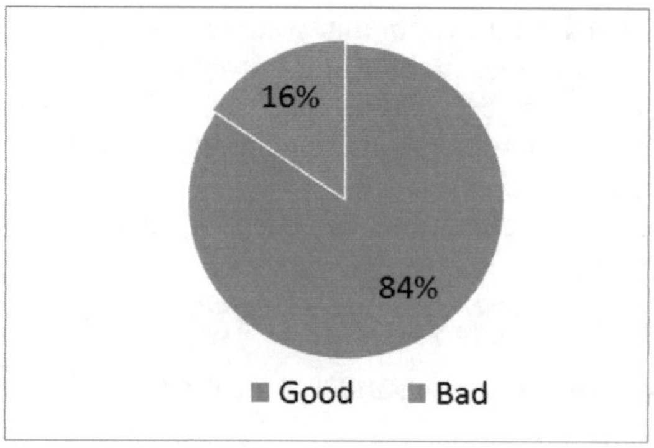

Figure 4. Water Quality According to Public Perception

Based on public perception, the degree of water sufficient described as follows:

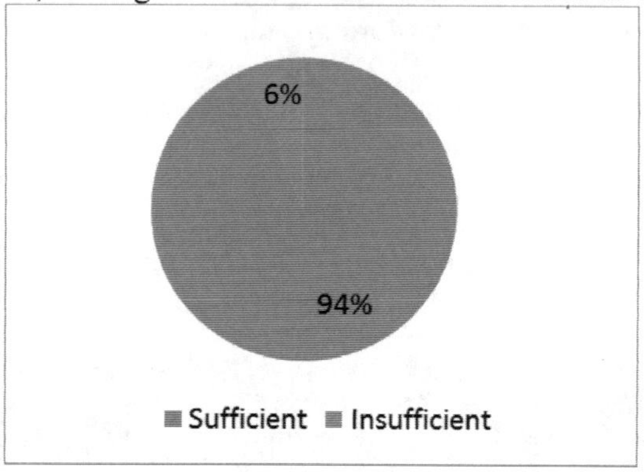

Figure 5. Water Quantity according to public perception

From the figure 5, most of the respondent consider that the quantity of clean water is sufficient for domestic and non-domestic needs. However, few people encounter insufficient water quantity distribution. Territorially, people who live close to the main well do not have insufficient water supply. Otherwise people who live far away from the main well have insufficient water supply. It is also related to water continuity which can be described as follows:

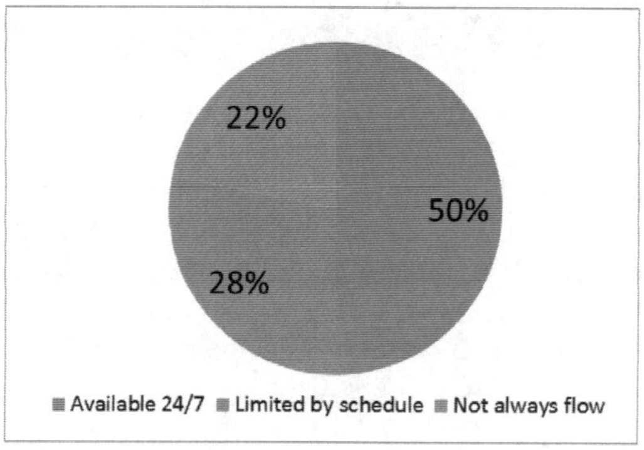

Figure 6. Water Continuity According to Public Perception

From Figure 6. it can be seen that 50% of people get clean water that flows every time and smoothly. While people who are farther away from the location of the main well complain about water that is limited by schedule (28%) and not smoothly distributed (22%). It proof that not all household have good quality, quantity, and continuity of water supply.

4.1.3. Condition of sanitation

Related to the ownership of sanitation in Kolorai Island, the following is a description of ownership of bathing (Mandi), washing (Cuci), toilet (Kakus) facilities (MCK):

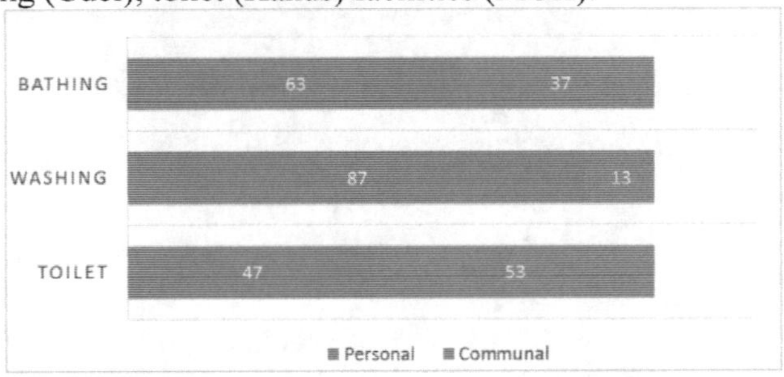

Figure 7. Mandi Cuci Kakus Ownership

Some households do not have a private toilet. There are 11 public toilet/MCK scattered over Kolorai Island for households who do not have a private toilet yet. For black water, each toilet is equipped with septictank, but there is no maintenance and regular suctioning faeces. For gray water, each household does not have a sewer, and discharge to the yard.

According to the statement of a nurse in n medical facility in Kolorai Island, open defecation (on the beach) is still high. Although there are private and public toilet available. Numbers of open defecation in Morotai Island reached 57% [7]. Open defecation free socialization have been frequently conducted by local medical officer, but public awareness is still low.

4.1.4. Waste management system

The garbage on Kolorai Island is managed conventionally. Here is an overview of waste management by communities on Kolorai Island:

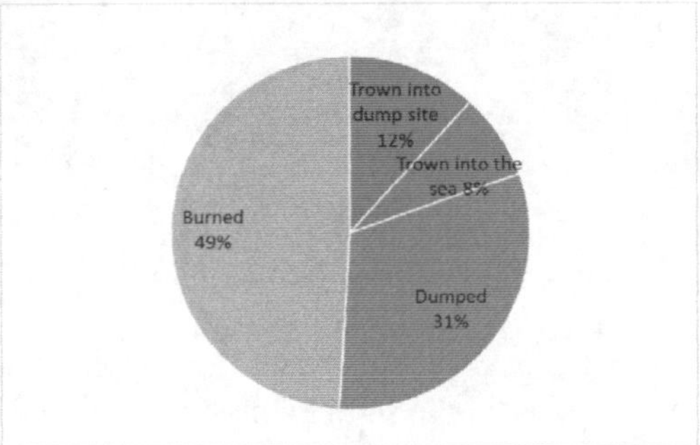

Figure 8. Waste handling by Kolorai Island Community

Figure 8 illustrate that most of household burn and dump their waste. Every household collect their waste and then disposed it to appointed location, and then the waste is dumped or burned. In small islands like Kolorai, the problem of waste management system becomes one of the environmental problems [2].

Figure 9. Piles of rubbish by the beach

4.2. Community Factors

4.2.1. Socio-economic conditions

Kolorai Island citizen is dominated by Galela, Tobelo, Ternate and Bajo tribes with Moslem religion majority. Most of them work as a fisherman, so their daily income is uncertain. Women on Kolorai Island mostly are housewives, and there are other tourism-related professions, such as pandan craftsmen and seaweed cultivator.

4.2.2. Settlement Patterns

The physical condition of settlement areas improved significantly after the Sail Morotai 2012 event. Homestay facilities was builted in several houses, as well as public toilets, solar cell-powered water reservoir and fences in front of the house. Most of the houses are built as permanent building, with concrete walls, zinc roofs, and cement floors.

Figure 10. Settlement in Kolorai Island

Their settlement pattern is oriented to the main docks. The homestay gathered on the main road to the dock. There is a mosque in the center of the settlement, it is also used as a public space. The citizen mobility are by walk, considering by the short space between houses, there are no vehicles on the Island.

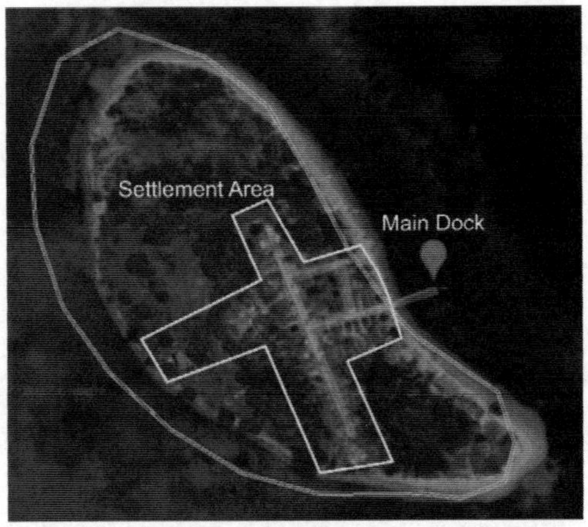

Figure 11. Settlement Pattern

4.2.3. Preferences of Technology Use
The adequacy of clean water and sanitation is seen from the community perception, which is described in the graph as follows:

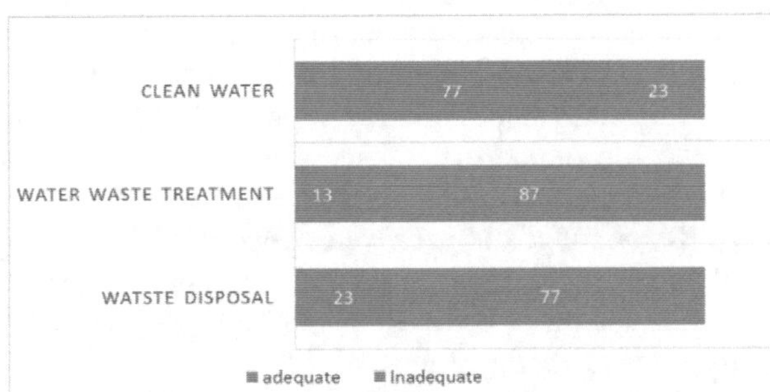

Figure 12. Public perception related to clean water and sanitation

From figure 11 can be seen that respondents have assumed that clean water (77%), as just waste water treatment facilities (13%) and waste disposal (23%) is adequate. This indicates that clean water facilities are adequate, but there are still 23% households have insufficient water supply. While waste water treatment and waste disposal is insufficient.

The community are welcome to water and sanitation technology. Exclude from the willingness to pay for technology, the potential of human resources to perform operational and maintenance for technology is exists. This can be demonstrated by a water reservoir with solar cell power that still exists and functioning well for use. One of the residents is appointed to be able to maintain and operate the the technology.

4.3. Financial Factor
Communities in Kolorai Island have not reached piping system by PDAM. The lack of electricity in this island causes the distribution of water use a generator set, both gasoline and diesel fuel. According to an interview with the community, to meet energy needs, including water distribution, each household requires Rp 20,000 to purchase gasoline or diesel fuel. But in holiday season when tourist come to the island, the need for energy and clean water becomes larger, reaching Rp 100,000 per day. Based on the results of interviews and questionnaires, the community is willing to pay more if quality of clean water and sanitation systems improved, with the following price ranges:

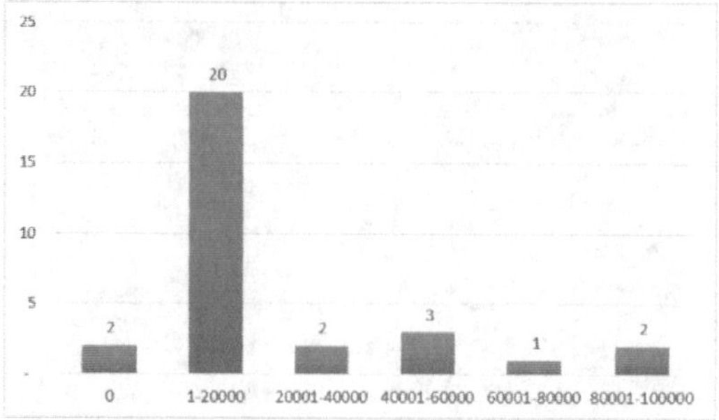

Figure 13. Willingness to Pay

Figure 12 shows most of the community is willing to pay more to get better water quality. The willingness to pay for the operation and maintenance is below price range Rp 20,000.

4.4. Institutional Factors

Along with the National Sanitation Settlement Development Program (PPSP), a program that aims to improve the quality of regional sanitation, in line with efforts to achieve Millennium Development Goals (MDGs) targets, the Government of Pulau Morotai has established a Working Group Team for the Acceleration of Sanitation Development Program of Pulau Morotai [8]. Its vision is "The realization of Morotai DALOHA in Development of Environmental Friendly Sanitation 2020". It is divided into support programs related to sanitation development in Pulau Morotai.

Bappeda Kabupaten Pulau Morotai (2017) provides recommendations of tourism support infrastructure in Kolorai Island, including water supply and sanitation. MCK are recommended to build near the main dock [9]. Recommendation for water needs use desalination of sea water and built near the village office or polindes. At the same time, recommendation for waste management system improvement is by providing garbage transporter using shuttle from Daruba.

Badan Pengembangan Infrastruktur Wilayah provides recommendations related to clean water and sanitation programs, such as drinking water treatment system, construction of rainwater catchment technology in homestay area, construction of waste management system scale area “TPS 3R” and settlement area development through the application of water-saving technology [2].

4.5. Technical Factor

Technically, Balitbang PUPR has technology that is ready to be applied and adapted to the needs. Its technology related to clean water are raw water source technology, processing system, and distribution. As for technology related to sanitation are waste water treatment and waste management system. It includes sorting and processing garbage that environmentally friendly. Advanced analysis of technology readiness level should be done. In 2017, technical advice by Balitbang PUPR only covers the study of technology needs in the main island of Pulau Morotai. It is related to the slum areas development on Daruba Beach, and has not cover the study of technology needs in small islands, including Kolorai Island.

4.6. Technology Requirement

From the analysis, the specific conditions or problems related to clean water and sanitation in Kolorai Island are:

1. Insufficient of water quantity
2. Brackish water quality
3. Limited water continuity (capacity and energy for water distribution)
4. There is no wastewater treatment system (black and gray water)
5. Traditional waste management system (dumped and burned)

Based on the problems above, can be formulated appropriate water and sanitation technology in Kolorai Island. For clean water technology, brackish water treatment into drinking water using a solar cell system. It designs with large capacity, which can be installed in 2 different locations, to ensure the affordability of the entire community in Kolorai Island. This technology aim to solve problems related to quantity, quality, and continuity of water at once. We should pay attention to water management and governance which plays an important role [10], including for water management on a small scale, aspect such as electricity or diesel fuel for groundwater pumps.

Waste water treatment is required communal wastewater treatment system to accommodate, collect, and process black water and gray water, so the effluent can be discharged directly to the sea. As for the waste management system, the current condition of the community is still collecting waste and processed with dumped and burned. However because of the geographical conditions of a small island, the condition will have a time limit. The burned garbage will accumulate because it can not burn out completely. As for the dumped garbage, the land used for hoarding is also limited. The sanira furnace (the burning furnace of non-toxic waste) is a technology from Balitbang PUPR. It can be a solution for waste management system in Kolorai Island, because zero waste, easy to use, using local furnace component, and energy saving [11]. Its operation use solar powered. Community participation in operation and maintenance is also crucial. In order to prove healthy and meaningful livelihoods for all of humanity, which is our major challenge in this century [12], this can be done, one of them by ensuring good service on drinking water supply and sanitation.

5. Conclusion

Potential and community mapping is needed in provision of clean water and sanitation facilities. The availability of clean water and sanitation is a key and primary element on tourism development in Kolorai Island. From research results, it can be concluded that the necessary technology for community are; the installation of processing and distribution of water with solar powered, communal wastewater treatment system, and sanira furnace for the processing of garbage. The chosen technology that suits to these needs, can be accommodated by technology products by Balitbang PUPR in accordance with the conditions and needs of Kolorai Island. In its implementation, further study is needed related to other technology application factors.

This research can be used to solve water and sanitation problems in other areas, especially on small islands, based on the application of community-based technologies

6. Acknowledgement

This research was supported by Research and Development Agency, Ministry of Public Works and Housing. We thank Nur Alvira, Arif Koeshernawan, Rivaldo Okono, for assistance with methodology, and Achjat Dwiatno, Ning Wiyatiningsih, and S.H. Aqsha for comments that greatly improved the manuscript.

References

[1] BPS Kabupaten Pulau Morotai 2016 *Kecamatan Morotai Selatan Dalam Angka Tahun 2016* Another reference
[2] Badan Pengembangan Infrastruktur Wilayah 2016 Rencana Pembangunan Infrastruktur Terpadu MOROTAI (Marine Heritage)
[3] Brikké, F., & Bredero, M 2003 Linking technology choice with operation and maintenance. WHO Press, 142. (Retrieved from http://www.who.int/entity/water_sanitation_health/hygiene/om/wsh9241562153.pdfaccesse, 2017, May 1)
[4] Tim Teknis Pembangunan Sanitasi 2010 Buku Referensi Opsi Sistem dan Teknologi Sanitasi
[5] Neville, S., Adams, J. and Cook, C. 2016 Using internetbased approaches to collect qualitative data from vulnerable groups: reflections from the field, Contemporary Nurse, 52:6, 657-668, DOI: 10.1080/10376178.2015.1095056

[6] Direktorat Pendayagunaan Pulau-Pulau Kecil 2012 Kolorai (http://www.ppk-kp3k.kkp.go.id/direktori-pulau/index.php/public_c/pulau_info/10695 accessed, 2017, July 1 Juli)
[7] PPSP Kabupaten Pulau Morotai 2014 Buku Putih Sanitasi Kabupaten Pulau Morotai. Kabupaten Pulau Morotai (2014)
[8] PPSP Kabupaten Pulau Morotai 2014 Strategi Sanitasi Kabupaten (SSK) Kabupaten Pulau Morotai (2014)
[9] Bappeda Kabupaten Pulau Morotai 2017 Rencana Induk Pengembangan Obyek Wisata Pulau Dodola, Pulau Sum-sum, dan Pulau Kolorai
[10] Wada, Yoshihide and Bierkens, Marc. 2014 Sustainability Of Global Water Use: Past Reconstruction And Future Projections, Environmental Research Letters Environ. Res. Lett. 9 (2014) 104003 (17pp) doi:10.1088/1748-9326/9/10/104003
[11] Balitbang Kementrian PUPR 2015 Katalog Layanan Badan Litbang PU, Pelayanan Uji Laboratorium, Sertifikasi, dan Advis Teknis (PULSA)
[12] Cosgrove, W. J., and D. P. Loucks 2015 Water management: Current and future challenges and research directions, Water Resource. Res., 51, 4823–4839, doi:10.1002/ 2014WR016869.

IOP Conf. Series: Earth and Environmental Science **152** (2018) 012008 doi:10.1088/1755-1315/152/1/012008

Reducing Heat Gains and Cooling Loads Through Roof Structure Configurations of A House in Medan

Irma Handayani Lubis[1] and Mochamad Donny Koerniawan[2]

[1]Mundinglaya Street No.4. Lebak Siliwangi Coblong. Bandung. 40132
[2]Labtek XI B Gedung Arsitektur Jalan Ganesha 10 Lebak Siliwangi Coblong Bandung 40132

[1]lubis.irma.handayani@gmail.com

Abstract. Heat gains and heat losses through building surfaces are the main factors that determine the building's cooling and heating loads. Roof as a building surface that has the most exposed area to the sun, contribute most of heat gains in the building. Therefore, the amount of solar heat gains on the roofs need to be minimized by roof structure configurations. This research aims to discover the optimization of roof structure configurations (coating material, structure material, inclination, overhang, and insulation) as one of passive design strategies that reduce heat gains and cooling loads of a house in Medan. The result showed that case four, white-painted metal roof combined with 45° roof pitched, 1.5m overhang, and addition of insulation, indicates the minimum heat gains production and the less cooling loads during clear sky day but not in the overcast sky condition. In conclusion, heat gains and cooling loads of a house in Medan could be diminished during clear sky day by the addition of roof coating with high reflectance low solar absorbtance, the slope roof, the extension of wider veranda, and the addition of insulation in the roof structure.

1. Introduction

"*The smaller the surface area of the house that is exposed to the sun, the smaller will be the total of the possible amount of radiation received*" [1].

Cities near the equator experiences the high amount of solar radiation throughout the year. Medan, located in 3.6°N and 98.7°E, receives large portion of daily solar radiation from January to December (see **Figure 1**). Thus, the amount of heat transfer in the buildings are possibly high.

Heat transfer in the form of radiation occurred between outside and inside surface of the buildings and it causes heat gain in the buildings. There are two sources of heat gains in the building, from external (walls, windows, roof, and floor) and internal (occupants and equipment) building [2]. Roof as an element in the building contributes to heat gains production.

Roof as one of building surface that dominantly exposed to the sun is obviously caused most of heat gains in the building [1]. Excessive heat gains usually result in discomfort; thus, the cooling loads is also high. Therefore, solar heat gains through roof need to be controlled to minimize the cooling demand in the house.

Passive cooling strategy is an approach to control heat gains with low or zero energy consumption. This strategy attempts to prevent or remove heat transfer in the building by means of natural cooling

[3]. Hence, passive cooling system through roof structure configuration can be a strategy to reduce solar heat gains in the building and cooling demand in a house.

This research aims to discover the optimization of roof structure configuration as one of passive design strategies that would reduce heat gains and cooling loads in the house. This research attempts to provide a solution for the climatic problem that could occur in the buildings in Medan.

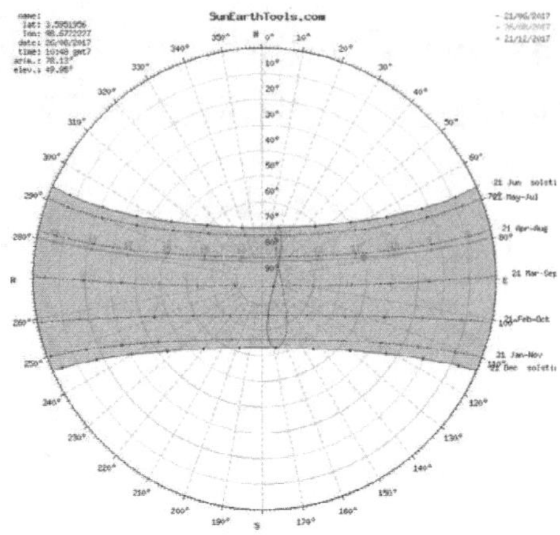

Figure 1. Sunpath Diagram of Medan

Source: Sunearthtools.com

2. Literature Review

2.1. Heat Gains Prevention by Passive Cooling Means

Antinucci, et al. (1992) explained four different techniques of heat dissipations (see Table 1). Radiative cooling applies the cool color (color with high solar reflectance) into the building surface to give heat loss effect. Evaporative cooling provides cooling by means of evaporation. Ventilation delivers cooling by means of air to remove heat. Ground cooling offers cooling from the ground temperature and the atmospheric environment [4].

Table 1. Heat dissipation techniques

Source: [4]

No.	Process	Heat Sink	Main Heat Transfer Mode
1	Radiative Cooling	Sky	Radiation
2	Evaporation Cooling	Air	Evaporation
3	Convective Cooling/ Ventilation	Air	Convection
4	Ground Cooling	Earth	Conduction

According to Santamouris and Asimakopoulus (1996), there are three different stages of passive cooling strategies in the building,

1. Heat gain prevention/ protection in the building
2. Heat gain modulation
3. Heat rejection from the interior building heat sinks (natural or hybrid cooling)

Source: [3]

2.2. Passive Cooling Strategies through Roof Construction

Oakley (1961) said that parasol roof (see **Figure 2**) could be one of building's solution in hot humid climates. Parasol roof serves as a large canopy towards the spaces and allowing air movement in sides between roof and ceiling [1].

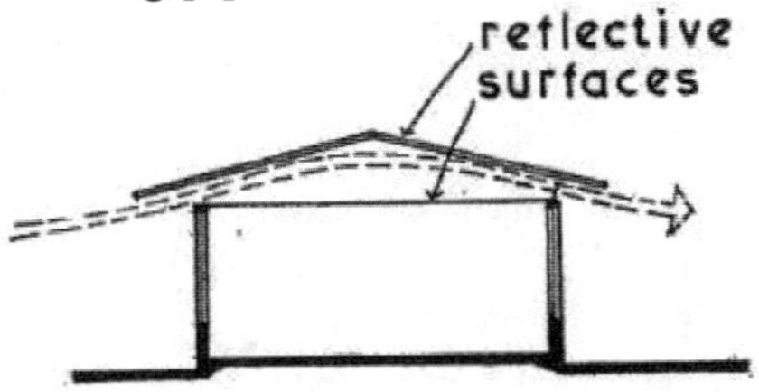

Figure 2 Parasol Roof Concept

Source: [1]

Al-Obaidi, Ismail, & Abdul Rahman (2014) stated that the use of reflective and radiative roofs gave a great reduction in building operational cost [5]. They suggested that the selection of colour and material properties should be taken into account as an approach to enhance the building thermal performance.

Chung, Park, & Ko (2015) proved that roof material with high reflectivity and emmissivity can decrease the urban temperature [6]. This cool roof also decreasing the cooling demand [6].

However, Sabzi, Haseli, Jafarian, Karimi, & Taheri, concluded that the most cooling reduction was provided by the application of water pond on roof, while the less reduction was shown by the use of water jacket on roof [7].

Arumugam, Garg, Ram, & Bhatia believed that two methods of making roof energy efficient are surface treatments (cool roof and radiant barriers) and thermal property modifications (roof insulation) [8]. They stated that the most suitable of roof in warm and humid climate were roof with R-value of 0.31 m2K/W, where it can be achieved by the addition layer such as Brick Bat Coba.

Roslan, Ibrahim, Affandi, Mohd Nawi, & Baharun argued that applying roofing passive design by reflective cooling with combination of optimum roof pitch and roof ventilated approach could contribute in the removal and reduction of heat gained on the roof [9]. They also added that the application of passive design in the roof could low the indoor temperature and the impact would less the energy demands by minimise the use of electricity for cooling purposes [9].

It is concluded that researches find some of passive strategies to reduce heat gains and cooling loads in the buildings. Researchers believe the application of cool roof, such as large canopy, cool coating colours, reflective roofs, evaporative roof, thermal properties modification, and combination of optimum roof pitch and ventilated roof can improve the building performance.

3. Methodology

This research included three stages in the process, consisted of solar radiation and sky cover range analysis of Medan, heat gains and cooling loads investigation of a house in Medan, and heat gains and cooling loads comparison by four different cases of roof structure configurations.

The research started with the analysis of solar radiation and sky cover range in Medan. This stage used Climate Consultant software to generate the weather data into charts and tables. This step determined the date for simulations. Because it was hard to determine the seasonal (rain and dry season) date in Medan, the date for simulations were selected based on the sky conditions, clear and overcast.

The second stage was the investigation of heat gains production and cooling loads of a house in Medan. This initial house was simulated by DesignBuilder. The heat gains simulation was based on two different sky conditions, clear and overcast. Then, the comparison between heat gains in the roof and cooling load in both sky conditions were discussed. This stage proved the performance of heat gains through building fabric occurred in the house.

The third step was the comparison of heat gains and cooling load improvement by four different cases of roof structure configurations. Four cases were simulated to observe the performance of heat gains and cooling loads. Then, the results were compared to analyse the reduction of heat gains and cooling loads from the original case.

In conclusion, the highest reduction of heat gains and cooling loads was indicated as the best performance of roof structure configurations. Then, this roof could be the parameters as the most suitable roof that could be applied in the house in Medan.

4. Results and Discussion

4.1. Solar Radiation and Sky Cover Range Analysis of Medan

Regarding solar radiation, the annual monthly average of daily hourly direct normal radiation was mostly dominated then its diffuse radiation (see **Figure 3**). Based on Energy Plus Weather Data, the highest monthly average of direct normal radiation was in July with 5315 Wh/m^2. While the lowest monthly average of direct normal radiation was 2398 Wh/m^2 in December. However, the maximum and minimum diffuse radiation was 3054 Wh/m^2 in October and 2313 Wh/m^2 in July, respectively.

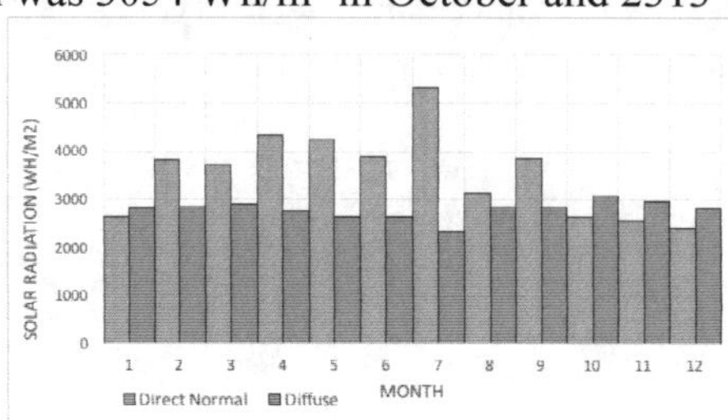

Figure 3 The Annual Monthly Average of Direct and Diffuse Solar Radiation of Medan

In terms of sky cover range of Medan, the monthly average high of cloud cover percentage was mostly near 100%, while the monthly average low was varied from about 35% to almost 70% of total cloud cover (see Figure 4). The average high demonstrates the overcast sky day, while the average low indicates the clear sky day. The huge gap between monthly average high and average low of sky cover percentage indicated that during one month the variation of sky condition could be occurred in Medan.

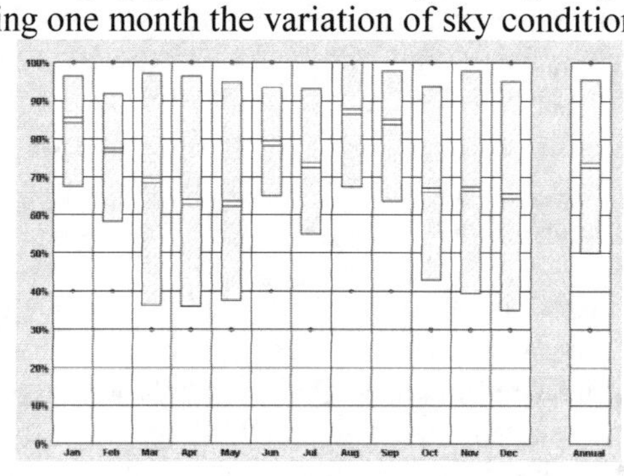

Figure 4 Sky Cover Range of Medan

Source: Generated by Climate Consultant

However, at some point, the larger direct solar radiation was not always identify the clear sky condition in Medan. For instance, in July, the monthly average of direct solar radiation was high but the mean average of sky range cover was also high. This was probably due to the amount of sky pollutant, such as, vehicle emission, industrial emission, and forest fire smoke caused the cloudy sky condition. Consequently, it was hard to determine the seasonal (rain and dry season) date in Medan.

Theoritically, tropical cities has two seasons, rainy and dry season. But, the the amount of solar radiation was not subjective to the season period in Medan. Therefore, this research classifies the sample date of simulation by clear and overcast sky day. To simulate the heat gains and cooling loads of the house, 15th December and 14th August were selected because they represented the most clear and overcast sky day.

4.2. Heat Gains and Cooling Loads of Existing House

In order to investigate heat gains and cooling loads of a house, computer simulation was conducted based on two different weather conditions, clear and overcast sky condition (15th December and 14th August). This research examined a landed house in Medan with the total ground floor area of 170.5m2, including four bedrooms, two bathrooms, a living room, a family room, a dining room, and a kitchen (see Figure 5). All the building fabrics (roofs, walls, windows, and floor) were set according to the real condition.

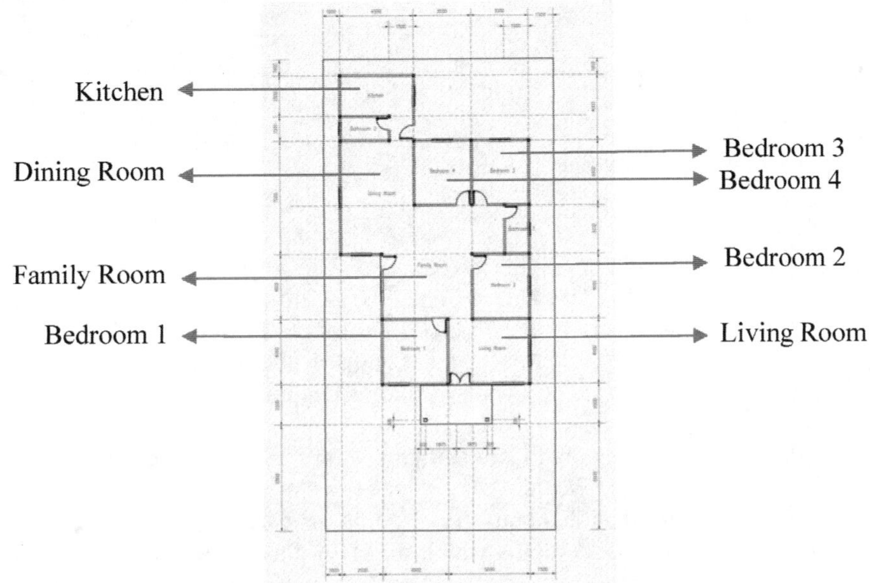

Figure 5. Layout of a house in Medan

Source: Generated by Autocad

The initial roof material contained 10mm of clay roof tile, 30mm of hard wood, 70mm of air gap, and 5mm of plywood as ceiling. The roof was constructed with 30° inclination, 0.6m overhang and without insulation.

The walls consisted of 120mm of brick and 15mm of plaster in both surfaces and the windows material was single glazing with painted wooden frame and the total opening was assumed as 30% of the walls.

The roof thermal properties are presented in Table 2. The total thermal resistance (R-value) was 0.36m2K/W, and the total thermal transmittance (U-value) was 2.7 W/m2K. The R-value is relatively high, and it is possible to reduced. However, the value of thermal absorptance and solar absorptance of clay roof tile were relatively high with 0.9 and 0.7, respectively. It is also noticeable that roof material is identified as lower heat reflection material, so it is easily absorbing heat from outside.

Table 2. The roof thermal properties of the house

Source: Generated by DesignBuilder

Fabric	Thick-ness	Conduc-tivity	Resistance (R-value)	Specific heat	Density	Emisivity	Solar Absorp-tance	Visible Absorp-tance
	m	W/m-K	m^2-K/W	J/kg-K	Kg/m^3			
Clay Tile (Roofing)	0.01	1.00	0.01	800	2000	0.90	0.70	0.70
Wood Hard	0.03	0.18	0.17	1255	721	0.90	0.78	0.78
Air Gap	0.07	-	0.15	-	-	-	-	-
Plywood (Lightweight)	0.005	0.15	0.03	2500	560	0.90	0.78	0.78
Total	0.115	R-value	0.36					
		U-value	2.78					

Regarding the heat gains production, the original roof construction presented the similar trend during clear and sky conditions. As described in **Figure 6**, the heat gains from roof material (red lines) was always dominated. It was reported that the roof contributed more than 4kW, while the rest of house elements produced less than 4kW of total heat gains in the house during both conditions. The second largest heat gains was generated by walls, where it reached maximum at 3.97kW and 3.77kW during clear-sky and cloudy-sky days.

Heat gains produced by roof was obviously started at 8 am and ended at 6 pm but the effect may remain at night. It began with 0.51kW and 0.47kW in clear and overcast day, respectively. Then, it was dramatically mounted into peak at noon with 4.76kW in clear day and 4.70kW in cloudy day. Last, it was significantly dropped into minus which means no heat gains was produced at 6 pm (see **Figure 6**).

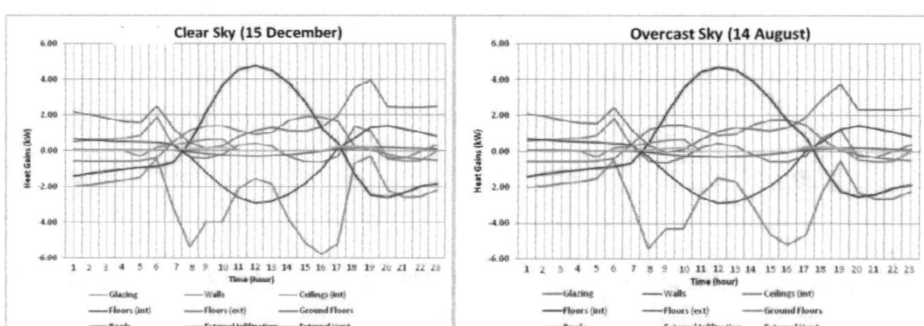

Figure 6. Total heat gains from building fabric and ventilation during clear sky and overcast sky condition. Source: Generated by DesignBuilder

4.3. Heat Gains and Cooling Loads Reduction from Roof Structure Improvement

To predict the reduction of heat gains and cooling loads by passive cooling strategies, four cases of roof structure configurations were simulated based on clear and overcast sky condition (15th December and 14th August). The roof structure configurations were selected based on the combination of coating and structure material, inclination, overhang, and insulation (see Table 3)

Table 3 Roof Structure Configuration

Case	Material	Slope	Overhang	Insulation
Original	Clay roof tile	30°	0.6m	-
1	Wood shingles, plain and plastic-film-faced	30°	0.6m	-
2		45°	1m	-
3	White-painted metal		1.5m	-
4				Wood fibre

As illustrated in Table 4, a gradual decrease of heat gains production was shown by four different cases. Case four is indicated the highest reduction of heat gains during both sky conditions. It is predicted that the maximum heat gains produced was decreased up to 3.68kW and 3.55kW during clear and overcast sky condition, respectively. It is also noticeable that the maximum heat gains of other building surfaces were also reduced but was not significant.

Table 4 Comparison of heat gains production

Source: Generated by DesignBuilder

Heat Gains	Original		Case 1		Case 2		Case 3		Case 4	
	Max (kW)		Max (kW)		Max (kW)		Max (kW)		Max (kW)	
	clear	over cast	clear	over cast	clear	over cast	clear	over cast	clear	over cast
Glazing	1.89	1.75	1.89	1.75	1.91	1.73	1.89	1.66	1.89	1.66
Walls	3.97	3.77	3.92	3.72	3.67	3.42	3.60	3.42	3.51	3.37
Floors (int)	1.39	1.44	0.89	0.92	0.86	0.87	0.78	0.79	0.36	0.36
Floors (eks)	0.2	0.20	0.12	0.12	0.19	0.20	0.33	0.34	0.19	0.20
Roof	4.76	4.70	3.62	3.63	3.97	3.92	3.38	3.34	1.08	1.15

However, it is predicted that the cooling loads of four different cases were also reduced during clear sky condition but not in cloudy sky day (see Table 5). This probably due to the roof improvement could not solve the high humidity during cloudy sky day result in discomfort, so the cooling loads was still high. Case four also obtained the best performance with the most reduction of cooling loads during clear sky days but not in overcast sky day. The maximum cooling loads of case four decreased 1.86kW during clear sky day but increased 1.32kW during cloudy sky day. Although the maximum cooling loads in case four was higher than the original case during overcast sky day, but the increase was least than three other cases of roof improvement.

Table 5. Comparison of cooling demand

Source: Generated by DesignBuilder

Original		Case 1		Case 2		Case 3		Case 4	
Max (kW)		Max (kW)		Max (kW)		Max (kW)		Max (kW)	
clear	over cast	clear	over cast	clear	over cast	clear	over cast	clear	over cast
11.96	8.94	11.85	11.10	11.22	10.74	10.31	10.34	10.01	10.26

Based on the result, whited painted metal roof presented the best performance among others. This was probably due to the value of solar absorptance, visible absorption, and solar reflectance (see Table 6). Whited painted metal roof has the minimum solar and visible absorptance and high solar reflectance. This material possibly absorp less solar radiation and reflect more solar radiation. Therefore, the value of heat gains and cooling load produced was also low.

The addition of insulation in the roof also provide a significant effect. Insulation provides high reduction of heat gains and cooling load. This is due to the U-value of wood fibre was minimum, with only 0.61 W/ m^2K (see Table 7). This means that the heat transfer was also low, so it produced less heat gains. Thus, the cooling loads was also low.

Table 6. Thermal properties of roof outer layer

Source: Generated by DesignBuilder

Outer Materials	Emissivity	Solar Absorptance	Visible Absorptance	Reflectance
Clay roof tile	0.9	0.7	0.7	0.3
Wood shingles, plain and plastic-film-faced	0.9	0.7	0.7	0.3
Whited painted Metal	0.9	0.4	0.4	0.6

Table 7. R-value and U-value of air gap and wood fibre as insulation

Source: Generated by DesignBuilder

Outer Materials	R-value (m^2K/W)	U-value (W/ m^2K)
Air gap	0.15	6.67
Insulation (Wood fibre)	1.63	0.61

Roof with higher inclination gave the big impact towards heat gains and cooling load reduction. In this case the maximum pitch was 45°, because the limitation of local material and the weather conditions. However, roof with higher inclination can give bad impacts to surrounding buildings. It gives glare and heat reflection, particularly in the metallic materials. Even though according to Antinucci et al. (1992) metallic materials gives the heat loss effect on a surface body due to its long wavelength radiation to the night sky, but it could reflect solar to other buildings. Hence, roof with high slope might not recommended and the suitable roof pitched for house in Medan should be less than 45°.

In addition, wider overhang was also obtained the most reduction of heat gains and cooling demand. This is probably because wide canopy can provide shading to house. Thus, the solar radiation would be decreased.

In general, the results assume that the best roof structure combination for house in Medan was whited painted metal roof with 45° roof pitched, 1.5m overhang, and addition of insulation.

5. Conclusion

In general, to solve the problem of heat gains and cooling demand in tropical cities, such as Medan, it can be solved by the arrangement of roof structure. Heat gains and cooling loads could be diminished during clear sky day by the addition of roof coating with high reflectance low solar absorbtance, the slope roof, the extension of wider veranda, and the addition of insulation in the roof structure.

6. Acknowledgment

Authors would like to dedicate gratitude to Dr. Robin Wilson, Department Architecture and Build Environment, The University of Nottingham for assisting during the research process.

References

[1] D. Oakley, Tropical Houses, London: Batsford, 1961.

[2] R. McMullan, Environmental Science in Building, Basingstoke: Palgrave Macmillan, 2012.

[3] M. Santamouris and D. N. Asimakopoulos, Passive Cooling of Buildings, London: James & James, 1996.

[4] M. Antinucci, D. Asiain, B. Fleury, J. Lopez, E. Maldonado, M. Santamouris, A. Tombazis and S. Yannas, "Passive and Hybrid Cooling of Buildings - State of The Art," *International Journal Solar Energy,* pp. 251-271, 1992.

[5] K. M. Al-Obaidi, M. Ismail and A. M. Abdul Rahman, "Passive Cooling Techniques Through Reflective and Radiative Roofs in Tropical Houses in Southeast Asia: A Literature Review," *Frontiers of Architectural Research,* pp. 283-297, 2014.

[6] M. H. Chung, J. C. Park and M. J. Ko, "Effect of The Solar Radiative Properties of Existing Building Roof Materials on The Energy Use in Humid Continental Climates," *Energy and Buildings,* pp. 172-180, 2015.

[7] D. Sabzi, P. Haseli, M. Jafarian, G. Karimi and M. Taheri, "Investigation of Cooling Load Reduction in Buildings by Passive Cooling Options Applied on Roof," *Energy and Buildings,* pp. 135-142, 2015.

[8] R. S. Arumugam, V. Garg, V. V. Ram and A. Bhatia, "Optimizing Roof Insulation for Roofs with High Albedo Coating and Radiant Barriers in India," *Journal of Building Engineering,* pp. 52-58, 2015.

[9] Q. Roslan, S. H. Ibrahim, R. Affandi, M. N. Mohd Nawi and A. Baharun, "A Literature Review on The Improvement Strategies of Passive Design for The Roofing System of The Modern House in A Hot and Humid Climate Region," *Frontiers of Architectural Research,* pp. 126-133, 2016.

Daylight performance of courtyard wall design at low-cost flat in the tropics

F Elsiana, A Juniwati and L S Arifin

Architecture Department, Petra Christian University, Siwalankerto 121-131, Surabaya 60236, Indonesia

E-mail: feny.elsiana@petra.ac.id

Abstract. Courtyard in a low-cost flat in the tropics provides daylight but it is usually protected with a corridor to control the solar radiation. Surrounded by a shaded corridor, daylight level of the adjacent dwelling rooms around the courtyard are reduced. Modification of courtyard wall by integrating a sloped light shelf is proposed to improve the daylight level on dwelling room and control excessive light penetration at the corridor. Experiment with simulation as a tool was used as a research method. Illuminance value, Daylight Factor and uniformity ratio of the base case, a courtyard wall design in a typical low-cost flat in Surabaya (latitude 7°15'55''S) using louver and case, a proposed courtyard wall design using sloped light shelf were compared, simultaneously with daylighting standards. The results demonstrated that courtyard wall modification reduced excessive average illuminance on the corridor in the range of 41% to 57.1% and improved average illuminance in dwelling room in the range of 0.9% to 19.2%. The proposed courtyard wall also increased illuminance uniformity ratio in both corridor and dwelling room. In order to improve daylight quantity and quality, a courtyard wall with sloped light shelf can be applied on a low-cost flat in the tropics.

1. Introduction

The courtyard is one of daylight enhancing techniques to bring light into the interior, with the aim of reducing active zones [1]. Compared with a glazed atrium, a courtyard has higher light available [2]. Several benefits of courtyard application such as provides direct links to the outdoor environment [1], regulates daylight, air movement and thermal interaction with the outdoor environment [3]. The courtyard is also energy efficient for low-rise building, below 13 floors, in hot-humid climate [1].

Previous research about courtyard had conducted by Freewan [3], focused on optimizing the courtyard's daylight performance by modifying its wall geometries. A predictive method for calculating the Daylight Factor (DF) of square courtyards under overcast sky conditions was developed by Acosta et al. [4], while a comparative analysis of energy performance between a courtyard and central atrium in buildings was investigated by Aldawoud and Clark[1]. Soflaei et al. [5] studied the impact of courtyard design variants on shading performance of traditional courtyard houses in a hot-arid climate of Iran.

In warm-humid regions, effective prevention of solar heating of the building is important for providing comfort [6]. Exposed to solar radiation, courtyard is usually protected from both sun and rain with cloisters or verandas [2]. Courtyards in tropical climate also have porches to provide a comfortable area being shaded while at the same time receiving natural ventilation [7]. This protection from sun and rain also found in many low-cost flats at Surabaya, at the form of corridor with louvre around courtyard. Surrounded by a shaded corridor, daylight level of the adjacent rooms around the courtyard are reduced. The surrounding walls will be in shadow and will not act as diffuse sources of daylight [2].

Modification of courtyard wall by integrating an advanced daylighting system that can block direct sun and admit diffuse light, at the same time redirect daylight into the dwelling area was proposed. A light shelf, a shading system [8] that can act both as a shading device and a light guiding into the depth of the room was selected.

A light shelf is one of a daylighting system designed to enhance daylight penetration into buildings. It can be mounted at the upper part of a typical window to provide solar shading and glare control to occupants near to the window while allowing daylight into the room [9]. Commonly in the form of horizontal or inclined projection attached to a window with a highly reflective surface, light shelf reflects sunlight to the ceiling and from there to the back of the room [10].

Previous researches about light shelf were focused on maximizing its daylighting performance by combining light shelf with a curved ceiling [11], combining light shelf with a horizontal light pipe [12] and using a dynamic internal light shelf without modifying external façade [13]. Claros and Soler, [14] investigated the influence of light shelf properties, i.e light shelf reflectance and model reflectance on daylight performance. Those researches showed light shelf's ability in providing shading while allowing daylight into the rear part of the room.

Different from previous research, the light shelf in this research is integrated on a low-cost flat's courtyard wall that has a distance of 1.5 m from the window of dwelling units. Integration of light shelf on the courtyard wall is expected to reduce the daylight level in 1.5 m wide corridor area and enhance daylight level in the dwelling unit area adjacent to the courtyard. The proposed light shelf had employed sloped reflector at its upper surface which was designed for Surabaya (latitude 7°15'55''S). The optimum angle of the reflector was determined based on the required angles of the incident and reflected solar rays.

The effect of proposed courtyard wall design on daylight performance was investigated in this study. Daylight level and uniformity of courtyard wall were evaluated not only on the dwelling unit but also on the corridor. Corridor in a low-cost flat is not only functioned as circulation area but also as a place where frequent social interaction between occupants occurred. Corridor occupation on low-cost flat also happened as a form of the household adaptation in order to expand their unit apartments so that their space requirement can be met [15]. Daylight adequacy and uniformity on both dwelling unit and corridor are important to achieve.

2. Courtyard wall design of low-cost flats

Several low-cost flats in Surabaya that utilize courtyard as their daylighting method are located at Grudo, Jambangan, and Siwalankerto (figure 1). They have a fully enclosed courtyard that placed at the centre of the building. Commonly consists of four stories and is designed in twin blocks, the low-cost flat building has two rows of dwelling units with corridors facing the courtyard. The width of the corridor is 1.5 m.

Courtyard wall design of existing low-cost flats in Surabaya consists of a white-painted brick wall at the bottom and horizontal louver at the top. The brick wall has 1 m in height while the louver has 0.6 m in width. The louvers are painted black and some of them are added with plantations (figure 1).

The adjacent dwelling unit located 1.5 m from courtyard wall was shaded by corridor and louver (figure 2 and figure 3). Utilization of courtyard as a secondary daylight source on dwelling units is limited as the surrounding walls will be in shadow and will not act as diffuse sources of daylight [2].

Courtyard wall modification was proposed to improve the daylight performance in corridor and dwelling unit. Proposed courtyard wall design utilize light shelf, a horizontal shading and redirecting devices [16], which was mounted at the upper part of the courtyard wall.

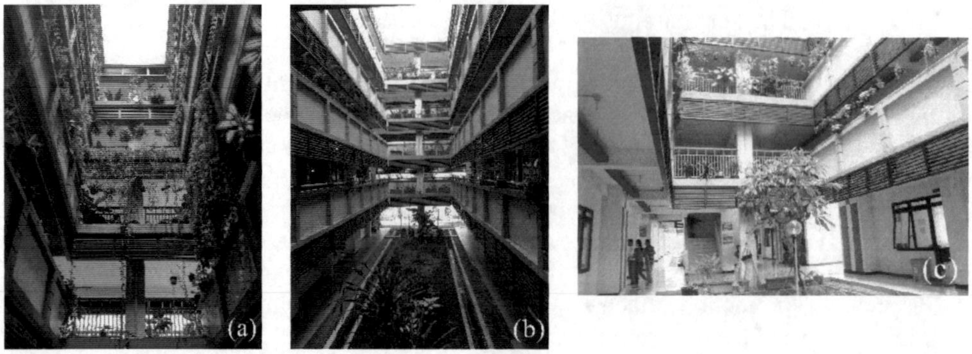

Figure 1. Courtyard wall design in low-cost flats at (a) Jambangan (b) Grudo and (c) Siwalankerto

The light shelf had totally 0.85 m in width, consists of 0.35 m internal and 0.50 m external light shelf (case in table 1). It had a sloped reflector which was developed by considering sun position of a specific location, i.e. Surabaya (latitude 7°15’55’’S).

The optimum angle of the reflector was determined based on the required angles of the incident and reflected solar rays. The reflector composed of a sloped and segmented surface to redirect sunlight with changing of solar altitudes (Case in Table 1). The material of the reflector was a highly reflective film (97.5%). In order to protect from dirt, the reflector was sealed and closed with clear glass on its both sides.

3. Methodology

To investigate the impact of courtyard’s wall design on daylight level and distribution in corridor and dwelling unit at low-cost flat, experiment with simulation as a tool was used as research method. A radiance-based computer simulation which had been validated in previous research [13] was employed. Radiance is a daylighting simulation program that uses a ray-tracing methodology to predict daylight’s behavior in space accurately [17].

A typical low-cost flat with fully enclosed courtyard, surrounded by single loaded corridor was studied. Illuminance value, Daylight Factor (DF) and uniformity ratio of the base case, a courtyard wall design in a typical low-cost flat in Surabaya (latitude 7°15’55’’S) and case, a proposed courtyard wall design were compared, simultaneously with daylighting standards (table 1). Existing ccourtyard wall design consists of 1m white-painted wall at the bottom and a black-painted horizontal louver at the upper part. A combination of 1 m white-painted wall and the sloped light shelf was studied as the case.

As a shading device, the light shelf was designed for a vertical shade angle of 60°, and had both internal and external part. The light shelf had 0.85 m in width, and was installed along the upper part of courtyard wall. A highly reflective material covered its top surface, with reflectance value as big as 97.5%.

The typical low-cost flat consisted of four stories and were designed in twin blocks. Its linear facades were orientated facing North and South. The building had two courtyards at its centre. Each courtyard had 4 m in width and 9 m in length (figure 2). Those courtyards were separated by a stair and corridor located at the centre of the building. Each building floor had totally seven dwelling units and had floor-to-floor height of 3 m.

Figure 2 shows the position of the dwelling unit and corridor investigated in this research. Located on the West side of the building, the unit had 4.2 m in width and 5.25 m in length. Ceiling height of the dwelling unit and corridor were 2.85 m.

Figure 3 describes the elevation and plan of dwelling unit and corridor on low-cost flat studied. The dwelling unit had four rooms, those were bath room, living room, kitchen and bed room. Located adjacent to the perimeter of the building, kitchen and bedroom were mainly daylighted by the side lighting. The courtyard provided daylight to the bathroom and living room through a top and bottom window in the living room and a top window at bathroom. Characteristics of the top and bottom window are described in table 2.

Daylight performance of courtyard wall at dwelling unit and corridor was measured on 21 June when the sun is lowest in the sky and the most critical time in the room facing North. Illuminance level, Daylight Factor (DF) and illuminance uniformity ratio of base and case then compared, simultaneously with daylighting standards.

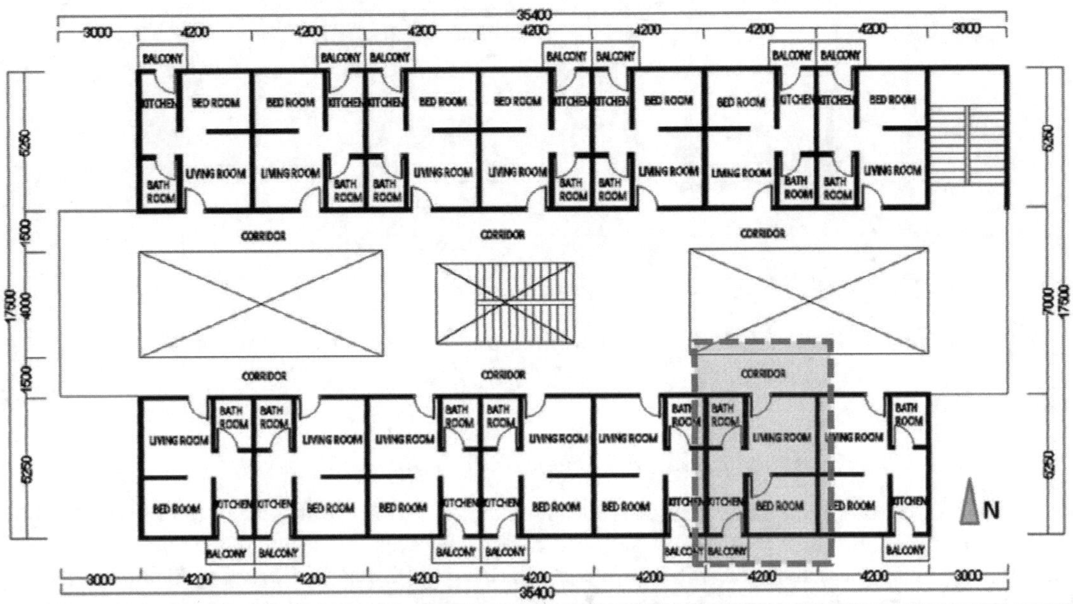

Figure 2. Location of dwelling unit and corridor studied on low cost flat

Table 1. Characteristics of dwelling unit, corridor and courtyard wall

	Dwelling Unit	
Surface Reflectance	Wall	62.8% (light grey)
	Floor	67.5% (off-white)
	Ceiling	74.99% (beige)
	Door	85.53%
Windows facing courtyard	WWR	9.5%
	Lower part: translucent glass	Transmittance 75.2% (frosted glass)
	Upper part: clear glass	Transmittance 88.5% (clear glass)
	Corridor and Courtyard Wall	
Surface reflectance	1 m wall	62.8% (light grey)
	Ceiling	74.99% (beige)
	Landscape at courtyard's floor	13.48% (RAL6007_Bottle_green)
	Louver	13.8 (dark grey)
	Light shelf	Upper surface: 97.5% (Galvanized metal LBNL) Covered glass: Clear glass 88.5% transmittance

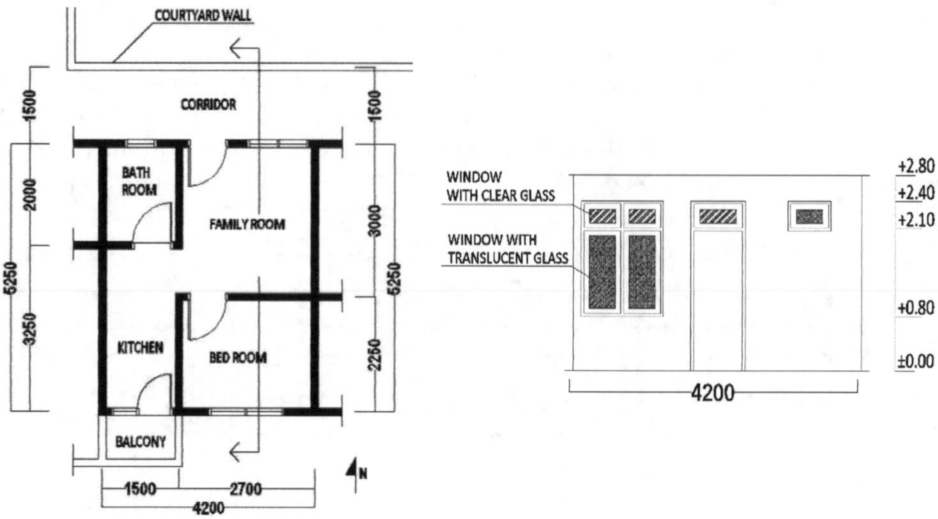

Figure 3. Plan and elevation of dwelling unit and corridor in low cost flat

4. Results and discussion

4.1. Illuminance level and daylight factor on corridor and dwelling unit

Figure 4, shows the daylight performance results of the existing courtyard wall (base case) and modification of the courtyard wall (case) at the corridor. The results showed that the base case, an existing courtyard wall consisted of 0.6 m louver and 1 m white-painted brick wall introduced high average illuminance level on the corridor. Average illuminance level on the corridor of the base case reached 513 lux, 267 lux and 261 lux at 21 June 9.00, 12.00 and 15.00, sequentially. Those average illuminance level were exceeded the IESNA recommended illuminance value for corridor where visual tasks are only occasionally performed (100-200 lux) [16]. High average illuminance level at corridor indicated that the louver could not shade well the corridor, especially at low sun altitude (at 9.00).

Different result occurred at the case, a courtyard wall consisted of a sloped light shelf at its top and 1 m white-painted brick wall at its bottom. Average illuminance level resulted by the case were lower than base case, as big as 220 lux, 157 lux and 154 lux at 09.00, 12.00 and 15.00, sequentially. Those average illuminance value were in the range of IESNA recommended illuminance value for corridor where visual tasks are only occasionally performed (100-200lux), except at 21 June 09.00.

Lower average illuminance value of case compared to the base case indicated the role of a sloped light shelf in providing shading and reducing the excess illuminance level at the adjacent area, that is the corridor. Reduction of average illuminance level generated by a sloped light shelf reached 57.1%, 41.2% and 41% at 09.00, 12.00 and 15.00, sequentially.

Figure 5 indicates the average illuminance of base case and case at dwelling unit. Analyses were mainly focused on two rooms inside dwelling unit that are adjacent to the courtyard, i.e. bathroom and living room. Bed room and kitchen area are also simulated but excluded in analysis considering both rooms receive daylight mostly from sidelighting facing South, not the courtyard.

The results showed that the base case generated low average illuminance level on the bathroom. Average illuminance level resulted by base case on bath room was 119 lux, 117 lux, and 117 lux at 9.00, 12.00 and 15.00 on 21 June. Those illuminance level had met illuminance target for a bathroom, as big as 100-200 lux [18]. Low illuminance value was caused by a small proportion of bathroom window (WWR 5.9%) and the use of translucent glass (VT 0.75) on the window to maintain privacy.

Modification of courtyard wall by integrating a sloped light shelf (case) increased average illuminance inside bathroom. Slightly improvement of average illuminance on bathroom occurred, as big as 3.4%, 0.9% and 0.9% at 09.00, 12.00 and 15.00 on 21 June, sequentially. Average illuminance of the case on bathroom was 123 lux, 118 lux and 118 lux at 09.00, 12.00 and 15.00 on 21 June, sequentially (figure 5). Those illuminance level had met illuminance target for a restroom, as big as 100-200 lux [18]. This result showed the role of light shelf in increasing daylight level on the rear part of the room, at the distance of 1.5 m from courtyard wall.

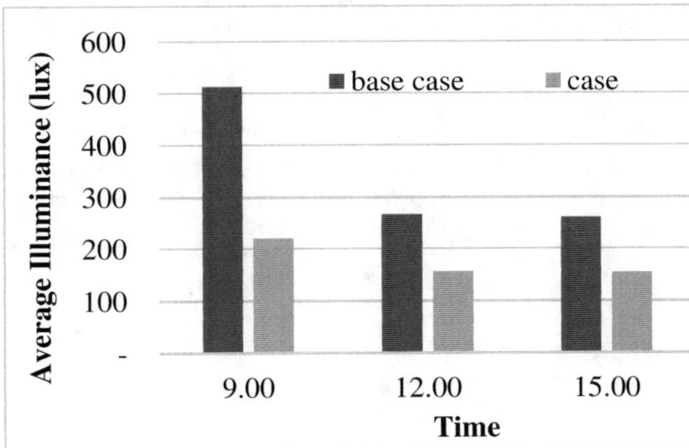

Figure 4. Average illuminance of base case and case on corridor

The results also showed that the base case yielded low average illuminance level on the living room. Average illuminance level of the base case on the living room was 120 lux, 118 lux and 114 lux at 9.00, 12.00 and 15.00 on 21 June, sequentially (figure 6).

Table 2. Experimental scheme

Base Case	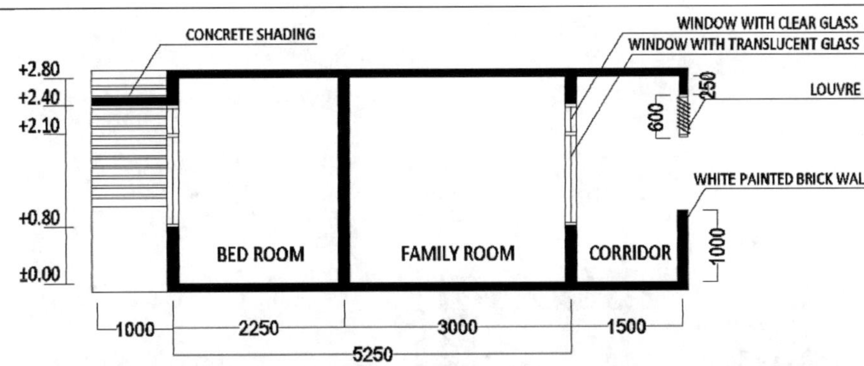 Courtyard wall consisted of black-painted louver and 1 m white-painted brick wall
Case	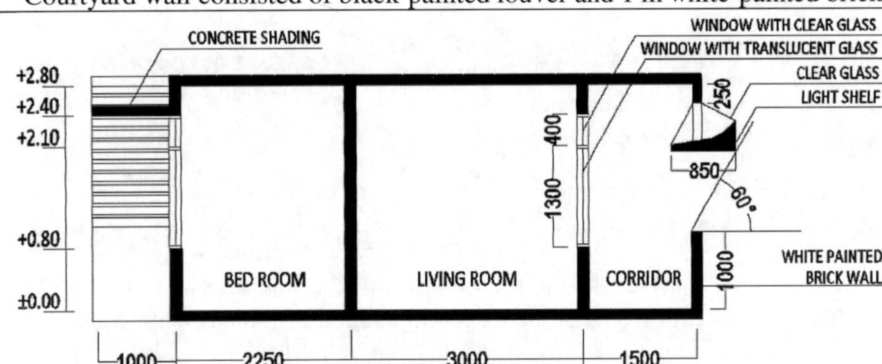 Courtyard wall consists of a sloped light shelf and 1 m white-painted brick wall

An improvement of average illuminance level was achieved in case. Average illuminance of the case on the living room was 143 lux, 116 lux and 117 lux at 9.00, 12.00 and 15.00 on 21 June, sequentially. This fact indicated that the biggest improvement of daylight level occurred at low angle sun (09.00), as big as 19.2%. Slightly improvement of daylight level occurred at high angle sun (12.00), as big as 1.8%. Improvement of daylight level at low angle sun at 15.00 were 2.6%. The low improvement of daylight level at 15.00 was caused by the presence of an obstruction, i.e the west side of low-cost building.

Those average illuminance value of both base case and case had met the illuminance target value for a living room where simple visual tasks are performed (100-200 lux). Considering that diverse activities took place at the living room, from a simple visual task (entertaining, playing, TV-watching) and an increase of visual task (ironing, studying), the improvement of daylight level by integrating a sloped light shelf is important.

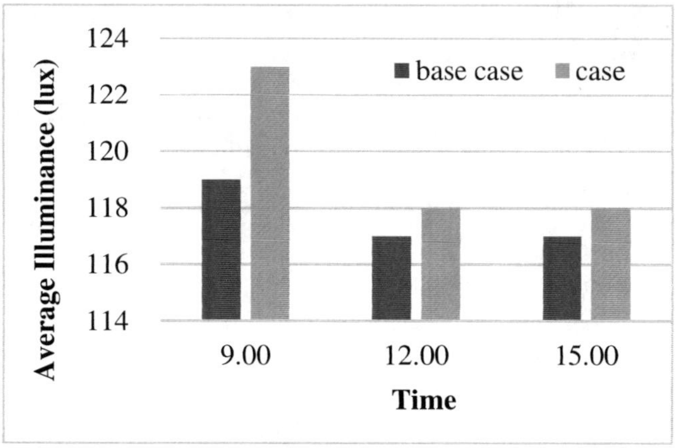

Figure 5. Average illuminance of base case and case on bath room of dwelling unit

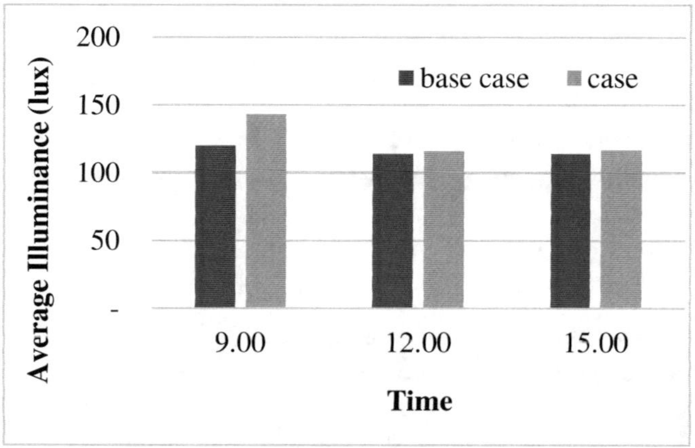

Figure 6. Average illuminance of base case and case on living room of dwelling unit

Figure 7 indicates the Daylight Factor (DF) profile at the centre of the corridor and living room. The case introduced lower DF value on the corridor than the base case. DF value of base case ranged from 2.9 to 5.8%, while the case ranged from 1.5 to 2.3%. DF value reduction on the corridor at the distance 1 m from courtyard wall reached 60.3%, 46.7% and 48.3% at 09.00, 12.00 and 15.00, respectively. These results showed that the sloped light shelf provided shading and reduced the excess daylight level at the area near the courtyard wall. Although all DF value for both base case and case were above the minimum DF for corridors (0.5% in Lechner [19]), but DF value of the base case, as big as 5.8% at 09.00 was over lit (above 5% which exceed the maximum 5% for avoiding glare and overheating).

Figure 7 also indicates that the case generated a higher DF value on the living room, at the distance of 2-4 m from courtyard wall than the base case. DF value of the base case ranged from 1.03 to 1.21% while the case ranged from 1.08 to 1.68%. The highest improvement of DF value by case reached 47.4% at low angle sun (09.00).

All DF value of both base case and case were above the minimum DF for a living room, as big as 1% [19]. Considering that diverse activities took place at the living room, from a simple visual task and an increase of visual task, the improvement of daylight level is useful.

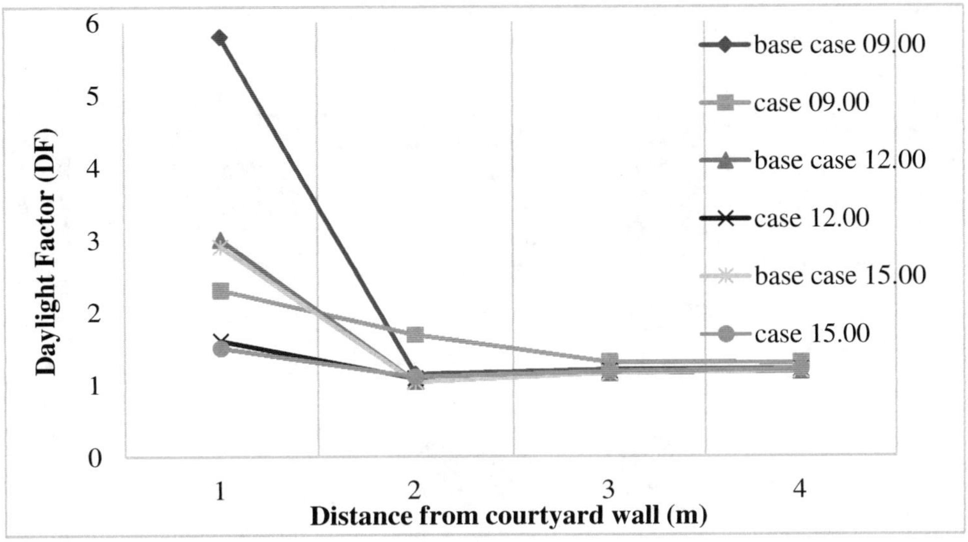

Figure 7. Daylight Factor (DF) profile on corridor and living room

4.2. Uniformity ratio

Table 3 summarizes the isolux plot, uniformity ratio of the base case and case. The existing base case recorded daylight uniformity ratio ranged from 0.32 to 0.50 at the corridor. Only at high sun altitude (12.00), daylight uniformity ratio of the base case had met the uniformity standard for corridor where a visual task was only occasionally performed, as big as 0.5 [18].

Modification of courtyard wall gave significant improvement on daylight uniformity, especially in the corridor. Uniformity ratio of the case on corridor was range from 0.79 to 0.89. Improvement of daylight uniformity ratio on the corridor by case reached 147%, 78% and 89.4% at 09.00, 12.00 and 15.00 on 21 June, sequentially. This improvement occurred because the light shelf shaded the corridor then decreased high daylight level near the courtyard (table 3). Daylight uniformity ratio of the case had met the target uniformity ratio for corridor [18]. Those results are in a good agreement with Lim et al. 2012 [13].

Table 3 also shows that both base case and case had high uniformity ratio at room adjacent to courtyard, i.e. bathroom and living room. Uniformity ratio of base case and case on bathroom was 0.99, 1.00 and 1.00 at 09.00, 12.00 and 15.00 on 21 June.

Modification of courtyard wall generated a higher uniformity ratio on living room than existing courtyard wall. An improvement of uniformity ratio by case was 2.4%, 1.2% and 3.5% at 09.00, 12.00 and 15.00 on 21 June.

5. Conclusions

The results demonstrated that modification of courtyard wall by replacing an existing a black-painted louver with a sloped light shelf at the upper part could provide an improvement in the daylight quality and quantity. The proposed courtyard wall could reduce excessive average illuminance on the area near to courtyard, i.e corridor, in the range of 41% to 57.1%. The proposed courtyard wall using 8.5 m light shelf also increase average illuminance level in dwelling room at the distance of 1.5m from courtyard wall. Improvement of average illuminance level in those spaces ranged between 1.8 to 19.2% on living room and 0.9% to 3.4% on bathroom.

The proposed courtyard wall design also had higher illuminance uniformity ratio on the area near the courtyard (corridor) and dwelling room at the distance 1.5 m from courtyard than existing base case. Improvement on illuminance uniformity ratio of the case reached 78% to 147% on the corridor and 1.2% to 3.5% on the living room. In order to improve daylight quantity and quality, a courtyard wall consists of a sloped light shelf at the upper part and 1m white-painted brick wall at the bottom can be applied on a low-cost flat in the tropics.

Table 3. Isolux plot, uniformity ratio of base case and case

		09.00	12.00	15.00
Base Case Courtyard wall consists of black-painted louver and 1 m white painted brick wall	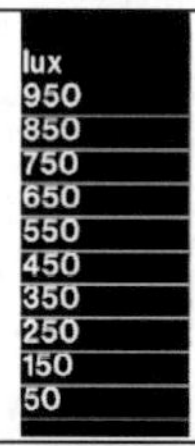	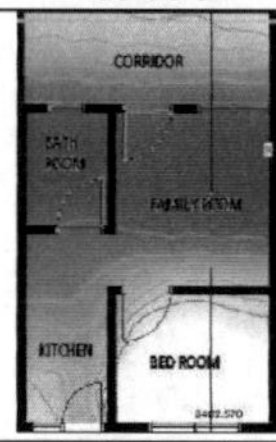	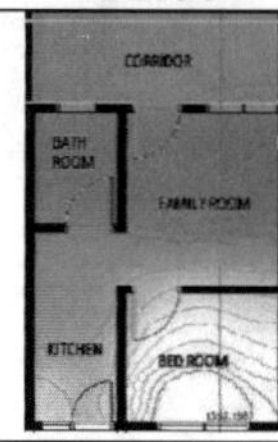	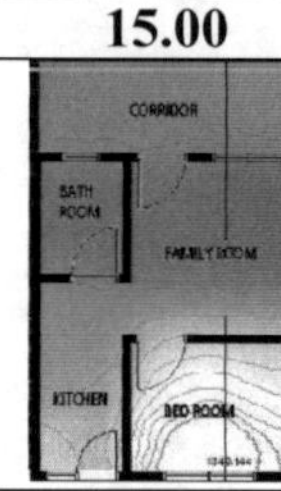
Uniformity ratio	corridor	0.32	0.50	0.47
	bath room	0.99	1.00	1.00
	living room	0.85	0.85	0.85
Case Courtyard wall consists of sloped light shelf and 1 m white painted brick wall	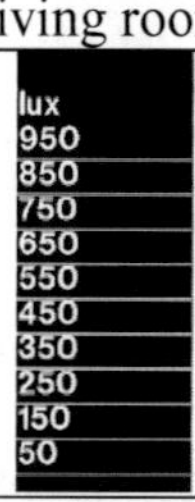	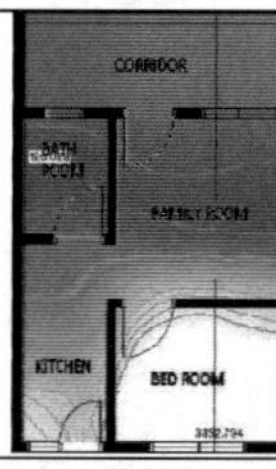	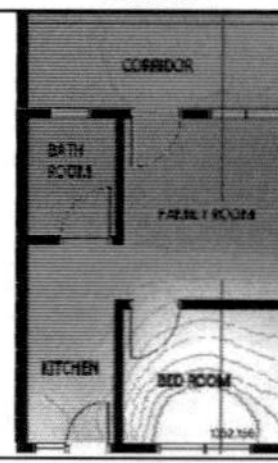	
Uniformity ratio	corridor	0.79	0.89	0.89
	bath room	0.99	1.00	1.00
	living room	0.87	0.86	0.88

6. Acknowledgements

This research is funded by LPPM (Lembaga Penelitian dan Pengabdian kepada Masyarakat) Petra Christian University through a research grant 01/G-RESEARCH/LPPM-UKP/XII/2016.

7. References

[1] Aldawoud A and Clark R 2008 Comparative analysis of energy performance between courtyard and atrium in buildings *Energy Build.* **40** (3) 209-214

[2] Baker N and Steemers K 2002 *Daylight Design of Buildings: A Handbook for Architects and Engineers* (London: James &James (Science Publishers) Ltd) p 50

[3] Freewan AA 2011 Modifying courtyard wall geometries to optimize the daylight performance of the courtyard *Smart Innov Syst Technol.* **7** 57-64

[4] Acosta I, Navarro J and Sendra JJ 2014 Lighting design in courtyards: predictive method of daylight factors under overcast sky conditions *Renew Energy* **71** 243-254

[5] Soflaei F, Shokouhian M, Abraveshdar H and Alipour A 2017 The impact of courtyard design variants on shading performance in hot- arid climates of iran *Energy Build.* **143** 71-83

[6] Givoni B 1998 *Climate Considerations in Building and Urban Design* (Canada: John Wiley and Sons, Inc) pp 379-383

[7] Taleghani M, Tenpierik M and Dobbelsteen A 2012 Environmental impact of courtyards- a review and comparison of residential courtyard buildings in different climates *J. Green Build.* **7** (2) 113-136
[8] Kischkoweit-Lopin M 2002 An overview of daylighting systems *Sol Energy* **73** (2) 77-82
[9] Wong I L 2017 A review of daylighting design and implementation in buildings *Renew Sustain* Energy *Rev.* **74** 959–968
[10] Warrier G A, Raphael B 2017 Performance evaluation of light shelves *Energy Build.* **140** 19-27
[11] Freewan A A 2009 Maximizing the lightshelf performance by interaction between lightshelf geometries and a curved ceiling *Energy Convers Manag.* **51** (8) 1600-1604
[12] Beltrán L O and Uppadhyaya K 2008 Displacing Electric Lighting with Optical Daylighting Systems *25th Conf. on Passive and Low Energy Architecture* October 20nd-24th Dublin
[13] Lim Y W, Kandar M Z, Ahmad M H, Ossen D R and Abdullah A M 2012 Building façade design for daylighting quality in typical government office building *Build Environ.* **57** 194-204
[14] Claros S T and Soler A 2002 Indoor daylight climate–influence of light shelf and model reflectance on light shelf performance in Madrid for hours with unit sunshine fraction *Build Environ.* **37** 587–598.
[15] Kisnarini R 2015 *Functionality and adaptability of low cost apartment space design: a case of Surabaya, Indonesia* (Eindhoven: Technische Universiteit Eindhoven) p 161
[16] Egan M D and Olgyay V W 2002 *Architectural Lighting* (New York: McGraw-Hill Company) p 34, 117
[17] Canziani R, Peron F, Rossi G 2004 Daylight and energy performances of a new type of light pipe *Energy Build.* **36** (11) 1163-1176
[18] Steffy G 2008 *Architectural Lighting Design* (New York: John Wiley & Sons, Inc.) pp 45-46
[19] Lechner N 2009 *Heating, Cooling, Lighting, Sustainable Design Methods for Architects* (Canada: John Wiley & Sons, Inc.) p 409

IOP Conf. Series: Earth and Environmental Science **152** (2018) 012010 doi:10.1088/1755-1315/152/1/012010

Thermal Comfort Assessment in The Open Space in Bandung Case Study Dago Street and Riau Street

M Sugangga[1], K I Janesonia[1], D F Illiyin[1], and M Donny Koerniawan[2]

[1]Major of Urban Design, Department of Architecture, School of Architecture, Planning, and Policy Development, Institut Teknologi Bandung, Indonesia

[2]Laboratory of Building Technology, Department of Architecture, School of Architecture, Planning, and Policy Development, Institut Teknologi Bandung, Indonesia

Abstract. Bandung's temperature has been higher since last years. This phenomenon affects the level of thermal comfort in open space. One indicator that determines the thermal comfort level is the type of activity performed by the open space user. Riau Street and Dago Street are corridors that are often used by the people for strolling, jogging, shopping. Dago Street has special event every Sunday namely car free day. Both corridors have different orientation; Dago Street is North to South corridor while Riau Street's is West to East. The goal of the study is to compare people's perception of thermal comfort in both corridors. This research uses two methods, namely qualitative method and quantitative method. Based on the results of qualitative analysis found that the thermal conditions in Dago Street more comfortable than the Riau Street. The result of quantitative analysis found that the average PET (thermal comfort indices) value of Dago Street was at 27.5 °C PET and Riau Street 28.6 °C PET. Dago Street is considered more convenient because it has a lower PET value than Riau Street. The people perception of thermal comfort is very important to start the steps for designing the orientation of street in urban design.

1. Introduction

The city is a gathering place in which humans live and perform various activities in a large collage [1]. A good city could provided comfort to every resident performing activities, one of which is thermal comfort. According to de Dear & Brager, thermal comfort is a condition in which there is satisfaction with the state of the existing environmental temperature so that it can be said comfort preference of every human and the thermal comfort level of every human is different [2]. Thermal comfort is closely related to air temperature, humidity, and wind [3]. These are factors that can affect thermal comfort. Some of these factors can be influenced by the development of the city such as building construction, energy generated by the building, and due to other factors such as the number of vehicles in a city [4].

Bandung is one of the fastest growing cities in Indonesia. Development of Bandung City is also accompanied by the construction of infrastructure such as streets with good pedestrian paths. One of the most well known corridors in Bandung is Ir. H. Juanda street or better known as Dago Street and L.L.R.E Martadinata street or known as Riau street. These two street corridors are notorious for their inner activity attracting the attention of the public both from within and outside the city.

Dago Street is one of the main streets in Bandung which is classified in secondary road collector level with a total length of 5.46 km, street width ± 12 meters, and pedestrian path ± 14 meters on both sides of the street. Dago street itself is one of the streets are very busy either through public or private vehicles, motorcycles, bicycles, and buses. On a typical day, the area is dominated by users who work in the area while on holidays, this area began to be visited by many tourists from out of town. Every Sunday, this street is used by the people's of Bandung to conduct Car Free Day activities with activities such as selling food, gymnastics, cycling, exercising and for the promotions.

Riau Street is a secondary collector road that has a total length of 2.6 km and a width of ± 10 meters and a pedestrian line ± 2 meters on two sides of the street. Land use in this area is similar to Dago Street, which is a commercial and commercial areas. In additions, Riau Street is famous for its factory outlet. This is because the many functions of building factory outlets that dominate. This street has a street orientation from the West to the East. The main activities on the street are shopping, and offices. In contrast of Dago Street, Riau Street only has a busy time at certain hours, starting at 09:00 until 22:00. Outside of that time, the condition of Riau Street area can be quite quiet especially pedestrian user.

The high development of the city of Bandung, especially on the street corridor can affect factors related to thermal comfort such as wind, air temperature, and humidity in this street corridor. This can certainly affect the thermal comfort felt by the people who use this street corridor for daily activities. This research was conducted to know thermal comfort in both corridors of this street.

2. Research Methods

2.1. Thermal comfort responder's preference

The Research method used in general is qualitative and quantitative methods. In the quantitative method using descriptive statistical analysis techniques. Descriptive statistical techniques are used to process data obtained by researchers from the results of questionnaires distributed to the public. Descriptive statistics are used that is by looking at the value of mode in the data obtained. In this analysis will describe quantitatively in the form of percentage of questionnaire results. These percentages explain the characteristics or profiles and preference of respondents regarding the level of thermal comfort according to their assumptions.

The data was collected during the Car Free Day (CFD) events on Ir. H. Juanda (Dago) Street at 08.00-09.00 and at the weekend at Riau Street. While the collection of data samples done by non-accidentally random sampling method is by bringing the form and conduct interviews on some people with certain gender, certain age range and certain activities.

2.2. Thermal comfort calculations

The data used to perform the thermal comfort calculations consist of Air Temperature (Ta), relatve humidity (Rh%), Heat Stress (TG) and wind velocity (v, m / s). Data collection using the measuring tool is done parallel with the data retrieval process through interview respondents, with the division of tasks as follows: first, two researchers who each operated TR74Ui Data Logger and Anemometer digital take data air temperature, humidity and wind speed at one point ; Second, the researcher who operated the WBGT measurements took air temperature (Ta), humidity (RH) and Heat Stress (TG) data simultaneously with the researcher who collected interview data on radius about 500 meters from the first researcher point.

The further data that has been obtained will be processed through software Rayman. The software functions to know the value of the thermal comfort index in PET and PMV units. PET is defined as the temperature of air needed to reproduce in a space with a standard of body heat production and surface

IOP Conf. Series: Earth and Environmental Science **152** (2018) 012010 doi:10.1088/1755-1315/152/1/012010

heat of a particular human skin [5]. PMV is a thermal comfort index introduced by Professor Fanger from the University of Denmark in 1982 and has a standard ISO 7730. Here is the formula for calculating PMV.

$$PMV = f(M, Rc, Ta, Tmrt, V, RH)$$

M is the rate of metabolism (e .g., Lying = 0.8 met; sitting = 1.0 met; run = 8.0 meth); Rc is clothing resistance (e.g., shorts and shirts = 0.1 clo; trousers and coat = 1.0 clo; winter clothing = 3.0 clo); Ta is Air Temperature; Tmrt = mean radiant temperature; V ie Speed wind; RH is Relative Humidity.

3. Observation Results

In this research, the comparison between variables that affect the thermal comfort in the outer space in both street corridors. The comparison is based on community preference and tool measurement, while the variables comprised include activity (metabolic rate), air temperature, wind speed, humidity and clothing resistance.

Dago Street and Riau Street have similar characteristics of land use in the form of trade and services include offices, hotels, cafes and restaurants. The width of the street and the width of the street sidewalk in the two corridors of this street has almost the same width, Dago Street has a width of ± 12 meters and the width of the sidewalk ± 4 meters and Riau Street has a width of ± 10 meters street and ± 2 meter wide sidewalk.

To support the existing activities in both corridors there are various types of supporting facilities such as street furniture that can be used by street users to move in the outer space. These two street corridors have different street orientations, in which the Dago Street corridor has a north-south street orientation and the Riau Street corridor has an orientation of the East-West street. This difference in street orientation affects the movement of the winds in each street corridor thus also affecting the comfort of the outer space.

Table 1. The comparison condition of Dago Street and Riau Street

Comparisons	Dago Street	Riau Street
Street Width	± 12 meters	± 10 meters
Width of sidewalk	± 4 meters	± 2 meters
Street Orientation	North - south	West - East
Land use	Trading and Services	Trading and Services
Building functions	Factory Outlet	Factory Outlet
	Café and Restaurant	Café and Restaurant
	Hotel	Hotel
	Offices	Offices
	Residential Home	Government Offices
Activities	Trading	Trading
	Offices	Offices
	Car Free Day	Social interactions (open space in the garden)

3.1. *Responders Characteristic*

Affect thermal comfort consists of two variables, the respondent's activity representing the metabolism value of the body and the clothing resistance that represents the respondents' adaptation index. At the time of the process of data collection, the most common activity of the community in both street corridors is sport activity. Data collection conducted at Dago Street coincides with the weekly event of Bandung city that is car free day so that dominating activities include the activities of sports streets, cycling, gymnastics and running. The type of activities that people do on Riau Street is slightly different from Dago Street because on Riau Street there is no weekly car free day event, but people also do sports activities such as walking casual, running, waiting and shopping.

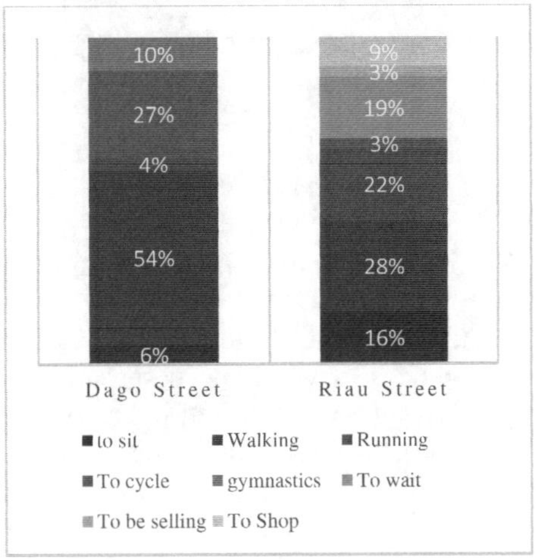

Figure 1. Respondents Acitivities

The clothing used by the respondents is also a variable that affects the thermal comfort preferences. The standard of clothing used in this research has a value of 0 clo to 3.0 clo which is where 0 clo means humans in the condition do not use clothes while 3.0 clo are in human condition wearing winter clothes. On Dago Street most of the respondents wear clothes with indexes of 1.5 are jackets or jackets and trousers, while for Riau street most respondents wear clothes with 0.5 clo indexes that are short sleeve and trousers.

Table 2. Comparison of clothing types

Comparisons	Dago Street	Riau Street
Not wearing clothes (0 clo)	0%	0%
Short shirts and Shorts (0.3 clo)	28%	29%
Short Shirts and Trousers (0.5 clo)	28%	38%
Kong sleeves and trousers (1.0 clo)	6%	33%
Coat and jacket and Trousers (1.5 clo)	38%	0%
Winter Clothes (3.0clo)	0%	0%

3.2. Thermal comfort preference

In this research, the measurement of thermal comfort preferences is divided into three variables: humidity, wind sensation and solar radiation. Preference about humidity sensation in this research consist of 3 scale that is Dry, Ok and Humid. Overall the humidity felt by the respondents between the two street corridors is relatively the same. In Dago Street 69% of respondents felt that the humidity in the street corridor was good enough, as did respondents in Riau Street 66% of respondents also felt that the humidity in Riau Street was in good condition.

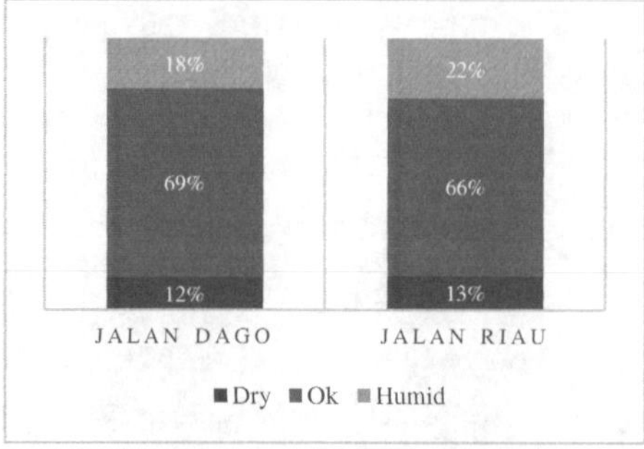

Figure 2. Humidity Condition Preference

The differences in street orientation in Dago Street and Riau Street affect the wind conditions in each of street condition. The preferences of wind speed sensation consists of 4 scales ie windy, Ok, Little Wind and still water, respondents who are located in Dago street, it is feel that wind speed in the street corridor is still less (little wind), different from Respondents who are in Riau street feel That wind speed is enough (windy). In addition to street orientation, the movement of the wind is also influenced by passing vehicles. Data retrieval located on Dago Street is done at the time of car free day so that vehicles passing in the corridor is only limited to the bike only so that the movement of the wind is not so great. On the other hand, on Jalan Riau, many vehicles cross the corridor causing high wind movement.

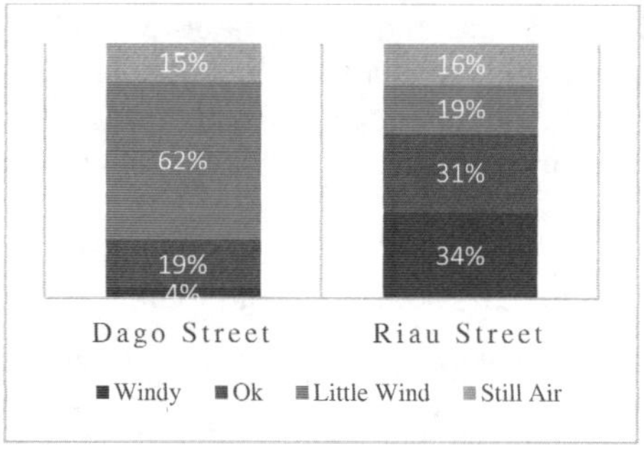

Figure 3. Wind Speed Sensation Preference

Solar radiation is the response of respondents related how the sun exposure in each corridor to the comfort felt by the respondent during the corridor. Sun exposure is influenced by tree or canopy imagery in each corridor. Dago Street Corridor has a shade of trees that are more shady compared to Riau Street so that respondents who are on Dago Street mostly not directly exposed to sunlight. There are 5 scales used to measure the sun exposure of Too Strong, Little Strong, Ok, Little Weak and Too Weak.

IOP Conf. Series: Earth and Environmental Science **152** (2018) 012010 doi:10.1088/1755-1315/152/1/012010

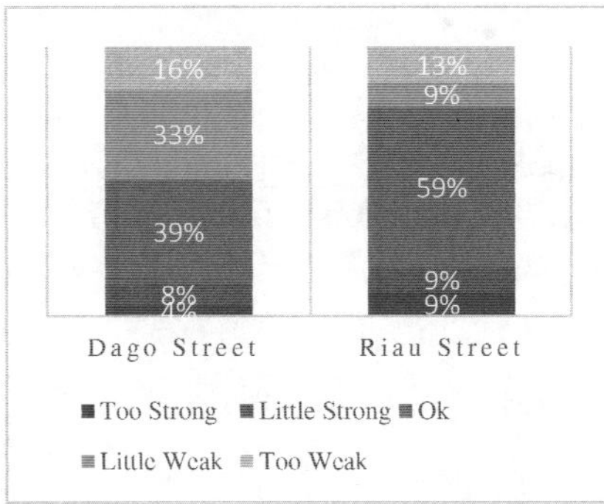

Figure 4. Solar Radiation Preference

From the condition of several variables affecting thermal comfort obtained comparison of respondent's preference to thermal conditions in each street corridor. To measure the thermal sensation there are 7 comfort scales including cold (-3), cool (-2), slightly cool (-1), neutral (0), slightly warm (+1), warm (+ 2) and hot (+3). The majority of respondents in Dago and Riau Street feels comfortable thermal sensation (neutral) with percentage of 65% and 44%. From the percentage, it is found that Dago Street has a more comfortable thermal sensation compared to Riau Street, this is because respondents in Dago Street are more comfortable compared to Riau Street.

Table 3. Scales comparison of ASHRAE

Scale	Jalan Dago	Jalan Riau
Cold (-3)	0%	0%
Cool (-2)	2%	22%
Slightly Cool	20%	22%
Neutral	65%	44%
Slightly Warm	12%	6%
Warm	0%	6%
Hot	0%	0%

Thermal comfort measured affects thermal acceptability felt by respondents in each street corridor. In the street of Dago 98% of respondents felt that the thermal comfort level was ideal, and as many as 50% of respondents wanted no change to the current thermal conditions. While for Riau Street, 78% of respondents felt that thermal comfort in the street corridor was ideal, and 69% of respondents wanted no change to the current thermal condition.

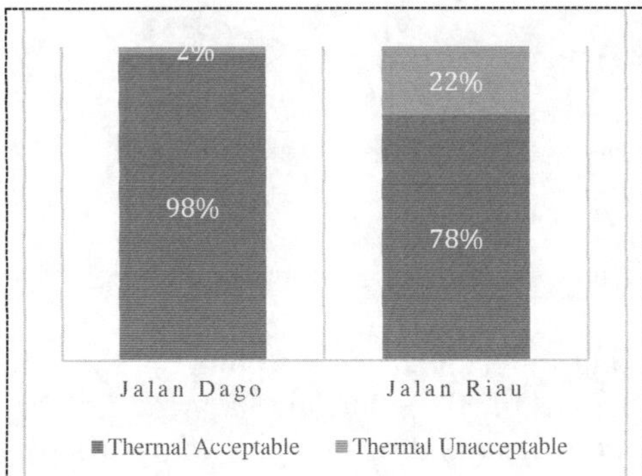

Figure 5. Thermal Acceptability

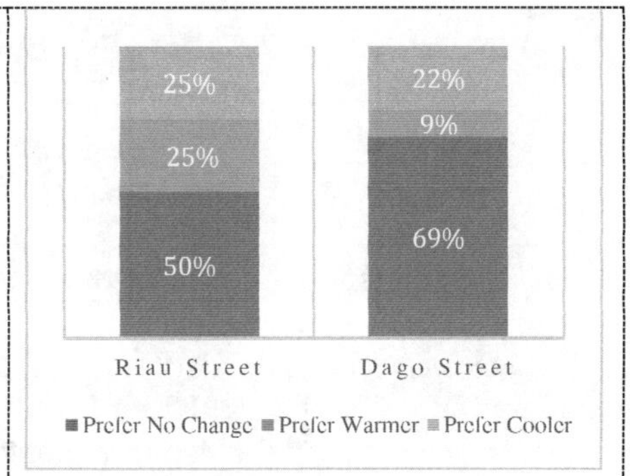

Figure 6. Thermal Preference

IOP Conf. Series: Earth and Environmental Science **152** (2018) 012010 doi:10.1088/1755-1315/152/1/012010

3.3. Processing data of tool measurement

The data obtained from the three tools are TR74Ui Data Logger, Digital Anemometer, and WBGT measuring instruments put together in the table to simplify the PET search process. The PMV and PET search process included four Air Temperature (Ta), Relative Humidity (Rh%), Heat stress (TG), and wind velocity (m / s) data.

Table 4. Data on the Dago Street

Time	TA	RH	m/s	TG	Tmrt
8:00	24.1	72.7	0	29.2	25.1
8:05	24.8	71.8	0.5	31.5	24.8
8:10	23.8	71	0.3	30.1	25.3
8:15	25.85	70.9	0.9	36.1	25.4
8:17	25.1	70	0.9	34.6	25.6
8:20	25.5	72	0	29.5	25.3
8:22	25.6	71	0	29.4	25.8
8:25	25.9	70	0.3	27.8	25.7
8:30	25.5	71	0.2	23.8	26.2
8:35	26.9	68	0.1	23.2	26.6
8:40	24.8	72.5	0.8	32.5	26.5
8:45	27.2	67	1.2	27.9	26.9
8:50	28.7	62	0.3	28.1	25.7
8:55	30.3	57	0	27.637	25.7
9:00	33.4	47	0.3	22.9	25.8

Table 5. Data on Riau Street

Time	TA	RH	m/s	TG	Tmrt
8.07	24.1	75.4	0,77	28.2	34.1
8.08	23.9	75.1	4,09	27.7	42.1
8.14	23.7	74.7	3,50	27.7	38.8
8.17	23.6	74.5	1,16	28.3	35.3
8.18	24.05	75.3	1,75	28.83	38.6
8.26	23.2	73.9	1,76	28.1	36.7
8.28	22.1	71.8	1,75	28.5	36.8
8.31	22.1	71.8	3,11	28.1	38.4
8.35	21.6	71	0,97	28.9	34.7
8.37	21.05	69.9	4,09	29.8	45.3
8.42	20.7	69.3	1,94	29.4	38.7
8.45	20	68.1	1,95	29.6	38.3
8.52	21.05	69.9	4,29	28.6	43.3
8.55	21.7	71.1	1,39	27.7	33.2
8.59	22	71.6	1,40	29.5	39.2

After the data is collected, the data is processed using Rayman software to find out the value of the thermal comfort index such as PET and PMV. PET Defined as the air temperature required to reproduce in a space with the standard production of body heat and surface heat of a particular human skin (Matzarakis, 2008). The PMV is a thermal comfort index introduced by Professor Fanger of the University of Denmark in 1982 and is standard ISO 7730. The equation uses a steady-state heat balance for the human body and links between the deviation and the minimum load on the mechanism.

The results obtained from the data processing using Rayman, the following is :

IOP Conf. Series: Earth and Environmental Science **152** (2018) 012010 doi:10.1088/1755-1315/152/1/012010

Table 6. PMV and PET results on the Dago Street

Time	PMV	PET
8:00	1.3	28.4
8:05	1.2	26.9
8:10	1.1	26.3
8:15	1.5	28.7
8:17	1.3	27.4
8:20	1.5	29.1
8:22	1.5	29
8:25	1.2	26.6
8:30	0.9	24.8
8:35	1.1	25.6
8:40	1.1	26.5
8:45	1.2	26
8:50	1.6	28.4
8:55	1.8	29.6
9:00	1.8	28.8
Average	**1.3**	**27.5**

Table 7. PMV and PET results on the Riau Street

Time	PMV	PET
8.07	1.3	27.6
8.08	1.1	25.2
8.14	1.1	25.4
8.17	1.3	27.4
8.18	1.4	27.9
8.26	1.2	26.8
8.28	1.3	27.3
8.31	1.2	26.1
8.35	1.5	28.2
8.37	1.6	28.4
8.42	1.5	28.4
8.45	1.6	28.7
8.52	1.3	26.5
8.55	1.2	26.4
8.59	1.5	28.9
Average	**1.6**	**28.6**

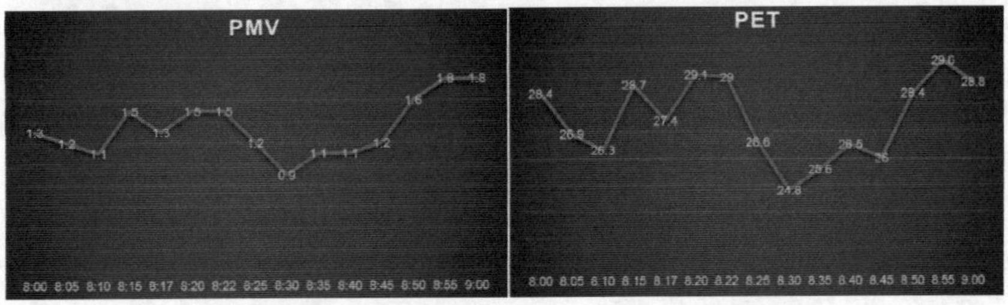

Figure 7. PMV and PET of Dago Street

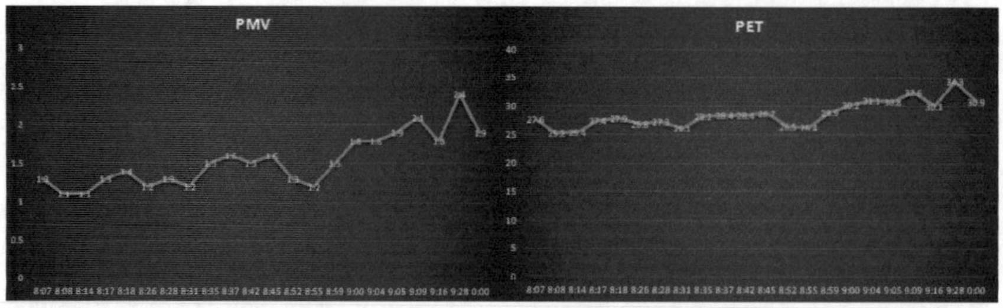

Figure 8. PMV And PET Riau Street

The averages PET on the Street of Dago corridor is 27.5 and the PMV is 1.3. This indicates that the thermal conditions of the Street Dago corridor are slightly above the thermal comfort standard index in hot-humid hotspots. While on the street of Riau found the average PET of 28.6 with a PMV of 1.6.

4. Conclusion

In generally, thermal comfort index values on both Dago Street and Riau Street are above the thermal comfort standard index for hot-humid climate countries. Some of the perceived preferences and sensations associated with the thermal conditions are influenced by the physical activity of the users of both street corridors. The difference in physical activity can cause differences that are influenced by clothing factors used during the move. On the Dago Street, thermal conditions are considered comfortable even though people do activities that are quite heavy such as running, gymnastics, and other sports. On the Riau street, thermal conditions are also considered to be comfortable even though the value of PET on Riau Street is higher than. This is due to the different activities that occur between Dago Street and Riau Street. The type of activity performed on Riau Street is lighter when compared to Dago Street.

Based on interviews, thermal comforts on Dago Street and Riau Street are considered to be sufficient to meet the comfort for the people who use both street corridors. With thermal comfort, the community can move outdoors optimally.

References

[1] Mulyandari, H. 2011. Pengantar Arsitektur Kota. Penerbit ANDI. Yogyakarta.Another reference
[2] de Dear, R.J., dan Brager, G.S. 1998. Developing An Adaptive Model Of Thermal Comfort And Preference. Macquarie University & University Of California Journal. California
[3] Matzarakis, A., Rutz, F. & Mayer, H. (2007). Modelling radiaton fluxes in simple and complex environments applicaton of the RayMan model. Internatonal Journal of Biometeorology, 51(4), 323-334
[4] Humphreys, M.A. 1976. Field Studies Of Thermal Comfort Compared And Applied. Oxford University. Oxford.
[5] Höppe, P. (1999) The physiological equivalent temperature – a universal index for the biometeoro- logical assessment of the thermal environment. Int J Biometeorol 43, 71–75

Wind Movement Comparison Between Student Dormitory 2 and 3 ITERA and The Correlation Toward its Indoor Thermal Comfort

Rendy Perdana Khidmat[1] **, M.Donny Koerniawan**[2] **, Suhendri**[2]

[1]Department of Architecture, Sumatera Institute of Technology, Lampung, Indonesia
[2]Department of Architecture, Bandung Institute of Technology, Ganesha street no 10, Bandung, Indonesia

Abstract. Student dormitory is a semi-private building that designated to occupies large number of habitats. This type of building mostly designated in simple type of vertical housing. In the context of utilization, dormitory surely requires indoor thermal comfort yet in the same way it requires the energy efficiency as well. Building in a tropical climate country is expected to be adequate to adopt a potention from its surrounding in order to switch air conditioner and gain efficiency in energy consume. One of its key factors is wind. This paper tries to describe and investigate wind movement that works on two different type of student dormitory in Sumatera Institute of Technology. The distinct difference between two blocks is one of the tower block utilizes void meanwhile the other are not. This research is conducted by using Computational Fluid Dynamic (CFD) based software. This study is expected to provide an overview of the wind movement and its effect on air temperature and its correlation to the indoor thermal comfort in both buildings.

1. Introduction

In recent years, the field research of thermal comfort has attracted the attention of many researchers around the world[1].Student dormitories or other vertical flat always requires the thermal comfort of the inhabitants. Some of these are buildings constructed by the government, many of these buildings do not use air conditioning and only utilize existing local micro climate around the site. Some standard government building designs typically adopt a local climate approach with natural ventilated building to conserve energy.

The thermal comfort in the building depends on the local microclimate. To start a plan in designing buildings, it is necessary to examine the detailed local climatic conditions especially for humid tropics [2], some of these factors are sun, wind, rainfall, and humidity.

In 2016 and 2017 Sumatera Institute of Technology (ITERA) has a grant for three dormitories from the government. The buildings designed with two different type. The differences in these two buildings is the utilization of void in one of the buildings. This paper will simulate the movement of wind by using Autodesk CFD software. It aims to see whether there is a difference in the wind movement will affective thermal comfort in both buildings.

The metodology conducted on this research is quantitative and exploration through simulation [3].

2. Computational Fluid Dynamic (CFD)

CFDs are branches of fluid mechanics. Initially CFDs were used for the aerospace industry and turbo engines. But today many CFDs are also used for wind driven rain, pedestrian wind comfort, surface heat transfer and urban air pollution [4].

CFD (Computational Fluid Dynamics) is a method that can be used for simulation of the indoor climate in buildings. With this technique the climate inside the whole building can be illustrated with a detailed picture of possible problems that may occur [5].
This research tries to examine the wind-flow around the two dormitories using Autodesk CFD.

3. Thermal Comfort

As it is proven that buildings account for 40% of the global energy consumption and contribute over 30% of the CO2 emission and the large proportion of this energy is used for thermal comfort [6], the design consideration is shifted to maximizing the local microclimate potential to reaching thermal comfort for the occupants.

According to ANSI/ASHRAE Standard 55-2010, thermal comfort is defined as "that condition of mind which expresses satisfaction with the thermal environment and is assessed by subjective evaluation." Also known as human comfort, thermal comfort is the occupants' satisfaction with the surrounding thermal conditions and is essential to consider when designing a structure that will be occupied by people.

There is some factors that can be considered related to the human comfort. The factors are include:

- Metabolic rate (met): The energy generated from the human body
- Clothing insulation: the amount of thermal insulation the person is wearing
- Air temperature: temperature of the air surrounding the occupants
- Radiant temperature: The weighted average of all the temperatures from surfaces surrounding an occupant
- Air velocity: Rate of air movement given distance over time
- Relative humidity: Percentage of water vapor in the air

Human thermal comfort is measured by the Predicted Mean Vote (PMV). PMV is presented in sensation scale between -3 which represent cold to +3 which represent hot. When the recommended scale that can be called as comfort is a range between -0.5 to +0.5.

4. Project Description

ITERA student dormitories number 2 and 3 were built in 2016 and 2017. Each building consists of 5 floors and 92 rooms. Dormitory 2 has 60 meters long and 19.2 meters wide. This building has a void surrounded by a 1.4 meter wide corridor. It has three main accesses the ladder in the center and two emergencies stairs on each side. On both sides of this building there is a void with areas of 106 m2. In this study the sample taken is typical floors 2 to 4. Where the ventilation contained in each room is a window on both sides that faces the outside of the building and toward the corridor. There is also a public pantry on each floor.

Figure 1. Dormitory 2

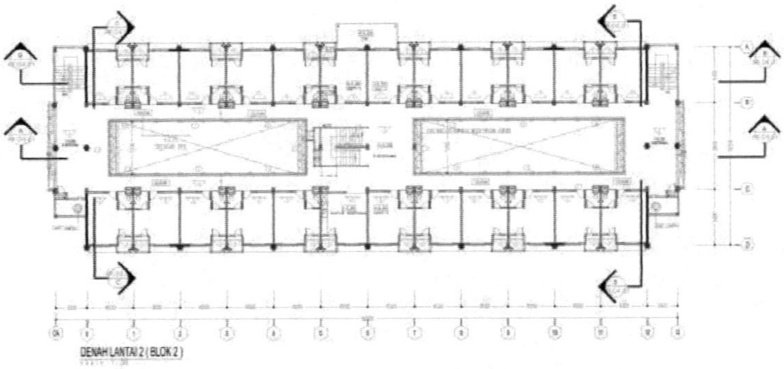

Figure 2. Dormitory 2 floor plan

While the dormitory 3, built 2017 has no void in the center of the building. The building has a length of 64 meter and 12.75 meter wide. On each floor there are typically 22 rooms that are not equipped with a balcony. Room with area of 17.3 meter square has windows overlooking the corridor only. There is also a public sunroom on each floor. The corridor in this building has a width of 1.8 m.

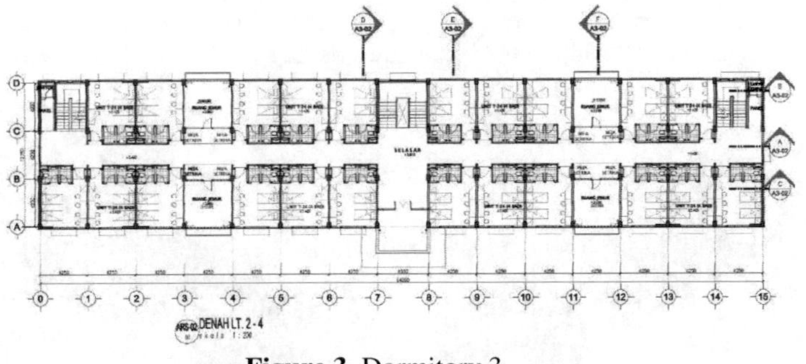

Figure 3. Dormitory 3

Figure 4. Dormitory 3 floor plan

This building has three main accesses, the ladder in the middle and two emergency stairs on each side. Both dormitories built with regular materials such as finished brick wall, aluminium in windows and door frames and ceramic tiles for flooring and gypsum board for its ceiling.

5. Bandar Lampung Climate

Bandar Lampung city is classified in zone that has a humid climate all over the year. Rainfall ranges between 2.257 - 2.454 mm/year. With air humidity ranged from 60% - 85% and average temperature ranges from 23 ° C -37 ° C. Wind velocity ranged from 2.78 - 3.80 knots with dominant direction from the west in November, from the north in March, from the east in June and from the south in September. And it located in latitude 700 mdpl.

6. Simulation

The simulation is conducted by using Autodesk Revit architecture software for modelling and Autodesk CFD to simulate wind movement that works on both buildings. The climatic data included for the analysis recipes in this simulation is microclimate data of Bandar Lampung city and the physical properties of the materials used in the two dormitory as well.

Firstly the geometry is constructed in revit acording to actual buildings yet with less components. The used components is limited to the floor, ceiling, walls and the openings including the void in dormitory 2. This aims to get the shorter calculation process in simulation.

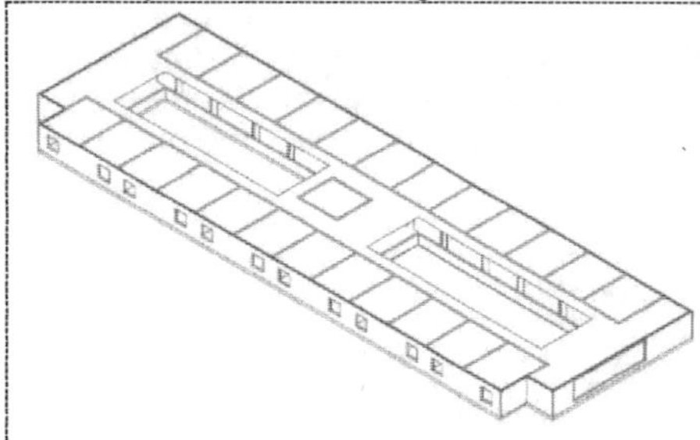

Figure 5. Dormitory 2 geometry

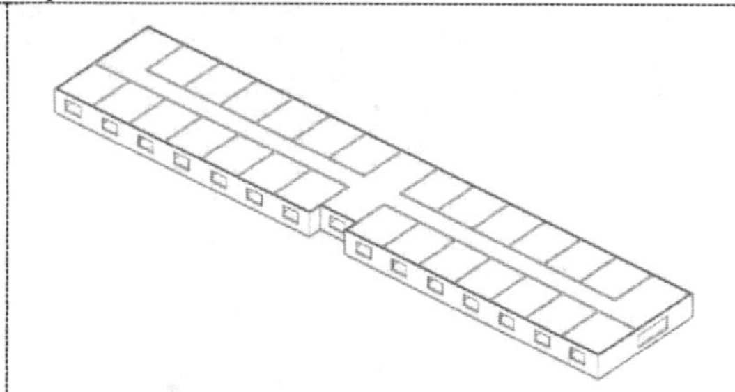

Figure 6. Dormitory 3 geometry

After the geometry is constructed, the next step is setting materials. The materials used in this simulation is regular materials that set to be similar for both buildings. Granite tiles is used for flooring and concrete is used for the walls. The ceiling again use the congcrete as materials assuming to be an open exposed construction. Besid the phsycal materials, the void on both building is filled by fluid air materials with fixed materials properties.

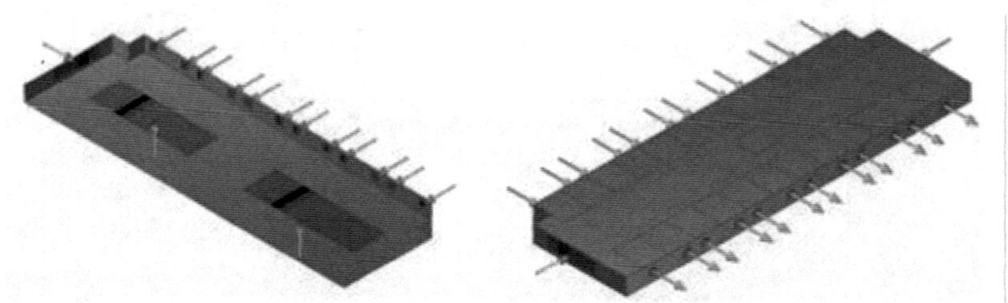

(a) (b)

Figure 7. Dormitory 2 (a) inlet and (b) outlet

The next step is to set the inlet and outlet of the wind flow. both building is set to get the force in same direction thatt is from the south west direction with velocity of 5 cm/s and temperature 27 ° C. And the outlet is located in the opposite direction, north east. The outlet is set to have a 0 Pa pressure.

IOP Conf. Series: Earth and Environmental Science **152** (2018) 012011 doi:10.1088/1755-1315/152/1/012011

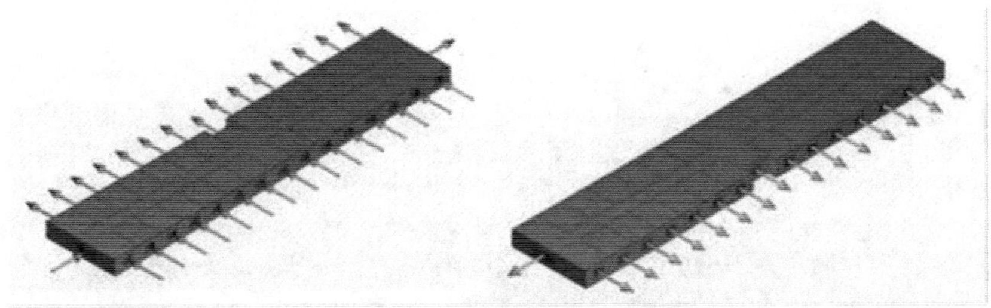

(a) (b)

Figure 8. Dormitory 3 (a) inlet and (b) outlet

The simulation running in different iteration. The simulation for dormitory 2 is stoped in itetation 500 while the dormitory 3 is stoped in iteration 371.

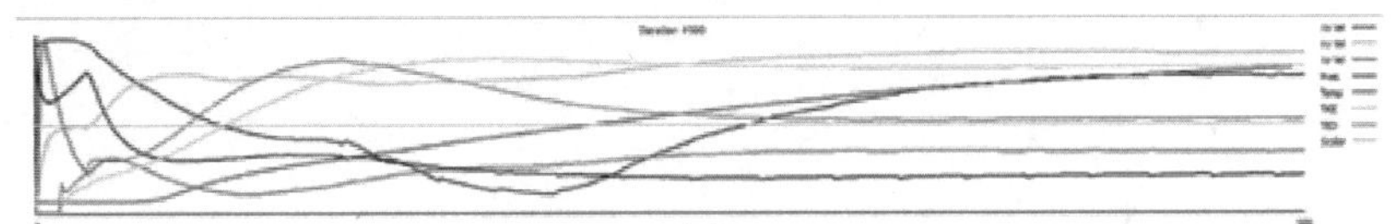

(a)

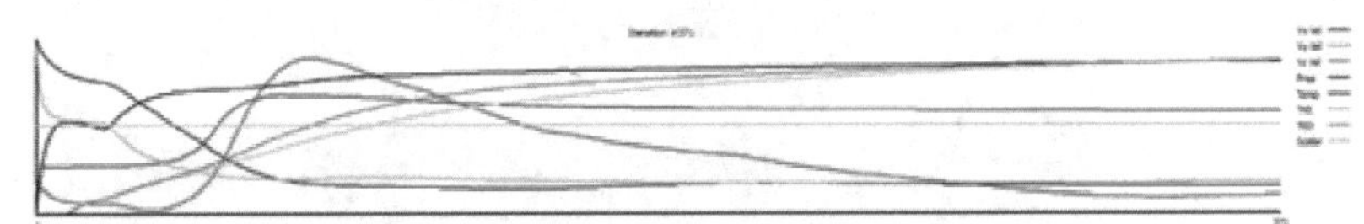

(b)

Figure 9. (a) (b) Iteration grafic of the two dormitories

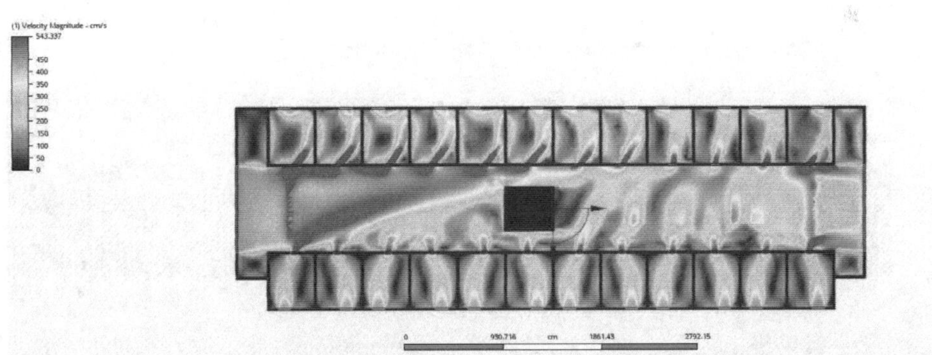

Figure 10. Wind velocity in dormitory 2

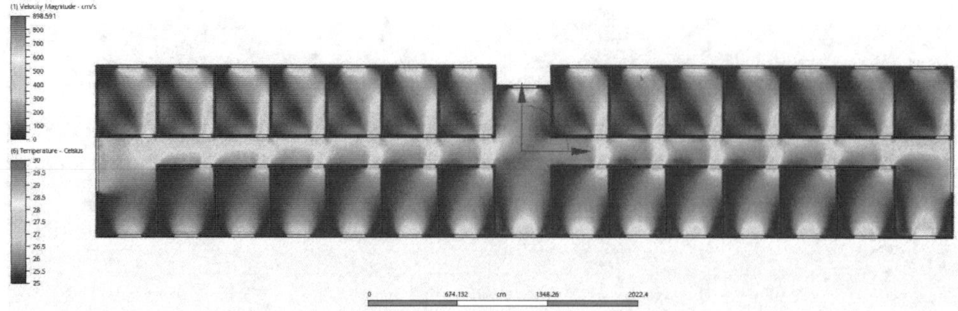

Figure 11. Wind velocity in dormitory 3

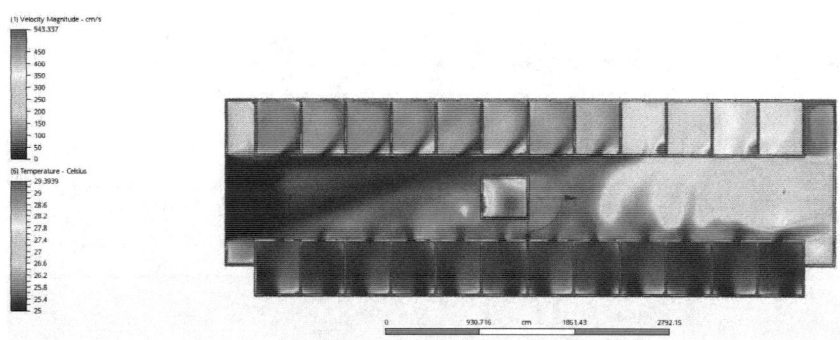

Figure 12. Temperature in dormitory 2

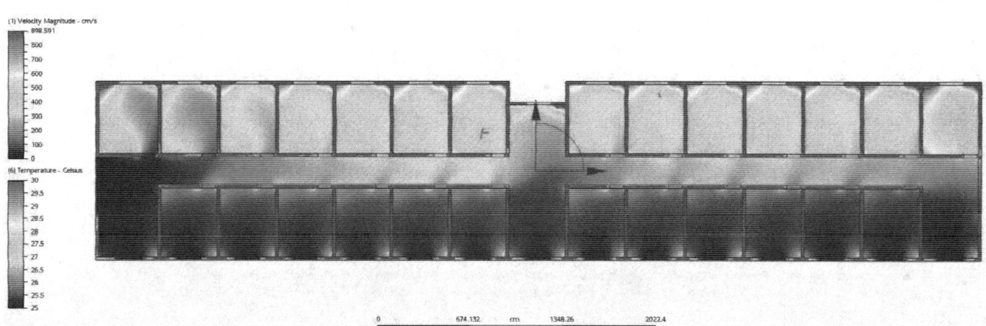

Figure 13. Temperature in dormitory 3

The results shows that there is two different wind flos works on the two dormitories. In the dormitory 2, wind flow is locally accumulated around the void. The wind and temperature are not well-distributed to the student rooms. And the wind entering through the coridor opening is affected bu the force that coming from the void. The result can be investigated in picture 10 and 11.

While the temperature that affected by the wind flow is works identicaly with the wind distribution. Here we can see that in dormitory 2 there is very hot area in the oposite side of the inlet while in the dormitory 3 the temperature in the outlet side seems tend to be reasonable. The distribution can be investigated through picture 12 and 13.

Picture 14 and 15 shows that the PMV distribution on the map is affected by the temperature as well. From the two picture below we can see that in dormitory 2 the distribution is concentrated in the void while in the dormitory 3 the better condition is spread around the coridor.

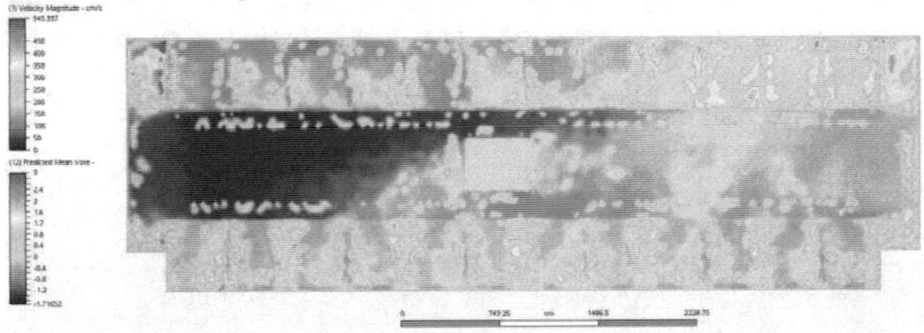

Figure 14. PMV in dormitory 2

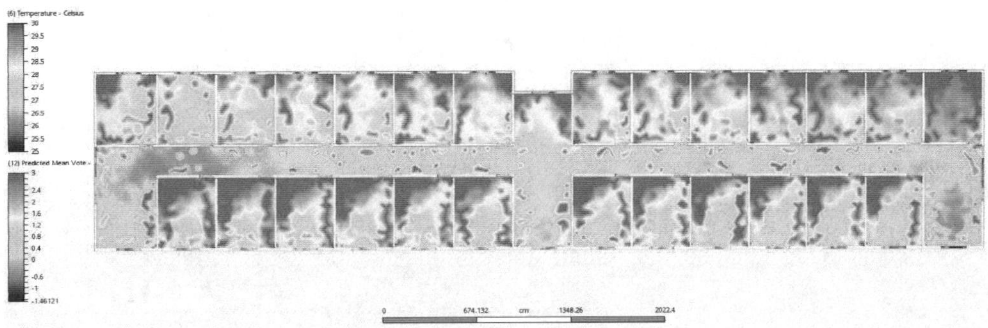

Figure 15. PMV in dormitory 3

7. Conclusions

This paper might not shows the real world condition yet it investigates the results through the simulation. This paper tries to compare the wind flow works in the two buildings. The different is lies on the used of void. From the results we can se that the wind flow works in dormitory 3 is shows better distribution compared to the dormitory 2. It can be conclude that the void does not fully affects the expected better wind distribution. The PMV also distributed according to the temperature carries bu the wind. The dormitory 3 shows the prefarable reason in PMV distribution rather than the dormitory 2.

8. Acknowledgments

This research is part of the "Penelitian Hibah Mandiri" funded by the Institute Technology of Sumatera and dedicated to the Architecture Department of Institute Technology of Sumatera.

9. References

[1] Rupp.R.F., Vasquez.N.G., Lamberts.R., 2015. *A review of Human thermal comfort in the built environment.* A field Study. Energy and Buildings, volume 105, p. 178-205

[2] Lippsmeier 1980. Tropenbau Building in the Tropics. Translated by Syahmir Nasution. Erlangga. Jakarta.

[3] Suhendri., Koerniawan.D, 2016. Investigasi Ventilasi Gaya Angin Rumah Tradisional Indonesia

dengan Simulasi CFD. Prosiding Temu Ilmiah IPLBI.
[4] Kaijima.S, Bouffanas. R, Willcox.K., 2013. *Computational Fluid Dynamics for Architectural Design. Proceedings of the 18th international conference of the assocoation of Computer-Aided Architectural Design Research in Asia CAADRIA 2013. CAADRIA.* Hong kong.
[5] Risberg.D. 2015. *CFD Simulation of Indoor Climate in Low Energy Buildings*. A Thesis. P. 1
[6] Yang. L., Yan. H., Lam. J., 2014 *Thermal comfort and building energy consumption implication-A review*. A field study. Applied energy, volume 115, p. 164-173.

Integrating Green Building Criteria Into Housing Design Processes Case Study: Tropical Apartment At Kebon Melati, Jakarta

V L Farid[1] and S Wonorahardjo[2]
[1] School of Architecture, Planning, and Policy Development, Bandung Institute Technology Bandung, Bandung, Indonesia
[2] School of Architecture, Planning, and Policy Development, Bandung Institute Technology Bandung, Bandung, Indonesia

E-mail: vanialarissafarid@gmail.com

Abstract. The implementation of Green Building criteria is relatively new in architectural practice, especially in Indonesia. Consequently, the integration of these criteria into design process has the potential to change the design process itself. The implementation of the green building criteria into the conventional design process will be discussed in this paper. The concept of this project is to design a residential unit with a natural air-conditioning system. To achieve this purpose, the Green Building criteria has been implemented since the beginning of the design process until the detailing process on the end of the project. Several studies was performed throughout the design process, such as: (1) Conceptual review, where several professionally proved theories related to Tropical Architecture and passive design are used for a reference, and (2) Computer simulations, such as Computational Fluid Dynamics (CFD) and wind tunnel simulation, used to represent the dynamic response of the surrounding environment towards the building. Hopefully this paper may become a reference for designing a green residential building.

1. Introduction

High-density residential buildings consume a huge amount of energy. The major cause of this phenomenon is the building's inability to adapt with the local climate. This buildings has a thermally uncomfortable rooms, thus force the tenants to use artificial air conditioning system.

To reduce and minimalize the use of energy on a high-rise residential building, a prompt design that can create thermally comfortable buildings is needed. To achieve this design, green building criteria needs to be included since the beginning of the design. To optimize the building's design, a guidance tools are needed for a better understanding of the building's actual condition. The guidance tools in this case study are simulator software that can simulate the thermal condition of a buildings and the airflow around the building. Several theories made by experienced professional who already applied in built projects are also used as a guidance tools.

This paper will describe the process of implementing green building criteria in a traditional design process.

2. The design process on tropical apartment project.

The case study is located at Kebon Melati, Central Jakarta. Jakarta is located on wet tropical area which has a constant high intensity of sun radiation and has moderate rainfall intensity, except for the winter months, when it becomes very heavy due to monsoon season. Most importantly, Jakarta has an average high temperature (28°~31° C). Consequently, the usage of artificial air conditioner becomes common. Through this design process, several green building criteria, based on tropical architecture principle, will be used to design a residential building which can minimalize air conditioner usage.

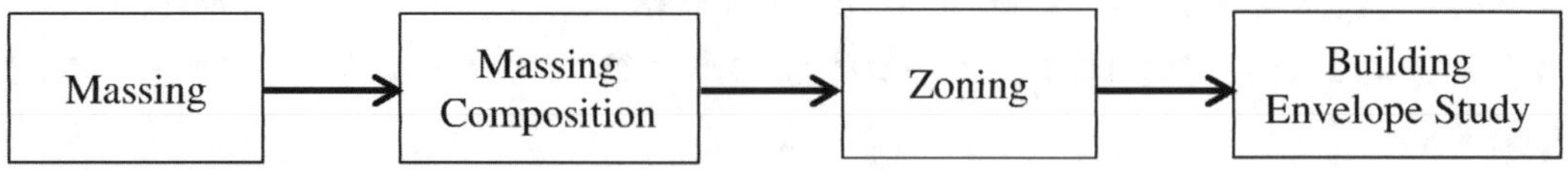

Figure 1. Traditional design process.

On this case study, the green building criteria will be integrated with traditional design process in order to change conventional process into a more advanced process to developed integrated green building. The traditional design process can be explained in below diagram:

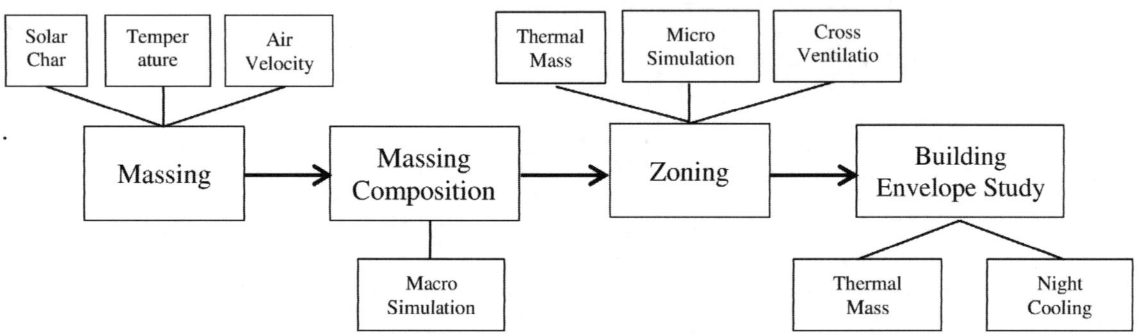

Figure 2. Traditional design process with green building criteria.

The green building criteria implemented on every stages of the traditional design process. These criteria enter as a requirement that must be fulfilled by the design itself. These criteria also cannot be detached from each other. It means that without fulfilling all the criteria, the green building system cannot be optimal.

On the massing stage, the green building criteria requires the design to adapt with existing physical situation of the site, such as solar chart, air temperature, and airflow. Following the initial design of the massing, a macro simulation, using Autodesk Flow Design software, is performed for a major comprehension of air circulation surrounding the massing. The results of this simulation will become a feedback in order to achieve the optimum design. These feedbacks will determine the ideal distance between the mass, and the massing composition.

Having the feedbacks implemented on the massing, the layout of the residential units inside is developed. Subsequently, these layouts will be taken to a micro simulation process to determine the horizontal and vertical zoning of the unit's layout to design the cross ventilation system inside the unit. In this stage, Thermal Mass theory is starting to be implemented on the unit's layout.

The last stage, facade study, is where the theories of Thermal Mass and Night Cooling will be implemented. The implementations of these theories are an important stage in this design process. This stage will show the result of green building criteria intervention towards a traditional design process.

3. Massing Stage

Primary and secondary data regarding the site and building's typology were collected during the early stage of massing process. These data contained government regulations, building codes, standards, precedent, and site analysis. The green building criteria at this stage help to design a mass that respond to existing site condition based on site analysis. These conditions were air temperatures, solar chart and airflow. These three conditions were the main variable within this stage. The air temperature inside the building must be lower than the outside and the airflow surrounding the building help to lower the air temperature inside the building, but still maintaining the comfortable condition for human activity.

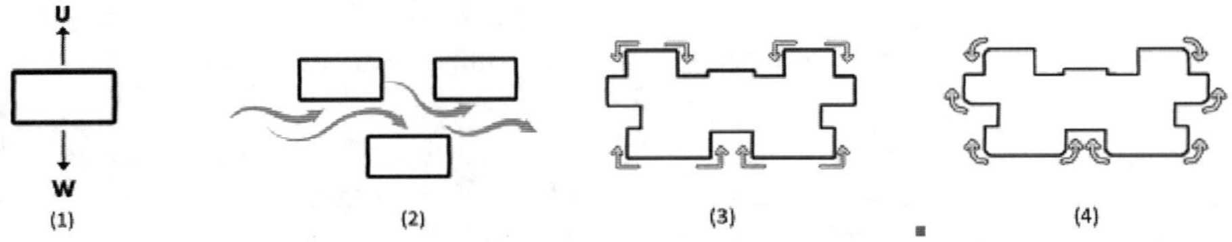

Figure 3. The initial stage on massing process.

The massing designed to minimalize solar radiation absorption through building's envelope while adapting with existing airflow around the building. According to the solar chart on Figure 4, the sun rise on east, and set on west. It indicates that the east and west side of the building's envelope will gets direct solar radiation. To avoid this situation, the building mass had to elongate from east to west to minimize the area of the east and west building's envelope.

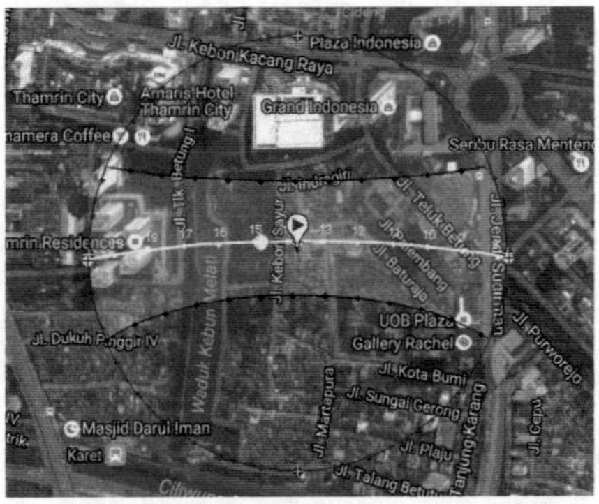

Figure 4. Solar chart on existing site at Kebon Melati.

Figure 3 describe the process massing to control the airflow around the building. Building's massing divided to three parts in order to create gaps between the massing, this gap will allow the air to move around the building, and enter the building. To control the air movement so the airflow will not be too high, the massing designed to be ridged. The massing's sides also made rounded to make the building aerodynamic, thus the air movement will be closer to the building's skin.

4. Massing Composition

The mass composition stage is an advanced process of the building's massing design stage. This stage determined the composition and the gaps between building masses, which divided into three in the previous stage. The distance between these massing will be decided based on the airflow around the building.

Airflow in an environment has a great influence to human comfort, especially on the physical environment 1.5 meters above the top of the terrain where the human activities occur. Table 1 shows the correlation between airflow and human comfort condition. Generally, a person feels comfortable with 1 ~ 5 m/s airflow [1]

Table 1. Airflow and human comfort condition.

Airflow (m/s)	Level of comfort
<1	Idle
1 ~ 5	Comfortable
5 ~ 10	Uncomfortable
10 ~ 15	Very uncomfortable

The airflow simulation ran on the initial building massing design and utilizing Autodesk Flow Design software, which offers the wind tunnel simulator tools and enabling flow analysis around the building mass. Below is the result of the simulation process:

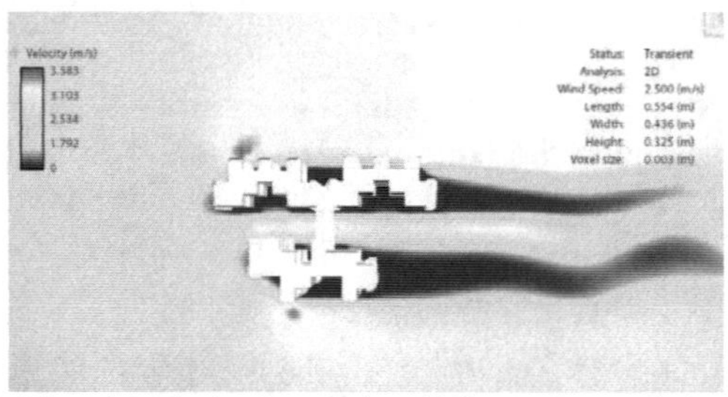

Figure 5. The simulation's result: airflows around the massing.

The result shows that the movement on the building's surface has deep blue color, which indicates the airflow there is below 1 m/s. This deep blue color mainly found on the indented side of the massing. This condition made the deep blue area has a poor air quality as the air scarcely has any movement there. Furthermore, to induce the surface cooling system, the airflow around the building's skin should be quite high. The yellow color on the massing gap indicates an area with the air flow around 3.1 m/s . This area will be connected with a sky bridge and will be used as an amenities area. Even though based on Table 1, the air flow rate 3.1 m/s still can be considered a comfortable condition, but to utilize this area for human activities, there must be an adjustment on the design.

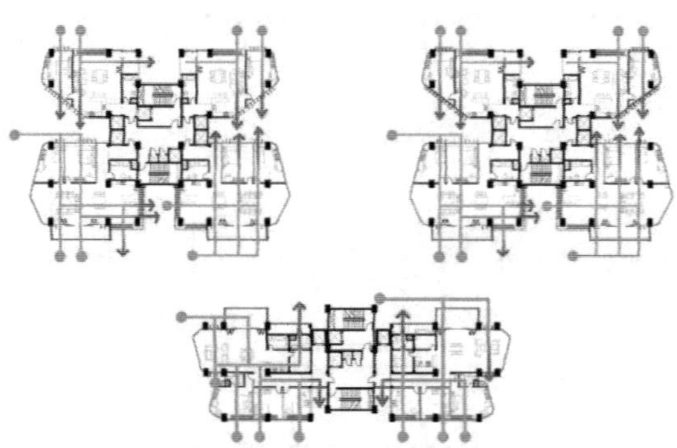

Figure 6. The revised design of the building massing.

These results were used as feedbacks for the design, and several adjustments were made based on these feedbacks. First, the initial massing were made to be less ridged and minimizing the cramped area which has a poor airflow rate. To lower the airflow rate in between the massing, the building surfaces that facing each massing were made to be more ridged and the habitable zone was located on the indented part of the massing. The distance between the massing was considered to be sufficient as the air movement can be detected on that area. Figure 6 is the revised building massing after adjustments based on the simulation feedbacks:

Figure 6 present the revised massing design based on the feedbacks from the airflow simulation. The outer surface of the massing, marked by red-lined square, is made of smooth surface without any indented part. This smooth building surface is expected to make the air flowing as near as possible to the building skin, thus can help cooling the building interior. This phenomenon is called surface cooling system.

Furthermore, the gap part of the whole building massing is made to be rougher, and has several indented part on the surface to slow down the airflow. By doing this, the gap part will have more comfortable airflow rate hence can create a comfortable condition for human activities. Balcony and openings are also located throughout this area. As a result, the air that will enter the residential units is expected to be a comfortable, breezy, fresh air with low airflow rate.

5. Zoning

A vertical zoning for the building massing is designed to equip the horizontal mass which has been done in the previous stage. Figure 7 shows the vertical zoning of the building mass.

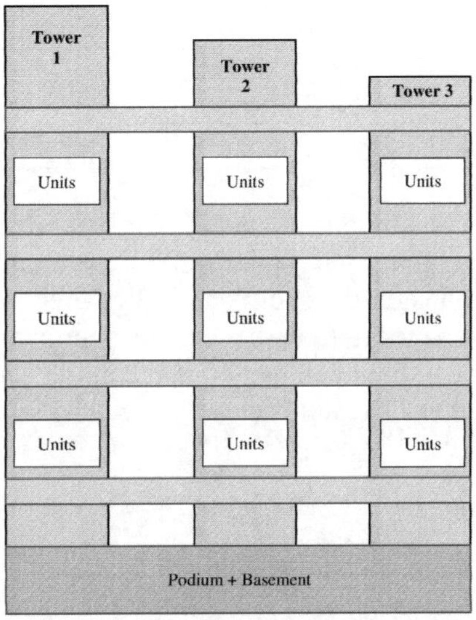

Figure 7. The building massing's vertical zoning.

The three massing of the tower will be connected by sky bridges, which function as a bridge between each tower, and an amenities area for the units above. Furthermore, this area is also equipped with a refuge area as a shelter for future emergency situation.

Following the completion of the vertical and horizontal zoning for the building massing, the interior of the residential area is developed through a horizontal zoning. This zoning was made on the schematic floor plan of the residential units. The important part of the zoning stage is to determine the concept of the cross-ventilation system. On the schematic floor plan, several openings is located on the opposite to each other to follow the passive design strategy. Having done that, the micro simulation for further understanding of the cross-ventilation design is done. The simulation is conducted using a software called Autodesk Computational Fluid Dynamics (CFD) which is a simulator to visualize how a liquid affects objects as it flows past.

This simulation will determine the openings placement, as well as to decide the location of each room. Prior to that, when the room schedule is arranged, it is distinguished by its needs to be equipped with air conditioner or not. The rooms which need to be air conditioned are the bedrooms. Formatting author affiliations

Please ensure that affiliations are as full and complete as possible and include the country. The addresses of the authors' affiliations follow the list of authors and should also be indented 25 mm to

Table 2. Room schedule of three bedrooms unit.

3 Bedrooms Unit	Area (sqm)
Master Bedroom	24
Bedroom 1	14
Bedroom 2	14
Dry Kitchen	9
Dining Room	20
Living Room	28
Master Bathroom	8
Bathroom 1	6
Balcony	9
Maid Room	8
Maid Bathroom	3
Total Area	143

The rooms that are highlighted with yellow are the one equipped with air conditioner. The rooms that are not equipped with air conditioner will be located on the area where fresh air can freely flow to the rooms so that the temperature inside the room can be lowered. Whereas the rooms which will be equipped with air conditioner will be located on the area with minimum airflow.

Prior to micro simulation process, the openings on the units are determined and is distinguished by the inlet and the outlet. The simulation's result is shown below:

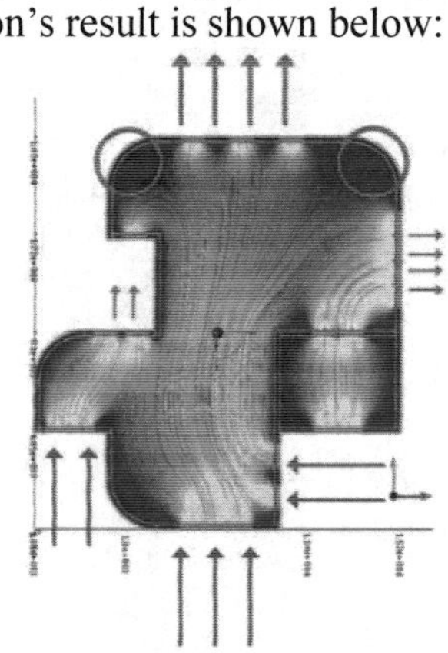

Figure 8. The micro simulation results using CFD simulator.

Figure 8 indicates the schematic floor plan with the inlet and outlet openings plan. The colour spectrum on the figure 8 indicates the airflow rate, and the curved line is showing the airflow according to the designed inlet and outlet.

The area with the deep blue colour showing that the airflow rate there is around 0 m/s. Thus, the area with the deep blue colour, which located mainly on the corner area, will be placed rooms with air conditioner system. The area coloured with lighter blue will be functionalized as rooms with cross-ventilation system since the lighter blue indicates the airflow with 2~3 m/s rate.

From the simulation result, the feedbacks for the schematic floor plan are applied to the design. Bedrooms are located on the north side of the unit, where the simulation result shows deep blue color. Living room and other less private rooms are located on the area with good airflow rate that indicated with lighter blue colour in the simulation result.

To help decrease the interior's air temperature, the thermal mass theory is applied on this design process by choosing a suitable wall material. Thermal mass is a material's ability to store and release heat energy. The principal of thermal mass can help to control the energy flow on the building to make the building more energy efficient. A material's thermal mass determine the material's ability to absorb and release heat. On this case study, which has the goal to design a residential building that minimize the usage of an air conditioner, a material which absorbs minimal amount of heat on the day and release maximal heat is the most ideal and beneficial for this case study.

A lightweight material, like aerated concrete blocks and gypsum, is a material which has a low value thermal mass, which means it has a low heat absorption capacity. Meanwhile, a heavyweight material, like brick and concrete, is a material which has a great capacity to absorb and release heat. This kind of materials performed as a thermal collector if it is facing east or west. They have relatively high heat-capacity value and automatically influence the thermal environment effectively as heat absorber or heat emitter [2]. Combining these two sorts of materials can built a system which can be utilized as a system to reduce the air temperature inside the building.

The east-west wall collects the heat from direct sunlight and release the heat to the air, while north-south wall may collect the heat indirectly from the air. Therefore, the lightweight material will be used on the east-west side building envelope.

Figure 9. Thermal mass absorp heats on the daytime (left) and release heat during nighttime [3]

Both the lightweight and heavyweight material will be used in this building. Aerated concrete block with aluminum composite panel as an exterior facade material is categorized as a lightweight material, whereas concrete and brick wall is used as a heavyweight material.

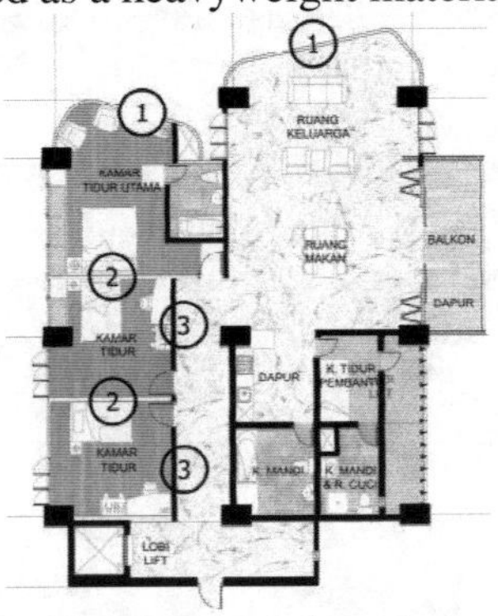

Figure 10. Floor plan of three bedrooms unit.

Figure 10 shows the location of heavyweight and lightweight walls. The solid black line shows the heavyweight walls, and the white one shows the lightweight wall. Rooms that are considered more private than the other rooms are located on the north side of the unit, where the openings are rather

IOP Conf. Series: Earth and Environmental Science **152** (2018) 012012 doi:10.1088/1755-1315/152/1/012012

small and equipped with air conditioner system. Less private room, such as living room, kitchen and dining room, is created as an open layout room which the future tenant has the freedom to adjust the layout. This also makes the rooms has a bigger space thus the air can flows freely as the room has no partition.

To expedite the airflow and maximize the cross-ventilation system on the unit, a large opening is placed as the main source of fresh air. Furthermore, this opening is already modified to adapt with weather changes. To ensure that the air can enter the unit and flows, an exhaust fan is installed as the outlet of the airflow. The scheme of the cross-ventilation system is shown below:

Figure 11. Inlet and outlet plan on the unit.

6. Study on Building Envelope Systems.

The green building criteria on the opening's design are based on the integration of thermal mass theory with a passive cooling system named night cooling system. The integration of these systems can help optimizing the passive cooling system on the unit.

Taking the advantages on thermal mass, the air temperature can be decreased by using the help from walls and floor slabs. The ideal condition to release the heat and cooling down happens on the night time. Therefore this system is called night cooling system. The basic concept of night cooling system involves cooling the building structure overnight in order to provide a heat sink during the occupancy period [4,5].

The night cooling system utilize the low temperature on the night time to cools down the walls and floor, thus on the next morning the room's temperature is lower, and can maintain the cool ambience. To establish this system, the openings and the air intake has to be designed. On the night time, the openings has to be opened maximum, thus the heat absorbed by the wall and floor materials during the day time can be released, and replaced with cool air. On the morning, the openings should be close and kept minimal openings so that the cool air absorbed the night before can be kept and the air temperature maintained to be low.

The implemented night cooling theory is followed by the modification of window's design. The night cooling and thermal mass system is a mutual and continuous system, and therefore the openings designs are adjusting with the integration of both systems.

The openings have two conditions depending on the time. On the night time, the openings can be opened maximum thus can encourage the fresh air to enter the unit and cools down the air. On the day time, the openings should be kept minimum so that the fresh air can flow inside without absorbing too much heat.

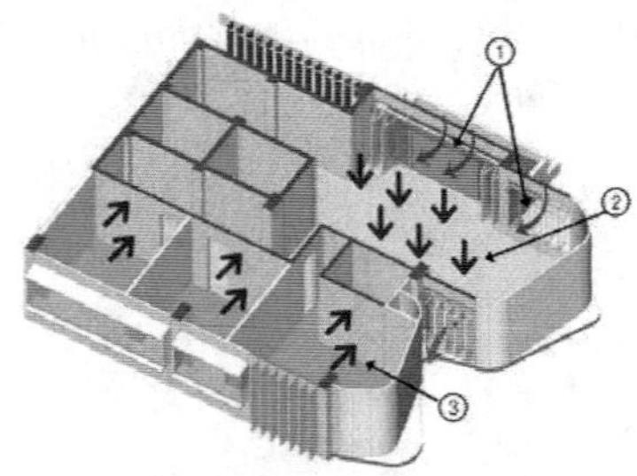

Figure 12. The passive cooling system work scheme on the night time.

Figure 12 shows the unit condition at night:

- Every openings are opened, thus the cool air with low temperature from outside can enter the unit.
- The low temperature air cools down the released heat from the wall and floor. Subsequently, the low temperature air absorbed by the walls and floors.
- The cool air released by the air conditioner also cools down the heavyweight walls and then transfer the low temperature towards the non-air conditioned rooms, thus can helps decreasing the air temperature.

On the day time, the low temperatures on the rooms should be kept by closing the openings and reducing the opportunity of heat absorption. Hereafter, the walls and floors will be absorbing the heat again. Thus, the openings must be kept open to a minimum degree for cross-ventilation to be happening.

7. Conclusion

The design process of Tropical Apartment at Kebon Melati, Jakarta, consist of conventional design process, which is commonly used, and then integrated with several green building criteria to answer the design's goals. The criteria of green building itself affect the design fundamentally. Since the building massing stage, the application of green building criteria has been used and the building masses are automatically formed due to the urgency to meet these criteria.

Therefore, in a building design process of a green building, the green building criteria itself is the most important part, and can be called as the core principle of the design and cannot operate optimally if the criteria are not considered and included from the early stages of the design process.

The rapid development of technology on the field of architecture, construction and material engineer, encourages architects to be able to design not only visually aesthetic buildings, but also has a good impact on the environment. In the future, the implementation of the green building concept can no longer be avoided. Nevertheless, this does not mean a limitation for an architects. In the end of the day, designing a building that is visually aesthetic and implementing the Implementing the concept of green building is very feasible.

8. Acknowledgement

This paper was supported by Dr. Ir. Woerjantari Kartidjo, MT. and Indah Widiastuti, ST., MT., Ph.D. who provided insight and expertise.

9. References

[1] Guo W, Liu X and Yuan X 2015 *A Case Study on Optimization of Building Design Based on CFD Simulation Technology of Wind Environment* (Guangzhou: Procedia Eng vol. 121) pp 225–231

[2] Halliday D, Resnick R and Walker J 2007 Fundamentals of Physics (Extended), (Asia: John Wiley & Sons (Asia) Pte Ltd) pp 476 –493.

[3] https://sustainabilityworkshop.autodesk.com/buildings/lighting-and-daylighting-design

[4] Artmann N 2009 *Cooling of the Building Structure by Night-time Ventilation* (Aalborg: Aalborg University)
[5] Wonorahardjo S 2012 *New Concepts in Districts Planning, Based on Heat Island Investigation. Elsevier* (Bandung: Procedia - Social and Behavioral Sciences vol 36) pp 235 – 242

IOP Conf. Series: Earth and Environmental Science **152** (2018) 012013 doi:10.1088/1755-1315/152/1/012013

A Vision of Daylight Technologies for High-Rise Residential Building in Tropic

R A Achsani [1], S Wonorahardjo [2], and F X N Soelami [3]
[1,2] Department of Architecture, Bandung Institute of Technology, Bandung, 40132, Indonesia.
[3]Department of Physics Engineering, Bandung Institute of Technology, Bandung, 40132, Indonesia.

[1]rizky.amalia7@gmail.com
[2]wonorahardjo@yahoo.com
[3]nugrohosoelami@gmail.com

Abstract. Sidelighting is still the primary lighting opening for most buildings in all part of the world. The technology allows only limited penetration of daylight that causes a high energy consumption for artificial lighting. Using a right daylight technology in high-rise residential building could saving more energy and have a good impact on productivity and occupant's health. This paper discusses the daylighting technologies for high-rise residential buildings in tropical area with literature review method. Some factors such function of systems, location, ability to change and transparency/ the view outside will be the factors that must consider in selecting daylight technologies for high-rise residential building in tropic. The diffuse light-guiding system is a practical daylight technologies for deep floor plan building in tropics. The studies shows that potential daylight technology is diffuse light-guiding system that composed from light shelf, fish sytem and anidolic system. The performance of light shelf influenced by direct sunlight. Its increasing iluminance near window area and causes uneven distribution of illumination in the room. Fish system shows better performance than ordinary blinds, however the result is from integration of fish system and external blinds at low internal illuminance. Application anidolic system in tropical area improving illuminance ratio by factor 3.3, reducing glare by 14%, under overcast sky could transfer daylight upto12.5 m and under intermediate sky up to 20 m. For future work, anidolic system expected to respond sky conditions and direct sunlight for improvement in tropical area.

1. Introduction

The process of designing daylighting in building must consider some factor in building, room, window, and daylighting system [1]. Tropical climate has a different sky conditions with other climates. CIE (International Commission on Illumination) and SNI 03-2396-2001 [2] divide the sky conditions into two types: overcast sky and clear sky, but the tropical climate has a sky condition between overcast and clear skies named cloudy sky [3]. Despite the dominance of the cloudy sky, the outdoor illumination levels in tropical climates reach of 10,000 to over 20,000 lux [3][4].

Sidelighting is still the primary lighting opening for most buildings in all part of the world. The building in the tropic has shading characteristics that became a challenge for designer to improve daylighting and leads to practical challenge of daylighting in tropic which is in deep floor plan buildings [5]. The demand in high-rise residential building is not synergize with the increasing of daylight technologies usage. High rise residential building has the characteristic of laying the apartment units on the side and circulation in the middle of the building. It leads to limited penetration of daylight, low illumination level in room especially in the rear room and increasing artificial lighting usage that cause increasing in high energy consumption. Using the right daylight technologies could saving more energy [6] and have a good impact on productivity, satisfaction and occupant's health [7][8][9]

2. Method and material

This paper discusses the daylighting technologies for high-rise residential buildings in tropical area with literature review method. Using daylighting in buildings does not just collect and install building components such as windows or skylights [10]. The application of daylighting is only one part of daylighting strategy. There are some factor that must consider when selecting the right technologies such as:

a. Function of systems
   - Multiple function
   - Single function
b. Location
   - Exterior
   - Window pane
   - Interior
   - Combined systems
c. Ability to change
   - Fixes, no change
   - Fixed, with change
   - Ajustable can be recessed
   - Sun tracking
d. Transparency / The view outside [1]

For this paper, the selecting of daylight technologies will accommodate:

a. The dominance of diffuse light that have high illumination level
b. High-rise residential building type that use sidelighting and have deep floor plan
c. The aim for this daylighting system is to even out the distribution of light in rear room

3. Daylight Systems For High Rise Residential Building In Tropic

In its application, the use of different types of daylight systems can be used simultaneously to achieve the required lighting requirements in space. Advanced Lighting Guidelines [11] calls this the 'advanced daylight system' which is a window and skylight design that regulates and shapes the distribution and level of room's lighting to meet the glare illumination requirements. The strategy is to

IOP Conf. Series: Earth and Environmental Science **152** (2018) 012013 doi:10.1088/1755-1315/152/1/012013

distinguish between a view window and a daylight window. Where a view window should be transparent, at the eye level of the user, leads to landscapes outside the building and minimizes glare. While a daylight window is designed to provide uniform illumination in a room with high visible light transmission and placed high above the wall.

Furthermore, IEA Task 21 [1] and Lopin [12] categorize daylight systems based on the presence and absence of shading devices. Daylighting systems with shading function to provide protection against glare and direct direct or diffuse light. The system is divided into two categories:

a. Systems that reject direct sunlight and using only diffuse skylight and such as: prism and venetian blinds
b. Systems that using direct sunlight, direct it onto the ceiling or above eye level such as louvers and blinds, light shelf, anidolic system

While daylighting system without shading is design to direct the light to areas far from windows or skylight openings. The system is divided into:

a. diffuse light-guiding system such as light shelf, anidolic system, fish system
b. direct light-guiding systems
c. light-scattering or diffusing systems
d. light-transport systems.

Many of this new daylight technologies termed “light redirecting devices”. Its aim to do: first, block or bend direct light that cause glare while allowing diffuse light. Secondly, uniformity in room. So, daylight systems divide to three categories:

a. Mirrored systems to reflect light such as light shelf, reflective louvers
b. Glazing systems to bend light such as glassblock
c. ‘Smart’ system that are dynamic in the sense that they adapt themselves to the conditions of the moment [13].

To obtain the uniform distribution of illumination from diffuse light on the deep floor plan, its requiring daylight technologies without shading device that rely on redirecting or guiding diffuse light. That make diffuse light-guiding system is a practical daylight technologies for deep floor plan building in tropics. Some daylight technologies in IEA Task 21 that fall in a daylight window category are:

3.1. Light Shelf

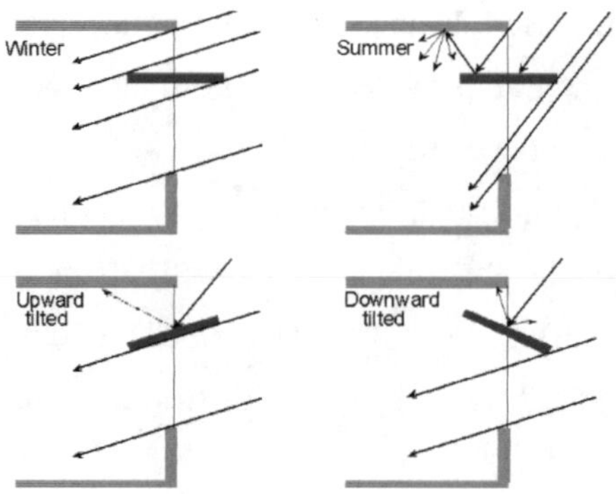

Figure 1. Light Shelves [1]

A lightshelf is a daylighting system designed to shade from direct sunlight (DSL) and then reflect light on its top surface [1]. Light shelves improved work plane illuminance by 24% at rear room, however the performance influenced by direct sunlight. Its show better result in overcast sky than under intermediate sky without direct sunlight. But, recent research show that light shelf increase iluminance near window area and causes uneven distribution of illumination in the room. So, designing light shelf must responding to change in sky conditions and direct sunlight. [14][15]

3.2. Fish System

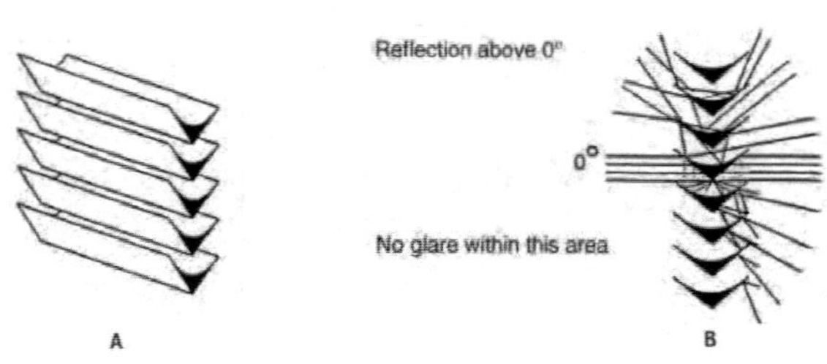

Figure 2. Fish System [1]

The advantage of louvers and blinds is to reduce luminance contrast, however it has weaknesses in daylight utilization [16]. Fish system is fixed horizontal louvers that have precisely aligned triangular section and designed for redirect light especially diffuse light and protect from glare. This system can transmitted light from upper part of the sky to ceiling to protect from glare. Its shows better performance than ordinary blinds, however the result is from integration of fish system and external blinds at low internal illuminance levels [1].

3.3. Anidolic System

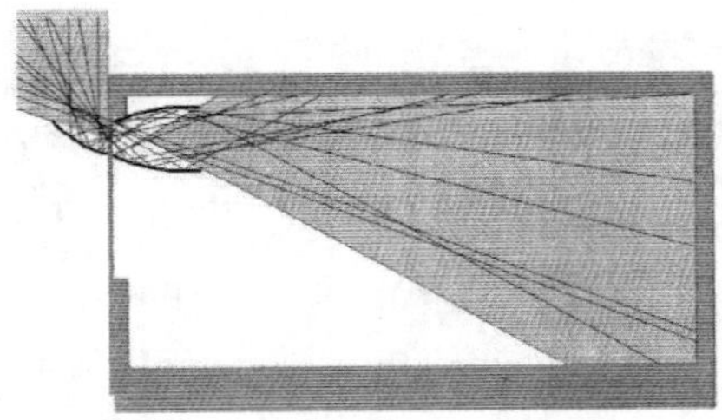

Figure 3. Anidolic System [13]

Anidolic systems is daylight system with compound parabolic concentrators that transport diffuse daylight from the sky to the rear rooms and consist of three component: collector, duct and distributor [1] [17]. The characteristic angle of the anidolic system, or θ, has an inverse relationship with the peak benefit's distance from the window façade; the wider the angle then the peak benefits will be closer [13]. This system aims to provide daylight availibity and better illumination in rear rooms under overcast conditions [1]. Application anidolic system in tropical area improving illuminance ratio by factor 3.3, reducing glare by 14%, under overcast sky could transfer daylight upto12.5 m and under intermediate sky up to 20 m [18] [19].

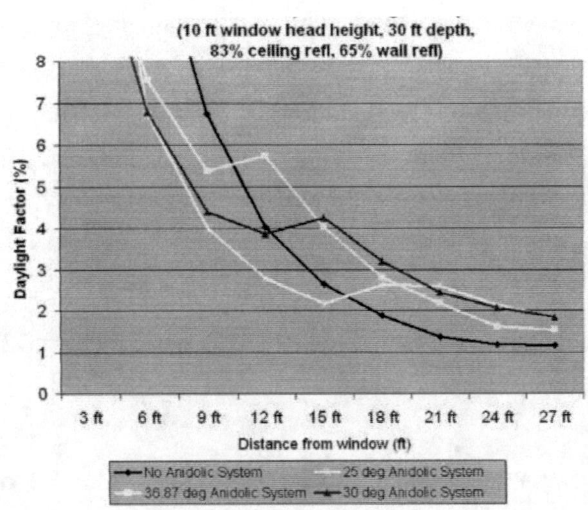

Figure 4. An illustration of the effects of anidolic system characteristic angle on the daylight factor profile [13]

4. Conclusions

From the literature, diffuse light-guiding system is a practical daylight technologies for deep floor plan building in tropic. The performance of light shelf influenced by direct sunlight. Its increasing iluminance near window area and causes uneven distribution of illumination in the room. Fish system shows better performance than ordinary blinds, however the result is from integration of fish system and external blinds at low internal illuminance. Application anidolic system in tropical area improving illuminance ratio by factor 3.3, reducing glare by 14%, under overcast sky could transfer daylight upto12.5 m and under intermediate sky up to 20 m. For future work, anidolic system expected to respond sky conditions and direct sunlight for improvement in tropical area.

5. Acknowledgments

This publication and research supported by Research Schema of desentralisasi ITB PUPT Risetdikti Indonesia 2017, contract No. 009/SP2H/LT/DRPM/IV/2017.

6. References

[1] IEA SHC TASK 21 2000 Daylight in Buildings- A Source Book on Daylighting Systems and Components.

[2] Badan Standardisasi Nasional (BSN) 2001 SNI 03-2396-2001: Tata cara perancangan sistem pencahayaan alami pada bangunan gedung. Jakarta

[3] Rahim R 2000 Analisa luminansi langit dengan metode rasio awan. Dimensi Teknik Arsitektur 28: 142-146.

[4] Lim YW dan Ahmad MH 2013 Daylighting as a sustainable approach for high rise office in tropics." International Journal of Real Estate Studies 8: 30-42.

[5] Edmonds I R, Greenup P J 2002 Daylighting in The Tropics. Solar Energy 73 (2): 111-121.

[6] Bodart, M., Herde, A.D 2002 Global energy savings in offices buildings by the use of daylighting" Energy and Buildings 34: 421-429.

[7] Heschong L 2002 Daylighting and Human Performance. ASHRAE Journal: 65-67

[8] Leather P, Pyrgas M, Beale D, and Lawrence C 1998 Windows in the workplace: Sunlight, view, and occupational stress. Environment and Behavior, 30(6): 739–762.

[9] Lam WMC 1977 Perception and Lighting as Formgivers for Architecture, McGraw-Hill, New York.

[10] Ander, G D 2016 Daylighting. Whole Building Design Guide. retrieved July 19, 2017, from https://www.wbdg.org/resources/daylighting

[11] New Building Institute 2003 Advanced Lighting Guidelines Project Team.

[12] Lopin MK 2002 An Overview of Daylighting systems. Solar Energy Vol 73, No 2, 77-82.

[13] Kleindienst SA 2006 Improving the Daylighting conditions of Existing Buildings: the benefits and limitations of integrating anidolic daylighting systems using the American Classroom as a model. Master Dissertation. Massachusetts Institute of Technology.

[14] Lim, Yaik-Wah, Ahmad, Mohd Hamdan 2014 The effects of direct sunlight on light shelf performance under tropical sky. Indoor and Built Environment 0 (0): 1-15

[15] Warrier G A, Raphael B 2017 Performance evaluation of light shelves. Energy and Buildings 140:19-27

[16] Lim Y W, Ahmad M H, Ossen D R 2013 Internal Shading for Efficient Tropical Daylighting in Malaysian Contemporary High-Rise Open Plan Office. Indoor Built Environment 22 (6): 932-951

[17] Linhart F, Wittkopf S K, Scartezzini JL 2010 Performance of anidolic daylighting system in tropical climate-parametric studies for identification of main influencing factors. Solar Energy 84: 1085-1094

[18] Wittkopf S K 2007 Daylight performance of anidolic ceiling under different sky conditions. Solar Energy 81:151-161

[19] Roshan M, Barau A S 2016. Assessing Anidolic Daylighting System for Efficient Daylight in Open Plan Office in The Tropics. Journal of Building Engineering 8: 58-69

IOP Conf. Series: Earth and Environmental Science **152** (2018) 012014 doi:10.1088/1755-1315/152/1/012014

Innovation of Iron Reinforcing Column of Partical From Frame of Light Steel

M R Ramadhan[1], A Faslih[2], M Z Umar[3]

[1] *Students of Department of Architecture, Vocational Education Program, Halu Oleo University, Kendari, Indonesia*
[2,3] *Lecturer of Department of Architecture, Vocational Education Program, University of Halu Oleo, Kendari, Indonesia*
H. E. Mokodompit Street, Kampus Bumi Hijau Tridharma, Andonohu, Kendari, 93232

Abstract. Almost half of houses in Indonesia are using lightweight steel roof truss today. The phenomenon in the field is that lightweight steel roof truss can blend with mortar mixture. Thus this phenomenon is captured for later applied dynamically, creatively, and innovatively with new idioms such as reinforcement for columns. This research aims to investigate the comparison of the way of making and the price of the materials between the column material made of the light steel and the column material made of the iron reinforcement which is the most efficient. Type of research is qualitative with a comparative causal approach. This research is divided into several stages, namely; Literature study, column creation, and validation. This study concluds that the manufacture of column material from reinforcement is more efficient, than the lightweight steel column material. The reinforcement column material is more efficient because of the more effective way of making and the price of the working materials more economical than the lightweight steel column material. Lightweight steel columns can be used for public housing on condition made by experienced craftsmen to make the process faster, and the dimensions of lightweight steel can be scaled down to make it more economical.

1. Introduction

Today, the world of residential construction has grown rapidly. The rapid development of housing construction also occurs in the world of construction in Indonesia, innovation is very important role in this regard. With the innovation, it is expected to create an end product of a fairly efficient construction. However, in reality today innovations in construction in Indonesia are considered immovable, this leads to inefficient construction work [1]. For small spans, the cost of lightweight steel roof truss is more efficient than wood [2]. Lightweight steel roof truss is a frame made of cold rolled steel silicon [3]. The use of mild steel as a roof truss more popular today. Almost half of the housing in Indonesia is already using lightweight steel roof truss, because the installation process is efficient [4].

Lightweight steel on buildings tends to be used only as a roof frame. The phenomenon in the field that lightweight steel roof truss can blend with mortar mix on the construction wall of horses made of red stone. This phenomenon is thus captured for dynamic, creative, and innovative application with new idioms such as reinforcement for practical columns. Practical columns serve to bind masonry pairs and funnel structural loads to buildings [5]. The purpose of this research is to investigate the comparison of the way of making and the price of the material between the column material made of the light steel and the column material made of the iron reinforcement which is consired the most efficient.

2. Literature Review

2.1. Practical Column Creation
The column practically serves as a binder and auxiliary wall of the main column. Maximum distance of practical column is 3 m. Practical columns are located on the wall, at meeting walls, and next to the door and window holes [6]. The practical columns occupy an important position in the building structure system. Practical column failures can have direct consequences on other structural components. In terms of the requirements for making practical columns, it should pay attention to cross-sectional strength [8]. The requirement of making practical columns has also been studied, it should also pay attention to the quality of concrete, and iron reinforcement. The column reinforcement consists of bending reinforcement and shear reinforcement. The bending hose is horizontally mounted and the shear reinforcement is mounted transversely [9].

At the practical manual column preparation stage, the compressive strength of concrete is made by means of low average manual (12,78 Mpa), so it does not meet the minimum requirement limits of compressive strength (20 MPa). The low compressive strength of the concrete column can not bear the combination of dead load, live load, and earthquake load. At the practical column preparation stage concluded that the practical column work stages are as follows: formwork stage, column reinforcement stage, and column casting stages. Thus it can be concluded that in the manufacture of practical columns should be considered cross section, concrete quality, reinforcement, compressive strength, and manufacturing procedures [8].

2.2. Dual Channel of C Profile
The current usage of channel profile C is not limited only for gording but also used as home structural elements such as columns [10]. The channel of C profile when used as a column and given a casting concrete cast, proved to prevent local buckling. To increase profile stability, reinforcement is used with reinforcing steel and add cast filler concrete within its profile [11,12]. The C profile of concrete filled with cracked patterns can prevent shear failure in concrete [13].

C canal profile beam with lightweight concrete filler can improve bending strength compared to beam C profile profile without lightweight concrete filler [14]. The channel profile C has a weakness in terms of stability. Therefore, in the application the profile needs to be given a stiffener and must be double installed. Thus it can be concluded that concrete C filled double channel C can prevent local buckling, increase load capacity, and add strength to the column, increase profile stability, and increase bending strength [15].

2.3. Efficient Principles on Building
Revenue is a description of the economic position of the household. If the income increases then the physical condition of the house will increase as well. The structure of the building must meet the requirements of stability, balance, strength, aesthetics, and economical. Terms of building construction in general should meet economic requirements, in the sense that there is no waste so that the financing becomes relatively efficient. The best structural system is attractive and cost saving. The results showed that the beams and columns of the house had smaller reinforcement dimensions than the installed reinforcement. This means that the reinforcement installed in the field is safe to use but can still be scaled to make it more economical [16].

To facilitate the implementation in the field should avoid the use of highly variable reinforcement with the aim of improving the efficiency of materials used. Commonly used wall-making materials are red bricks and concrete blocks. Red bricks and concrete bricks are the most economical wall materials. To save the cost, it is better that the brick installation should be flat so that the plaster has the same surface [17]. Thus it can be concluded that the principles for building structures require principles that are fast, economical, and effective in order to be efficient.

IOP Conf. Series: Earth and Environmental Science **152** (2018) 012014 doi:10.1088/1755-1315/152/1/012014

3. Research Methods

This research is a qualitative design with a comparative causal approach. This research is divided into several stages, namely; literature study, column creation, and validation. The literature study stage is presented in descriptive form.

The design stage consists of: making a column of lightweight steel material and a column of reinforcing steel material. The Validation stage is done by comparing the way of making and calculating the price of materials work in the form of data tabulation.

4. Results and Discussion

4.1. Comparison of How to Create

Table 1. Comparison of How to Make

No.	Category	How to Make	Effectiveness How to Make
1.	Column material of Lightweight steel (3.00 m x 0.15 m x 0.15 m)	Stage of column framing	Effective
		Stage of making column molding	Effective
		Stage insert the column frame in to the mold	Ineffective
		Stages of making a mixture of columns	Effective
		Stage of column casting	Ineffective
2.	Column material of reinforcing steel (3.00 m x 0.15 m x 0.15 m)	Stage of column framing	Effective
		Stage of making column molding	Effective
		Stage insert the column frame in to the mold	Effective
		Stages of making a mixture of columns	Effective
		Stage of column casting	Effective
3.	Equation	All ways to make it effective except the stage of inserting mild steel in the mold and the pool casting stage	

Table 1. above explains that the steps of preparing the column material of light steel and the column material of the reinforcement have an equation. The first stage is the preparation of column frames. The second stage is making column molds. The third stage is to insert the column into the mold. Columns of reinforcement material are easily and rapidly performed at this stage. Meanwhile, lightweight steel of column material is difficult because an anchor is mounted on the column mold by drilling from side edges to sideways. At the stage of installing the anchor need carefulness because the anchor hole on the molding board must be translucent straight on mild steel to penetrate the side.

The fourth stage is the creation of a mixture of columns. The fifth stage is the column casting. The column material of iron reinforcement is easy and fast at this stage. Whereas, lightweight steel column material is difficult because the mixture is inserted from the top with a ladder. The cast mixture is not only filled on the outside of the mild steel, but on the hole between the lightweight steel, so to insert the cast mixture need to be pierced with reinforcement. When inserting the mixture into a mild steel cavity should be done with great care because of the shape of a small hole, in the hole there is an anchorage iron, and the shape of large pebbles. Thus, the inner light steel is feared to be unfilled with the cast mixture, although it is pierced with reinforcing steel to mix the filling up to the bottom of the column mold (shown in Figure 1).

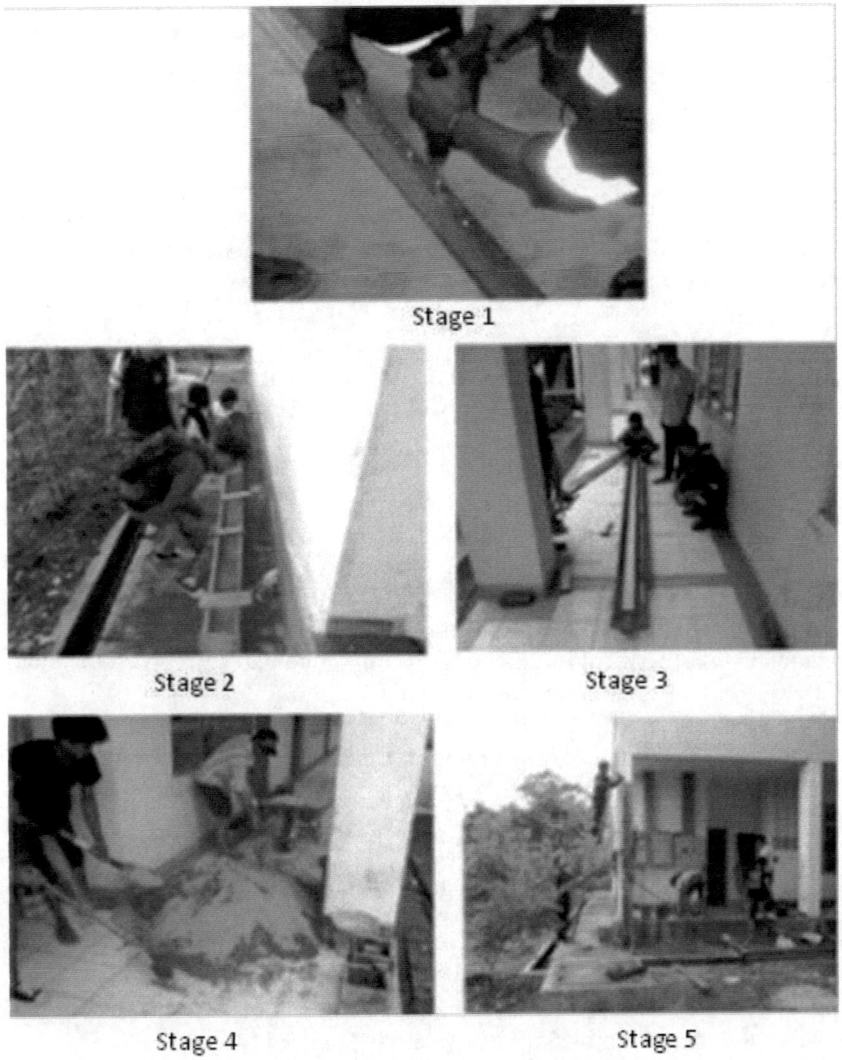

Figure 1. The steps of making column materials of mild steel
Source: Result of documentation 2017

4.2. Material Price Comparison

Table 2. Comparison of How to Make

No.	Work Materials	Column Material from Light Steel (3.00 m x 0.15 m x 0.15 m)	Column Material from Iron Reinforcement (3.00 m x 0.15 m x 0.15 m)

1.	Iron reinforce with dia 6 mm	1 bars	26.000	2 bars	52.000
2.	Iron reinforce with dia 8 mm	-	-	1 bars	36.000
3.	Cement	½ sack	35.000	½ sack	35.000
4.	Sand	2 bars	10.500	2 bars	10.500
5.	Gravel	3 bars	11.000	3 bars	11.000
6.	Light steel	1 bars	75.000	-	-
Total Price		Rp. 157.500,-		Rp. 144.500,-	
Diffrence			Rp. 13.000,-		

Table 2 above explains that the price of working materials to reinforce column material is more economical (Rp 144,500,-) than the lightweight steel column material (Rp 157,500, -) and price difference is Rp. 13,000, - (thirteen thousand rupiah) per column.

5. Conclusion

This study concludes that the manufacture of column material of reinforcing steel is more efficient than lightweight steel column material. Column material of reinforcing steel is more efficient because the way of making is more effective and the price of working materials is more economical than lightweight steel column materials. Material of lightweight steel columns can be used for public housing on condition made by experienced craftsmen to make the process faster, and the dimensions of lightweight steel can be scaled down to make it more economical.

6. References

[1] Buku Manual National Steel Truss Innovation Contest "Ritter" 1 2016 In National Steel Truss Innovation Contest Civil of Engineering Faculty Tadulako University
[2] Irianto 2013 *J. Ilmu Pengetahuan dan Teknologi* **3** 50
[3] Rahayu S A and Manalu D F 2015 *J. Fropil* **3** 118
[4] Wildensyah I 2010 *Rangka Atap Baja Ringan* (Bandung: Alfabeta)
[5] Nisa I J, Mustikawati T, and Yatnawijaya B 2014 *Student J. Architec.* 4
[6] Supriani F 2011 *J Teknik Sipil Inersia* **2** 23
[8] Solikin M and Setiawan B 2015 in Simposium Nasional Teknologi Terapan (SNTT) 3 S132
[9] Tunggul G M 2015 Thesis (Undergraduate) Civil Engineering Studies Program Faculty of Engineering Muhammadiyah University Surakarta p 2
[10] Lisantono A, Santoso B H and Sugianto R 2013 In Konferensi Nasional Teknik Sipil 7 (KoNTekS 7) Sebelas Maret University Surakarta October 24-26 p S-171
[11] Jiwandono D P P 2010 Thesis (Undergraduate) Civil Engineering Studies Program Faculty of Engineering Atmajaya University Yogyakarta Chapter **II** 1
[12] Laksono D B 2009 Thesis (Undergraduate) Civil Engineering Studies Program Faculty of Engineering Atmajaya University Yogyakarta p xvi
[13] Wigroho H Y 2008 *J. Teknik Sipil* **8** 265
[14] Lisantono A and Sari M R 2009 In Konferensi Nasional Teknik Sipil 3 (KoNTekS 3, Jakarta, 6-7 Mei) p S-99
[15] Lisantono A and Jiwandono P P 2011 In Prosiding Peningkatan Peran Riset dan Pendidikan Teknik Sipil dalam Penyelenggaraan Infrastruktur yang Efektif, Efisien, dan Berkelanjutan Sumatera Utara University Medan October 14 **2** S-271
[16] Rantung C M, Windah R S, and Sumajouw M D J 2014 *J. Sipil Statistik* **2** 301-309
[17] Sabaruddin A 2008 *Membangun Rumah Sederhana Sehat Tahan Gempa* (Jakarta: Penebar Swadaya)

Alternative Housing System & Materials Criteria For Land Subsidence Area (Case Study: Bandarharjo, Semarang)

N S Saharom, S C Diana, and D Kusyala

Institut Teknologi Bandung, Jl. Ganesha No.10, 40132, Kota Bandung, Jawa Barat, Indonesia.

syakirinsharom@gmail.com, claussiedibbetz@gmail.com, dibjakusyala@ar.itb.ac.id

Abstract. The critical land subsidence at rate 10cm per year happening in Bandarharjo district of Semarang is a big concern for the local community. With the increasing number of population, the building load increases as well thus could be one of the factors that hasten the land subsidence process. This phenomena causes the locals to spend a lot on house renovation and repairment. To keep up with the road level increased by the government, they had to increase their house floor level. Some demolished and rebuilt their houses. This paper seek to study the ideal criteria of alternative housing systems and materials that is adaptable for land subsidence and flood situation in Semarang by reference research methodology which various papers and guidelines of construction system based on the context of Bandarharjo. The existing housing material and system needs to be strategised for more affordable and adaptable housing system. Although there are already adaptation responses towards these natural disasters by the local communities but the responses are without the awareness about the ideal house system and materials that that helps them to adapt and live with natural disaster in a sustainable and more durable system.

1. Land subsidence and Flash Flood in Semarang

Semarang, located in the central of Java Island of Indonesia, with the second highest density of population is one of the participating cities that joined the 100 resilient cities network that works together in tackling tangible and intangible issues of land subsidence and the flood that results to damages to houses, infrastructures and also economic loss.

In the 6th century, the coastline of Semarang was still at the boundary of Bergota area, which was not too far from Simpang Lima area which is now is the current city center. However, the sedimentation that coming from upper Semarang to the coastal area, is shifting the coastline further to the Java Sea. Until 2007 the coastline has extended as far as ±6,5km [1].

Two main factors that influences the occurance of land subsidence and tidal flood which is also known as 'rob' in Indonesian language are; through excessive ground water extraction and load of building and infrastructure [2]. According to *"In* Monitoring Land Subsidence in Semarang, Indonesia" [3] the causes of land subsidence is categorised as naturally occurred and man-made voids which relates to the massive withdrawal of groundwater in Semarang but this paper will focus more on the load of building infrastructure that causes the land subsidence.

Figure 1. Semarang coastal line shift of 6.5km [1].

Another factor of the land subsidence in Semarang is the soil of Northen part of Semarang that is composed by very young alluvium with high compressibility [4] and currently the topography level of Semarang is at the elevation of 0-3. from sea level. Although this area is prone to flooding and land subsidence issues, this area has become a strategic area for settlements for its economic value of port, industrial and commercial area.

Due to the increasing population and urban development in Semarang area, excessive ground water has been extracted for industry and domestic use and load of development also causes increase in building and infrastructure load. This has led to serious land subsidence and tidal flood.

2. Bandarharjo District, Semarang

Figure 2. Bandarharjo. District, Semarang

Bandarharjo district which is located in the upper North of Semarang, neighbouring with the Tanjung Mas Semarang Port area is comprised of housing settlements and also industrial area. The main environmental issues that is being faced by Bandarharjo is the land subsidence which is at the rate of 7-8cm subsidence per year based on PS InSAR simulation or 6-7 cm subsidence per year based on GPS simulation calculation [4]. At this rate, a single storey house could be completely sunken in 30 years. In response, the government raised of road levels by 50cm every 5 years to catch up with the original elevation of the land. This does not only affect the infrastructure and houses but also causes economic loss to the local community which majority are in the category of low income community.

IOP Conf. Series: Earth and Environmental Science **152** (2018) 012015 doi:10.1088/1755-1315/152/1/012015

The land subsidence also gives negative impacts to the two types of floods that happens in Bandarharjo which is the tidal flood and the flash flood. However, the recent installation of flood control pumps at few points in site had stopped the flooding for two years.

Figure 3. Sunken houses in Kampung Kalibaru, Bandarharjo.

Although the flood has been stopped, land subsidence is a continuous process and will grow worse if the groundwater extraction of the whole Semarang is not stopped and also urban development that increases the building loads which eventually will causes the tidal flood again in the future. Therefore, criteria of house system and materials that is suitable for this condition should be studied to prepare the local community towards awareness about these natural disasters and the strategies for them to adapt and be resilient.

3. Methodology

In order to develop more understanding about the material and construction system that suits best in adapting land subsidence and flood in the context of Bandarharjo, data and information were gathered and studied through literature studies and focusing on the literatures based on housing system and materials in respond to land subsidence and tidal flood in the context of Semarang or other regions with similar context. Besides that, to develop deeper understanding about the issues of housing that affected by the land subsidence and flood, the site visit was carried out. The purpose of the site observation is also to study the local communities responds towards the natural disaster issues, and to collect information and documentation about house construction system, material, condition and damages of selected houses. Moreover, a deep interview with locals who has lived more than 30 years in the area is carried out to further investigate the condition of the land subsidence and flood and also their responds towards the natural disaster. Last, an interview with an oceanography researcher was also carried out to develop more understanding about flood and land subsidence scientifically that is happening right now and the possibilities in the future.

4. Construction system and material criteria for land subsidence and tidal flood area

Damaged caused by the land subsidence and flood disasters had come to studies of ideal types of house construction and materials which is more adaptable and flexible in adapting these issues. Ideally, the type of house construction system is a design system whereby the building could be lifted by phases [5], which the ground is left as is and the construction is built in suiting the present condition of the ground which means no road or ground level rise. As for building load factor, lightweight building materials are to be considered for a house system in context of unstable land which is prone to land subsidence and flood [5].

4.1. House construction system

IOP Conf. Series: Earth and Environmental Science **152** (2018) 012015 doi:10.1088/1755-1315/152/1/012015

There are three potential house construction system for areas that experience both land subsidence and tidal flood; stilt house system, floating house system, and raising house level system and these three potential systems are evaluated and analyzed based on 5 criteria as shown in the table below [6], which was then further evaluated to get the house type which best fit for the condition of land subsidence and flood in Bandarharjo.

Table 1. Criteria score [6]

Aspect	Value (%)
Ability in adapting tidal flood	40
Feasibility of construction process	20
Material availability	15
Financial	10
Community preference	15
Total	100

4.1.1. Stilt House System (Rumah Panggung)
Floor is elevated from the ground surface. It is already common in the Indonesian traditional architecture which was meant to protect the inhabitants from danger; nature and wild animals. The space underneath can be utilised as house barn, or as a communal area for the family. Since wood is widely available and easy to assemble and dissemble, it is common to be used as the material for this kind of construction system.

The advantages of stilt house system are its flexibility of knock-down for adjustment and repairmen, extra space underneath that can be utilise as a multiuse space, and protects its inhabitants from danger of flood and so on [6]. The space underneath the stilt house can reduce the humidity from the ground to the floor and also the increment of house level can be done by phases according to the user's financial ability.

4.1.2. Floating House (Rumah Apung)
Focusing on not to touch the ground, the whole structure is to be floating which usually on site with water. There are variety of floating system that can be used such as drum, kayau or bamboo. The advantage of this system is the flexibility in adapting to the water level. However, the technical aspect of this system is very complicated.

4.1.3. Raised Floor House (Rumah Urug)
This house construction type is whereby the level of house floor is being raised in response to the land subsidence by filling more soil to the ground. This is the most popular construction system chose by the locals. The advantage of this system is that it is easy and feasible for the locals but on the other hand, it comes with high cost and also the need of retaining wall construction.

Aspect	Score (%)	Point	Score (%) x Point Value (%)

		RP	RA	RU	RP	RA	RU
Flood Adaptibility	40	50	20	30	20	8	12
Construction Feasibility	20	30	20	50	6	4	10
Local Community Preference	15	50	20	30	7.5	3	4.5
Financial	15	35	25	40	5.25	25	4
Material Availability	10	40	20	40	4		6
				Total	42.75	20.75	36.5

RP: Rumah Panggung (Stilted House)
RA: Rumah Apung (Floating House)
RU: Rumah Urug (Raising of floor level)

Table 2. House system evaluation based on criteria score system [6].

Thi from the matrix of house construction type evaluation criteria and score values in Table 2 by [6] in the paper Sistem Konstruksi Sederhana Pada Perbaikan Rumah Warga di Kalurahan Kemijen, it is found that stilted houses has the highest score for its adaptability towards flood, construction feasibility, local community preferences, financial capability and also material availability.

In conclusion, every each of house types has its advantages and disadvantages for the condition of land subsidence and flood in Bandarharjo. The types as housing system and construction responds from the locals are to be further studied in chapter 3.3.

4.2. House construction system

According to Muhammad Ali, decision making of building construction material in flooding are, the aspects of durability in terms of waterproof and also corrosion are needed to be considered especially for the structural members that has direct contact with ground or the ones that are exposed to floo d [7]. The alternative material should be adaptable and suitable with the condition of the existing house type and also the condition of the land subsidence and flood disaster [5].

In choosing the ideal material for houses in land subsidence and flood prone area, the materials chosen should be considered with its functionality which functions as different components of structure of the house construction which are the foundation, column, floor beam, floor panels, perimeter beam, roof structure members, roof covering, and wall panels and finishes [6]. On the other hand, in respond to unstable land that has land subsidence issues lightweight material should be considered [5].

Of these studies about ideal house construction material criteria's, it can be concluded that materials for houses in land subsidence and rob prone area should consider the following criteria [6];

1) Durable
2) Waterproof
3) Corrosion proof
4) Lightweight

However, in the context of Bandarharjo which involves the local community, material choice decision should also consider the following criteria [6]:

1) Structural strength
2) Economic capacity
3) Material availability
4) Construction feasibility and skill availability
5) Sea water durability

Considering the ideal material criteria, there are variety of materials than can be considered in both bio based and non-bio based materials such as wood, bamboo, lightweight steel and lightweight concrete respectively.

5. Bandarhajo Houses condition

5.1. Existing Houses System

In 2010, there are 1614 units of permanent concrete houses at 40.97% (refer figure 4) , 1018 units of semi-permanent at 25.65% and 1311 units of non-permanent or wooden houses 33.66% (refer Figure 5) in Bandarharjo district which is prone to land subsidence and flood [7].

Figure 4. Sunken Brick House in Kampung Kalibaru, Bandarharjo.

Figure 5. Wooden and Bamboo House in Kampung Kalibaru, Bandarharjo.

Most of the existing houses in Bandarharjo are landed house with direct contact to the ground which uses concrete as main structural material. From site visits, no stilt house type was found except for few blocks of 5 storey apartments which are lifted up and ground floor serves as open public area for community activities and also community market. The type of foundation system commonly used by the existing houses is the strip foundation, buried underground.

In responding to the flood condition, this landed house system is not a suitable house system for its direct contact towards the ground that could allow flooding in both short-term and long-term condition of land subsidence in Bandarharjo district. Majority of the houses are lower than the road level whereby this condition could lead to more serious flood damages due to the ground level differences that allows more water to flood the house compared to a house which its ground floor is the same or higher than the road level.

5.2. Existing Houses Material

The condition of land subsidence and flood could be worsened with the big number of houses in Bandarharjo district that uses brick and concrete as the main structure which increases the construction load towards the ground.

Although wood and bamboo are used as non-structural material which are mainly as roof structure or secondary structural member, such as column for porches and smaller structural members for roof, there are still room for the local lightweight material to be further developed and innovate to attract the local community to use it as the main structure in reference of a small number of good quality existing old wooden house in the area.

Today there are variety of choices for lightweight and durable materials for houses in land subsidence and flood prone area. However, it is also important to consider the other aspects of decision making in relation to Bandarharjo context involving the 5 criteria mentioned in chapter 3.2 [6]; structural strength, economic capacity, material availability, construction feasibility and skill availability and also sea water durability.

5.3. Adaptation to Flood and Land Subsidence

Living in an area with the extreme condition of flood that happens almost every day and land subsidence at the rate of about 10cm per year had given no other opportunity for the local community of Bandarharjo other than adapting to the condition by renovating their houses.

Reason	%
Moving to other area	9
Increasing House Level	74
Do nothing	13
Others	4
Total	100

Table 3. Local community Wants in Rob Area [7]

According to a survey of what the local community wants in respond to land subsidence and rob, 74% of the community wanted to raise up their ground floor level and 9% of the community chose to move to another area [7]. The house system adaptation behaviour of the local community of Bandarharjo district is influenced by both internal and external factors as follows [8]:

a. Internal Factor:

Internal factors that influence the adaptation behaviour in housing adjustment in respond to extreme condition in Bandarharjo district are education background, knowledge and experience for the community to adapt, economic capacity in affording house adjustment, community network and also land and house ownership status.

b. External factor

External factor that influences the adaptation behavior in housing adjustment in adapting the land subsidence and flood condition in Bandarharjo are environmental condition, infrastructure network system that plays important role in flood management such as drainage system and also other infrastructure and facilities of the area.

5.3.1. Current Adaptation System. House adjustment raising the floor level or renovation towards elevated house type are usually done according to economic capacity of the home owners. Based on the House Adjustment in Flood Area Guideline Theory [8], there are four types of adaptation system introduced which are:

Type A : Raising up house structure
Type B : Raising up ground floor level
Type C : Waterproof fence construction
Type D : Using waterproof material

In relation to the guideline, referring to a thesis titled "Housing Building Loss Cause by Rob and Direction of Management Policy" by Muhammad Ali, 74% of the community in Bandarharjo district which is facing land subsidence and flood disaster choose to adapt the condition by raising their house's ground floor level by adding concrete layer. This system of adaptation is chosen for its construction feasibility and also material availability.

This method of raising up the ground floor level by adding concrete layer is not the best solution in responding to rob and also land subsidence. The addition another layer of concrete onto the existing floor results to accumulation of volume of concrete that results to accumulation of building load towards the unstable ground. This will hasten the sinking process of the house into the ground yet the road level is raised up to 50cm every five years by the government.

.

Figure 6. Floor level is raised higher by adding more concrete layer to avoid flood.

Sunken house in catching up the raised road level had raised the urgency to keep raising up the floor level therefore the head clearance between floor to ceiling will be reduced. In 20 to 30 years period, the ground floor space will be no longer functional as a living space. Thus, the house needs to be renovated or even demolished to be functional again as a living space creating an unsustainable cycle of house adjustment which not only cost more money but also not environmental sustainable in terms of material usage.

This indicates that the local community is not aware or does not prefer the ideal housing system suggested by some studies which the building could be lifted by phases, according to the condition and also affordability of the house owner [5]. The ideal house construction system is further described as elevated stilt house system that uses lightweight material in reducing load towards unstable ground [5].

In addition to the load reduction, this house construction system has more flexibility for adjustment and also repairmen. The elevated ground floor provides protection for the house residents from flood danger in compared to a landed house system which is more prone to flood. Besides that, space underneath the elevated floor could be serve as a multifunctional space, providing additional living space and also space for storage or as small barn for living stock.

6. Conclusion

Based on recommendations and suggestions from various references, in can be concluded that stilt house construction system fulfills most of criteria suggested specifically in the context of Bandarharjo district which has the feature of adaptability and flexibility that enables the house to be lifted by phases according to the flood condition. This will provide protection for both residents and their properties from flood. The flexibility also considers the economic capacity of the Bandarharjo residents which mostly are under the low-income category.

Lightweight material either bio based or non-bio based, agreed as the ideal material in most references for its capability of reducing building load but in context of Bandarharjo, it should be durable, waterproof, corrosion proof, affordable and widely available for construction. However, it is also important to consider the local community preferences which usually determined by their knowledge and familiarity.

The introduction of lightweight elevated house on stilts may not be the silver bullet for this land subsidence and flood issue for its drastic requirement of demolishing the current permanent concrete house for a new one which is totally different than the common house type in Bandarharjo district but it can be slowly implemented and further developed with the available small construction industries in the area.

The possibilities of house construction system and materials could be anything as long as it fulfills the concluded criteria that specifically refers to the context of Bandarharjo.

Acknowledgement

Community Service Program 'Development of Applied Technology of Housing Unit and Neighborhood in Kampung Kota' under P3MI, Building Technology Research Group Institut Teknologi Bandung.

References

[1] Semarang City Government. (2016). *Resilient Semarang Strategy Book.* Semarang: Semarang City Government. [Online]. Available: http://www.100resilientcities.org/wp-content/uploads/2016/05/Semarang20Resilience20Strategy20-202016.pdf [Accessed: 05-July- 2017].

[2] Radjawane, I. M. (2017). Memahami Banjir, Kekeringan dan Penurunan Muka Tanah Studi Kasus: Kota Semarang. Bandung.

[3] Muh , A., & Lorenz, K. (2007). Monitoring Land Subsidence in Semarang, Indonesia. . [Online]. Available:https://link.springer.com/article/10.1007/s00254-007-0680-3 [Accessed: 05- July- 2017].

[4] Abidin, H. Z., Gumilar, I., Sidiq, T. P., Andreas, H., Maiyudi, R., Gamal, M., & Fukuda, Y. (2013). Mapping And Evaluating The Impact Of Land Subsidence In Semarang (Indonesia). *Geospatial Vol. 2*, 26-40. [Online]. Available: http://journals.itb.ac.id/index.php/ijog/article/view/2185/1120 [Accessed: 05- July- 2017].

[5] Krisprantono, & Ch, K. (2012). Pemecahan Masalah Rumah Ambles Di Semarang Utara. . *Vitasphere*, 45-48. [Online]. Available: http://journal.unika.ac.id/index.php/vit/article/view/366/323 [Accessed: 05- July- 2017].

[6] Pribadi, S. B., Indriastjario, Wulandri, A. R., Wibowo, Y. T., Janatin, B., & Muzammil, M. (2011). Sistem Konstruksi Bangunan Sederhana Pada Perbaikan Rumah Warga di Daerah Rob (Studi Kasus: Kelurahan Kemijen, Semarang Timur). [Online]. Available: https://ejournal.undip.ac.id/index.php/modul/article/view/1461/1226 [Accessed: 05- July- 2017].

[7] Ali, M. (2010). Kerugian Bangunan Perumahan Akibat Rob dan Arah Kebijakan Penangannya di Kelurahan Bandarharjo Kota Semarang. [Online]. Available: http://eprints.undip.ac.id/23702/1/MUHAMMAD_ALI.pdf [Accessed: 05- July- 2017].

[8] Ariandini, D., Utami, S., & Yatnawijaya, B. (2016). Adaptasi Fisik Bangunan Rumah Tinggal di Permukiman Rawan Banjir (Studi Kasus: Kelurahan Bandarharjo, Semarang Utara). Vol 4, No 2.[Online]. Available:http://arsitektur.studentjournal.ub.ac.id/index.php/jma/article/view/263/255 [Accessed: 05- July- 2017].

IOP Conf. Series: Earth and Environmental Science **152** (2018) 012016 doi:10.1088/1755-1315/152/1/012016

Effectiveness of Double Skin Façade in Controlling Indoor Air Temperature of Tropical Buildings

Akhlish Diinal Aziiz.[1], S. Wonorahardjo[2], M.D Koerniawan[2]

[123] Architecture Department, Institut Teknologi Bandung, Ganesha Street No 10, Bandung, Indonesia

[1]akhlishdiaz@gmail.com

Abstract. Double Skin Facade (DSF) application on buildings envelope in tropical area aims to decrease cooling load of buildings. Generally, the system is applied on building in an extreme outdoors air temperature, so it could achieve a comfortable air temperature for occupants. This paper discusses effectiveness to control indoor air temperature of some DSF factors such as; 1) air gap between DSF and building façade, 2) glass type. For these purposes, collect outdoor, indoor and air gap air temperatures, surface temperature of all glass surfaces and indoor-outdoor globe temperatures. The initial results show an optimum setting of air gap and glass type of the DSF, as well as some guidance on glass type selection for building surface and DSF.

1. Introduction

Air conditioning system as an important part in building cooling strategies in tropical area consumes a lot of energy during the year [1]. Cooling load of the Aircon system comes from several aspects such as home appliance [2], human activities [3], however the biggest heat source is from outdoor-indoor heat flow at building envelope [4][5].

Energy consumption of building has been proved by many researchers that show big amount each year [5][6][7], the design of double skin facade on the building is construct to decrease this sector of its demands [7][8]. This system of facade is taken as an effective strategy to response dynamic thermal condition outside of the building and caused comfort thermal in indoor temperature [9]. This system defines by involving 2 layers between occupied space of interior and exterior environment which separated by an air gap [9] and one of the heat avoiding strategy, which this strategy is more effective than cooling strategy [10]. In certain size, this air gap could hold back the radiation which consist extreme temperature for entering indoor space and let it out [11].

Several researches referred in this paper has the states to define the role of double skin façade on the building. Aslim in her research explain that this system could play as retrofit strategy that relate to decreasing energy consumption over the building [12]. Dewi also states that double skin façade use air flow between the layers as a stack effect to decrease indoor temperature [13]. Darmayanty in her thesis report wrote based on the research that double skin façade could role as modifier which has design to confront heat outside of the building [14] Iyati also explain that double skin façade as an envelope for the building should use the wind as the potential for decreasing temperature on building surfaces [4]

Meanwhile the previous research by Aziiz shows that double skin façade revealed a larger distance managed to lower the temperature better than smaller distances because it delays the heat to entry the building, even the wind flow inside the gap have smaller wind velocity [15]

1.1 Aplication of Double Skin facade
Many types of double skin facade system are created to applied for responding various thermal condition in a specific site around the world. One of the biggest percentage of DSF project has done in Northern Europe and Japan. This data shows limited research on another place such some tropical humid climate countries which has warm tendencies in thermal environment. This environment highly requires better cooling system to decrease energy consumption over the year.

However, the project in double skin facade has done several times, researcher found that in the summer time, overheating commonly occur in the air gap part of this system and cause increase of cooling load in the building [16]. Tropical-humid climate such as Indonesia has mostly high temperature throughout the year and there are only two season that is dry and rainy season. And those are cause its sun light intensity in all over the year. Research also states that the overheating risk on the system that applied in tropical country are high [17]. This air gap which design to prevent the extreme air temperature from outside environment must be configurate specifically base on the environment case of each place.

The configuration of its size has been confirmed to define the behave of air flow for entering indoor space [18]. Research recently explain that the more bigger air gap is, the better it would prevent the extreme temperature from coming in. Research proved that this gap one of the vital part of the system which could defines the performance [15]. Many configurations such as this, and also the angle or the material we should provide are important to create best performance on this system [13].

In some cases, architects include shading devices in the air gap of double skin facade to responses solar exposure on the outer skin, so it would reflect out the radiation that brings heat to the indoor occupied space. In the other hand, configurate the color of the shading devices that commonly called blinds is also considered recently to influence the radiation. This setting has a great result for decreasing the heat from coming in [19].

This research on the paper took Bandung as in Indonesia's humid climate city to begin the experiment and to get the aims to control the temperature inside the building.

2. Model Testing Double Skin Facade
This paper involving a real scale model of double skin facade (figure.1) with adjustable air gap to investigate the ideal level of temperature indoor within its width 0.5, 1.0 and 1.5.

Figure 1. Research Aziiz, Wonorardjo, Koerniawan (2017) using real scale of model double skin facade with adjustable air gap.

This air gap easily adjustable only by user doing for research purposes and has a certain size of the existing part of the building in Bandung environment.

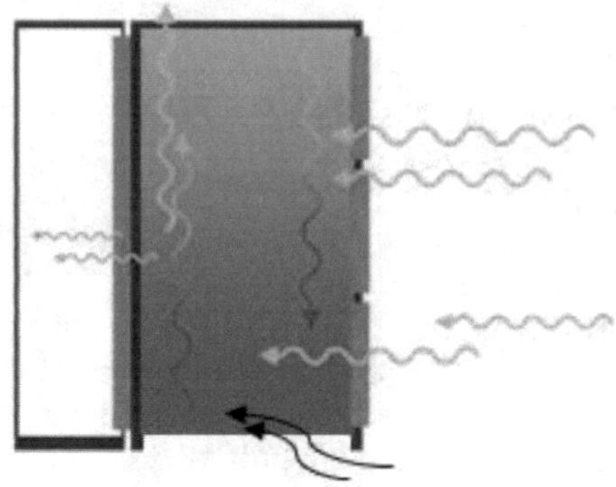

Figure 2. Illustration on research testing model with natural ventilation

In some cases, architects include shading devices in the air gap to response solar exposure on the outer skin, so it would reflect out the radiation that brings heat to the indoor occupied space. In the other hand, configurate the color of the shading devices that commonly called blinds is also considered recently to influence the radiation. This setting has a great result for decreasing the heat from coming in [20]. In this research on the paper provide one-way glass material as second layer of the system to make reflection effect, otherwise the color of this layer is dark bronze with reflection characteristic (figure. 3). So hypothetically it could decrease the heat in solar radiation that received by this layer which facing directly outside environment.

Type of ventilation for the system also allow to pull out the heat from the air gap before it entering inside [21]. This certain ventilation cause buoyancy effect which naturally driven the heat out of the system. On this paper researcher used open ventilation below the air gap system to make natural ventilation in every scenario (figure. 2).

Figure 3. One-way glass in second layer in the research as outer skin of the testing model.

3. Measurements Methods

This research present 3 kind of scenario that has the same material of outer material, that is one-way glass 5mm in dark bronze colour. To begin the testing, it is also involved model scenario for indoor environment to gain the temperature data in interior space by every scenario (figure. 4).

In this research, thermocouple devices were involved to measure the temperature value in certain spot (figure. 4). The measurements took 8 spot which consist 3 value of air temperature, 3 value of globe temperature, and 2 value of surface temperature. 3 air temperature will place on indoor, air gap and outdoor environment. So are the other 3 of temperature globe. 2 values of surface will take place on first skin of the system which seize clear glass and the other on the second skin that is one-way glass dark brown. Double skin façade generally must be ready-designed to facing the sunlight intensity over the year with certain of humidity and wind condition. Some of the research are using simulation to use the wind as the potential to decrease the temperature surface on double skin façade.

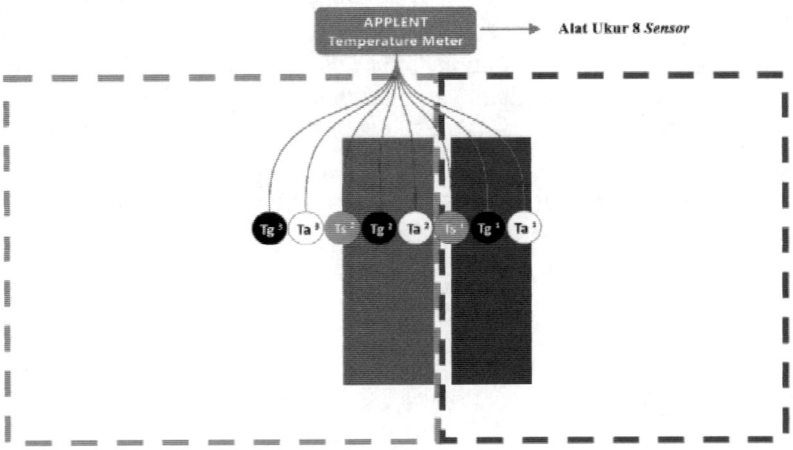

Figure 4. measurements scenario in the research. Tg represent Globe Temperature, Ta represent Air Temperature, while Ts is Surface Temperature.

This several spots were defined to investigate the changes of temperature within every scenario and result comparison to see the most effective width with outer skin material one-way glass as a reflective layer.

Figure 5. measurements device in the experiment

The measurement device that is thermocouple will be used simultaneously to gain the data over the scenario (figure. 5). This device allows to measure 8 spot of temperature in one time, so that the authors would gain the same value with the same load condition in the field.

4. Research Analysis and Discussion

In this research, analysis of chart is presented to see the temperature changes and the differences with one spot of measurement and another. Researcher also involving software JMP 7.0 to investigate the latent pattern that could have not been show on the previous analysis, and produce analysis of correlation and analysis of variant.

4.1 Analysis Distribution

The first analysis in this paper will be presented in a charts mode of diagram. This diagram directly inform the changes and the differences about the temperature value of each spot in measurements. This data contains value of Ta as the air temperature, Tg as globe temperature to see the value of heat

IOP Conf. Series: Earth and Environmental Science **152** (2018) 012016 doi:10.1088/1755-1315/152/1/012016

radiation in air temperature, and also Ts that is surface temperature that inform about the temperature of each layer of the skin.

The diagram below generally shows that there is indeed a temperature changes by the application of double skin facade system (Fig 6.). The changes are a decrease temperature in mostly each spot of measurements. Researcher use temperature globe (Tg) to see this decrease pattern because of its capability to show the value without others variables that might disturb the real value of the heat. And apparently the value of the temperature surface in every section and width of the air gap is also have te same decrease pattern tendencies.

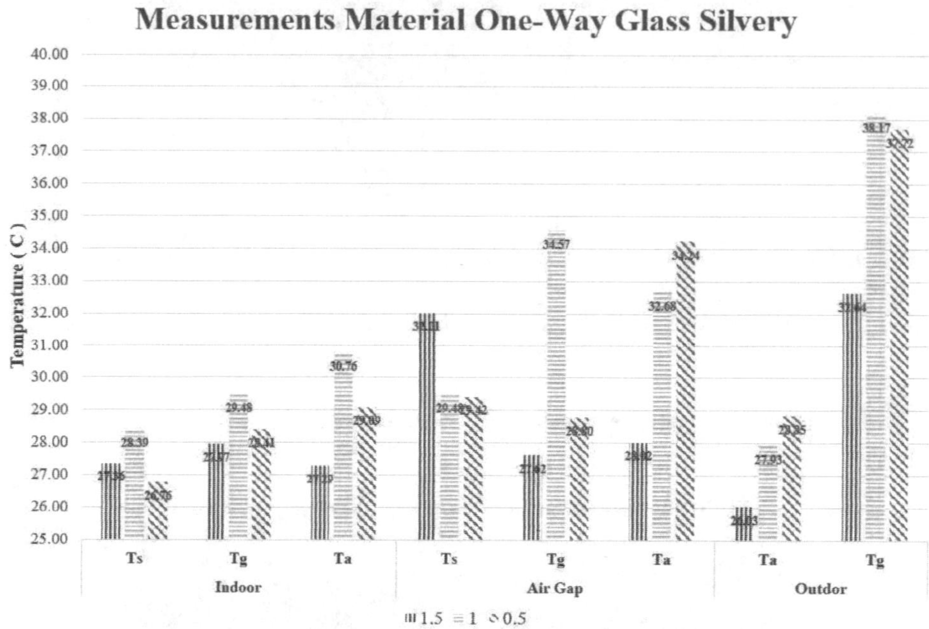

Figure 6. Comparison result in every section of the system, outdoor, air gap, and indoor. Tg represent Globe Temperature, Ta represent Air Temperature, while Ts is Surface Temperature.

However, the decrease of the temperature is not quietly same in pattern. For example, the decrease pattern seems likely was not happening in temperature air in every width of the air gap. The width 1.5 and 0.5 has changes of increase temperature to the indoor section. In the other hand width of 1 meter is decreasing. This value of temperature in Ta spot in every section shows different pattern because of its variables that might happen in the measurement such as wind flow and air humidity that for exact it could make a change of its temperature.

The diagram also shows pattern of the 1-meter width of air gap. This width generally inform that 1 meter is happen to show the highest temperature of all. This pattern caused by the condition of the outdoor temperature when the measurement. However, if we look at the comparation of the differences between the outdoor temperature and air gap temperature (figure. 7), 1-meter width has the smallest changes. In the other word, these smallest changes of temperature mean that the capability of 1-meter scenario in the research is lowest than the other to prevent over heating in the air gap of the system.

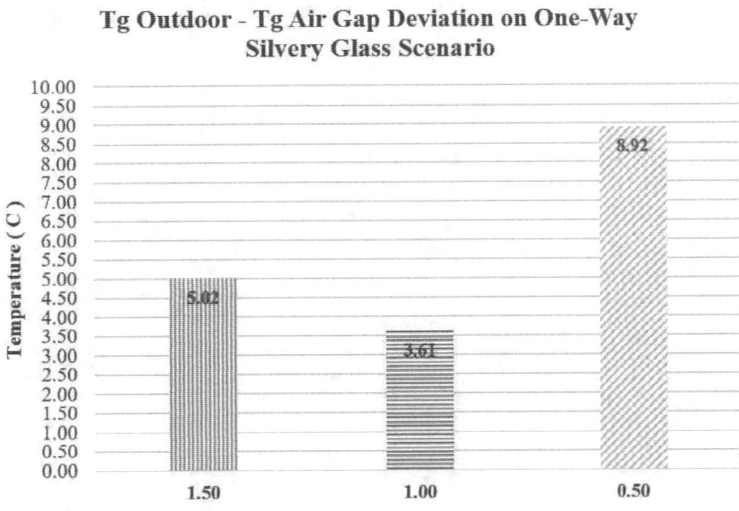

Figure 7. Comparison result deviation in every width scenario on section Outdoor Tg to section Air Gap Tg. Tg represent Globe Temperature, Ta represent Air Temperature, while Ts is Surface Temperature.

In the data above also show precisely the smaller width produce highest changes which mean this scenario is so far, the best to prevent over heating in the air gap.

On the previous diagram (figure. 6), the changes of temperature are not showing some particular pattern, while 1 and 1.5 meter shows an increase changes of temperature to the indoor space, the 0.5-meter scenario shows a decrease change. So, from this data researcher could say that the scenario 0.5 meter of air gap could help cooling the indoor space.

In this paper researcher use one-way glass in every scenario as the outer skin or second layer that directly facing the outside environment. This layer will be rolled as reflectance layer from the solar radiation which contains heat.

From the diagram, it's informing that the value of temperature in spot Ts (figure.6). this value shows the temperature of the surface and the changes between the second layer to the first layer by the temperature outside.

Generally based from the data, this material one-way glass as the second layer (outer layer) could help decrease the heat from outside environment. Its reflection characteristic helps to reflect the heat from sun light as electromagnetic wave that trying to transmit to the system.

Researcher using differences to show the capability of each width scenario to see which one has the biggest amount of decrease changes. It is again informing that scenario 0.5 has the biggest amount of decrease change in the surface temperature (figure. 8).

This result of analysis strengthens the argument that the 0.5 with one-way glass material scenario has the potential to be better in the double skin facade system. This biggest changes in the scenario 0.5 meter of air gap represent that the width allows material to decrease more than the other width.

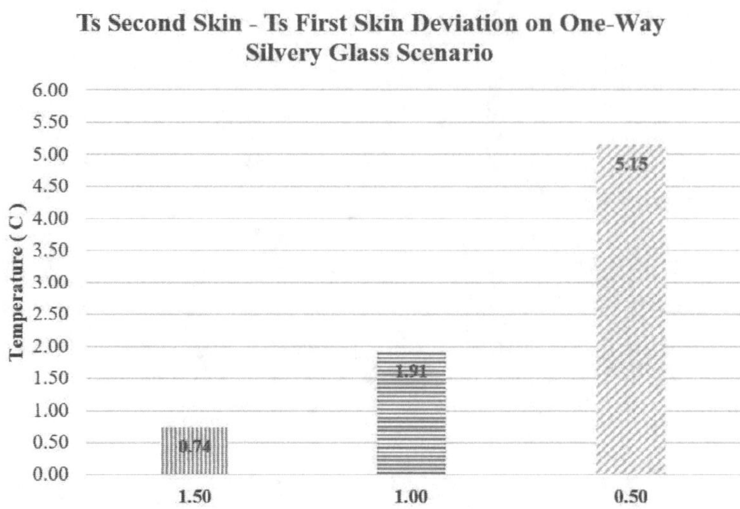

Figure 8. Comparison result deviation in every width scenario on section Ts Second Layer Skin to section Ts First Layer Skin. Tg represent Globe Temperature, Ta represent Air Temperature, while Ts is Surface Temperature.

Researcher assume that this caused by the width itself. The bigger the gap, the smaller the changes to the other skin will be. The gap in the system allow the flow to move out to the other place. While the smaller gap applied, it will not have the same time to move out, and smaller gap makes the flow go faster than the other. That is why the decrease changes were not so much in bigger gap.

4.2 Analysis Correlation and Analysis of Variant

The second analysis would be using software JMP 7.0 to see the correlation between parameter which shown on each spot of measurement. For example, this analysis would investigate if there is any correlation between Ta outdoor to the Ta indoor, or if there is any direct influence between the width and the Ta indoor. The previous discussion in the previous analysis, research present several correlations between two same parameters or not. Same parameter would present in analysis of correlation, and the different parameter such as width and the temperature indoor (ta) would be present in analysis of variant.

Diagram below informing about the correlation between parameter Tg Outdoor and Tg Air Gap. This diagram value indicate significance which mean that the Tg in this spot as temperature outdoor given an effect for the Tg in air gap.

Diagram below also informing the trend of each measurement on the polynomial green. It is explaining that there is certain trend happen in these measurements

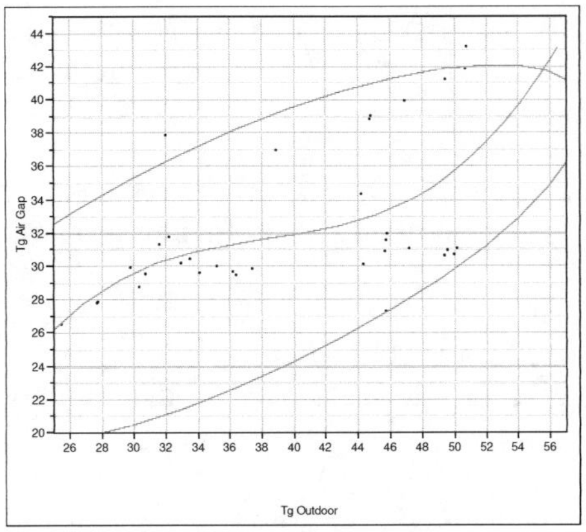

Source	DF	Sum of Squares	Mean Square	F Ratio
Model	3	751.7922	250.597	20.1807
Error	39	484.2894	12.418	Prob > F
C. Total	42	1236.0817		<.0001

Figure 9. analysis correlation Tg air gap by Tg outdoor. Tg on figure above represent Globe Temperature.

Generally, if there is an increasing value in Tg outdoor, the increasing also happens in Tg air gap. This trend happens until certain value of temperature. Based on the data, this trend will be slowing down and come back to increase.

Another information indirectly explains that most of the data are classified in one particular pattern which shown by blue circle. This blue circle is representing the reach of 90% data of all the measurements.

The next analysis would be presented on diagram of analysis variant. This analysis allows to compare two different parameters in the measurements. This analysis will include parameter the width and indoor temperature (Ta). these two parameters have different value which the width would represent 3 scenarios, and the other one represents value of temperature. So, this analysis would be informing the pattern of these parameter.

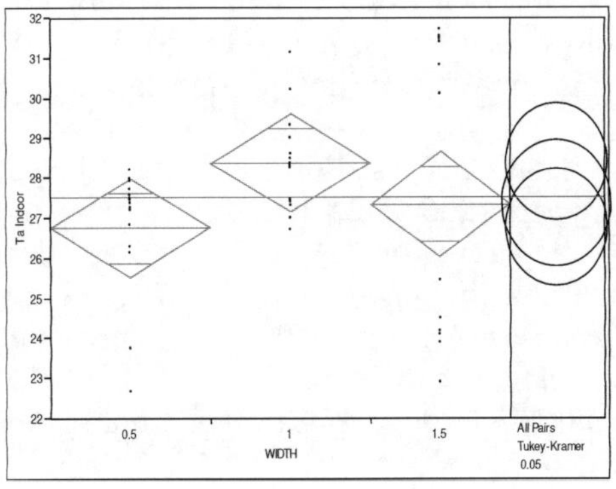

Source	DF	Sum of Squares	Mean Square	F Ratio	Prob > F
WIDTH	2	20.38544	10.1927	1.8499	0.1705
Error	40	220.39661	5.5099		
C. Total	42	240.78205			

Figure 10. Tg Indoor by width of air gap. Tg on figure above represent Globe Temperature.

The analysis above indicate value of insignificant by the value Prob of 0.17 which it should be smaller than 0.1 to be significance. But the diagram informing about the result of the scenarios in the research. That is width of 0.5 meter has the best result for decreasing temperature indoor, and 1 meter of width shows opposite result.

Researcher assume that this significance value in the software caused by the width parameter would not affecting directly into indoor value. But experimentally, this result made differences.

5. Conclusion

The research came up with the ideal of how to controlling indoor temperature by using DSF system in the specific site in humid-climate country such as Bandung-Indonesia. The research providing real scale of model to be tested and resulting that this system also applicable to be use in tropical humid country to decrease heating inside the room. The recommendation in this research says that width 0.5 meter is the best scenario to reduce indoor air temperature. But this result only could be gain by providing outer skin one-way glass as a reflectance layer that directly facing the outdoor environment.

6. Acknowledgments

This publication and research supported by Lembaga Pengelola Dana Pendidikan (LPDP) and Research Schema of Desentralisasi ITB PUPT Ristekdikti Indonesia 2017, contract No. 009/SP2H/LT/DRPM/IV/2017

7. References

[1] S. Aisyah, S. Ahmad, H. Bahadur, and S. Wonorahardjo, "Field study on adaptive thermal comfort in of fi ce buildings in Malaysia , Indonesia , Singapore , and Japan during hot and humid season," Build. Environ., vol. 109, pp. 208–223, 2016.

[2] R. Lindberg, M. Korpi, and J. Vinha, "Factors affecting energy consumption of buildings," 2008.

[3] C. Deg, H. Comfort, and I. M. S. Kam, "25.5 Deg C and Human Comfort," 2006.

[4] W. Iyati, "Double Facade ENgineering with The Use of Air Flow on The bUilding Envelope SUrfaces of Multi STorey Office, Case STude: Wisma CIMB Niaga Bandung," Institut Teknologi Bandung, 2012.

[5] H. Poirazis, "Double Skin Facade : A Literature Review," Lund, 2006.

[6] A. L. S. Chan, T. T. Chow, K. F. Fong, and Z. Lin, "Investigation on energy performance of double skin façade in Hong Kong," Energy Build., vol. 41, no. 11, pp. 1135–1142, 2009.

[7] M. A. Shameri, M. A. Alghoul, K. Sopian, M. F. M. Zain, and O. Elayeb, "Perspectives of double skin facade systems in buildings and energy saving," Renew. Sustain. Energy Rev., vol. 15, no. 3, pp. 1468–1475, 2011.

[8] M. Taleghani, M. Tenpierik, S. Kurvers, and A. Van Den Dobbelsteen, "A review into thermal comfort in buildings," Renew. Sustain. Energy Rev., vol. 26, pp. 201–215, 2013.

[9] F. Marques and M. Glo, "Gap inner pressures in multi-storey double skin facades," Elsevier ; Energy Build., vol. 40, pp. 1553–1559, 2008.

[10] S. Wonorahardjo, "New Concepts in Districts Planning , Based on Heat Island Investigation," Procedia - Soc. Behav. Sci., vol. 36, no. June 2011, pp. 235–242, 2012.

[11] G. Baldinelli, "Double skin facades for warm climate regions: Analysis of a solution with an integrated movable shading system," Build. Environ., vol. 44, no. 6, pp. 1107–1118, 2009.

[12] N. Aslim, "Effectiveness of DOuble Skin facade as Retrofit to Reduce Energy Operational Consumption of Rent Office Building," Institut Teknologi Bandung, 2015.

[13] N. I. K. Dewi, "Active Performance of DOuble Skin Facade as an Effort to Control Air Flow on The Building Envelope Surfaces," Institut Teknologi Bandung, 2014.

[14] N. Darmayanty, "The Effect of Double Skin Facade's Glass Tilt ANgle on Internal Heat Gain: Bandung as Case Study," Institut Teknologi Bandung, 2014.

[15] A. D. Aziiz, "Passive Strategy to Control Air Temperature Indoor by Configurate Air Gap Width and Material on Double Skin Facade. Case Study: Bandung Cinty," Institut Teknologi Bandung, 2017.

[16] C. Dewi, R. Huang, and A. Nugroho, "Strategi Double Skin Fasade pada Bangunan Kampus National Central University dalam Menurunkan Kebutuhan Energi Pendinginan," J. Ruas, vol. 11, pp. 51–59, 2014.

[17] T. M. Boake, K. Harrison, D. Collins, A. Chatham, and R. Lee, "Understanding the General Principles of the Double Skin Façade System," no. November, pp. 1–18, 2003.

[18] T. Başaran and T. İnan, "Experimental investigation of the pressure loss through a double skin facade by using perforated plates," Energy Build., vol. 133, pp. 628–639, 2016.

[19] B. Rahmani, M. Z. Kandar, and P. Rahmani, "How double skin facade's air-gap sizes effect on lowering solar heat gain in tropical climate?," World Appl. Sci. J., vol. 18, no. 6, pp. 774–778, 2012.

[20] E. Gratia and D. A. Herde, "The most efficient position of shading devices in a double-skin facade," Energy and Buldings, vol. 39, pp. 364–373, 2007.

[21] A. Pappas and Z. Zhai, "Numerical investigation on thermal performance and correlations of double skin fac ¸ ade with buoyancy-driven airflow," Energy Build., vol. 40, pp. 466–475, 2008.

Wind-Driven Natural Ventilation Design Of Walk-Up Apartment In Coastal Region North Jakarta

Fathina I Nugrahanti[1], P E Yasin[2] and A Nurdini[3]

[123]Architecture Department, Institut Teknologi Bandung, Ganesha Street No 10, Bandung, Indonesia

Email: [1]fathinaizmi@gmail.com, [2]praseffendy@gmail.com, [3]allisnurd@gmail.com

Abstract. Housing has been the second most energy-consuming sector in Indonesia nowadays. According to the data released by government, the biggest consumption in housing sector is the use of air conditioning. This consumption will significantly rise in metropolitan-high density city like Jakarta along with the increase of vertical housing supply. This research focus on design iteration to achieve optimum model of wind-driven naturally ventilated housing. Cilincing District, North Jakarta, known as industrial and settlement area is used as case study. Since the location by the bay area, Cilincing represents the characteristic of tropical coastal area. This research utilizes the tropical coastal characteristic especially wind to design a naturally ventilated housing. Various building elements are determined as variables and tested using Ansys Fluent CFD simulator to achieve thermal comfort stadard by SNI 03-6572-2001. Preliminary results shows that unlinear (zig-zag) building layout and combination of various building distances give big impact to airflow movement around the buildings. Narrowing building distance in the middle of the site can create a kind-of tunnel / trap that strengthen the wind along the site. Inlet and outlet area should be balance to avoid uneven airflow distribution inside the room and located in different level to maximize cross-ventilation.

1. Introduction

Energy consumption of South East Asia is increasing nowadays. Indonesia, country with the highest population number in South East Asia, spends energy the most among other countries [1]. Household is the second most energy-consuming sector after industry in Indonesia. The Ministry of Energy and Natural Resource releases data which states that in household sector, energy is mostly used for air conditioning (ventilation).

Meanwhile, this energy comsumption will keep rising along with the increase of housing needs, especially in big city like DKI Jakarta. Therefore, we need a housing solution that can save energy especially for ventilation.

To fulfill housing demands, the goverment of DKI Jakarta sets a program to build vertical housing, mainly for middle-class society which is still dominating the income distribution of Jakarta [2]

Location is an important issue in housing design. Development of Jakarta is slowly moving toward the north area. One of the example is massive reclamation plan along North Jakarta coast. In Nort Jakarta, Marunda Village in Cilincing District can be a perfect location because it suits with goverment's plan (RDTR DKI Jakarta) to develop this area into vertical housing zone. This area has lower density than other districts and the land price is affordable, but already has good infrastructure

IOP Conf. Series: Earth and Environmental Science **152** (2018) 012017 doi:10.1088/1755-1315/152/1/012017

and public facilities. This location is also predicted will continue to grow as industrial centre along the reclamation of O, P, and Q islands as the *Port of Jakarta*.

Adjacent to Java Sea, this location has a strong character of coastal region. Based in local data from BMKG Tanjung Priok, this area has monsoon climate with average wind velocity 2 m/s mostly comes from north-east, average temperature 28°C, and humidity between 64.58 – 85.17% around the year.

The availibility of sunlight and wind in coastal tropical area is a potential that can be utilized to produce a thermally comfortable, energy-effient building design using natural ventilation.

1.1. Walk-Up Apartment

Walk-Up Apartment is a vertical housing which use stair as the main vertical transportation system. Elevator is not mandatory and each block usually consists of 12 – 40 units [3]. Other characteristics of walk-up apartment are: walkable; units usually accessed by interior hallway; each unit has private outdoor space (such us balcony / patio); off-street or basement parking.

In Indonesia, walk-up apartment can be categorized as *rumah susun (rusun)*. There are some classification of *rusun* but this project focuses on the design of low-rise, middle-class, commercial *rusun*. This typology has been commonly applied in other countries but to fit with Indonesian habitational behaviour, there are some adjustment needed, such us the unit size (1 bedroom 24 m^2 unit and 2 bedroom 36 m^2 unit are chosen), the existence of private open space (balcony in every unit) as a substitution of garden in landed housing, also spacious corridor (corridor width 1.8 metre) and community room in each floor.

1.2. Thermal Comfort

Air conditioning, the main issue of housing energy consumption in Indonesia, is an artificial way to achieve thermal comfort using mechanical utility. The key to achieve thermal comfort is mantaining the body temperature between 37°C [4]. Human preference of thermal comfort may vary but SNI 03-6572-2001 has set a standard as written below.

Table 1 Thermal comfort standard.

Theory	Temperature (°C)	Wind (m/s)	Humidity (%)
SNI 03-6572-2001	Comfortable cool20.5-22.8	0.1 - 0.35	40-60
	Optimum comfot 22.8-25.8		
	Comfortable warm 25.8-27.2		

There are some environmental factors of thermal comfort: air temperature, radiant temperature, air movement, and humidity. The most effective way to achieve thermal comfort in tropical area is optimization of wind movement because the breeze allows heat dissipation on the surface of your body [5].

This project located on the coastal area which has dominant wind characteristic, so the design iteration and simulation will focuse on thermal management through wind-driven natural ventilation.

1.3. Wind-Driven Natural Ventilation

Wind-driven natural ventilation occur when higher wind velocity hits the building surface and creates higher air pressure (+) on the inlet and lower pressure (-) on the other side. This pressure differences on the building's surface will initiate air to move across the building, bring the fresh air and also reduce room temperature [4]. From the literature study, we conclude that atrium or courtyard can penetrate sunlight and improve the quality of natural ventilation. Chosing atrium-shape mass also allow the application of double-side or cross ventilation where inlet and outlet are located in the opposite direction with each other.

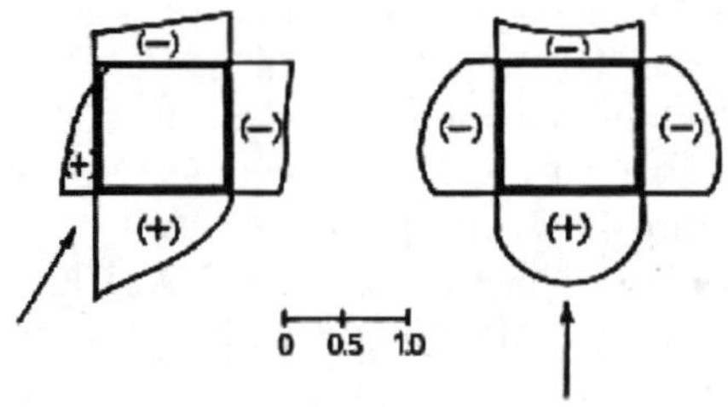

Figure 1 Wind pressure around a building.

Air movement around building is affected by its environment. There are some variables that will affect air movement around the building: 1) Building form [6, 7]; 2) Orientation [8, 9]; 3) Building layout [7]; 4) windscoop [6].

Inside the room, hot air moves from the bottom and flows near the ceiling, waiting to be removed and replaced by fresh air. There are also some variables that will maximize air movement inside the building: 1) Opening dimension [6] and [10]; 2) Opening orientation [11]; c) Inlet & outlet location [6].

2. Design Case

This project is located near North Jakarta beach, in Marunda, Cilincing District, Indonesia. This are is proposed as vertical housing zone in RDTR DKI Jakarta and surrounded by green area, schools and industrial zone. The site has 1.4 Ha area and will have maximum 35% built floor area. Density that will be achieved is 460 persons/Ha, so we will have 6 blocks 5-storeys buildings with 288 room units/block.

Figure 2 Site Location (indicated by bold yellow box).

3. Design Simulation

Simulations are done using **Ansys Fluent** Simulator in order to test the effect of variables mentioned before to get optimum design model of wind-driven naturally ventilated housing.

3.1. Outdoor Simulation

The datas assigned in the beginning are wind velocity 2 m/s, temperature 28°C, and wind direction from north-east. The first simulation is a step to check the general wind behaviour around the site effected by the surrounding buildings in radius 250 metre. The site (inside rectangle) is barely shaded by other building so wind can still move freely inside the site.

Table 2 Model 1 and 2 in simulation.

(1.a) Rectangular	**(1.b) Square**	**(2) With Pilotis**
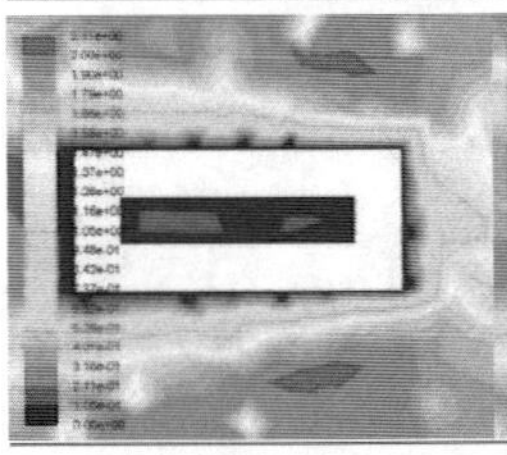	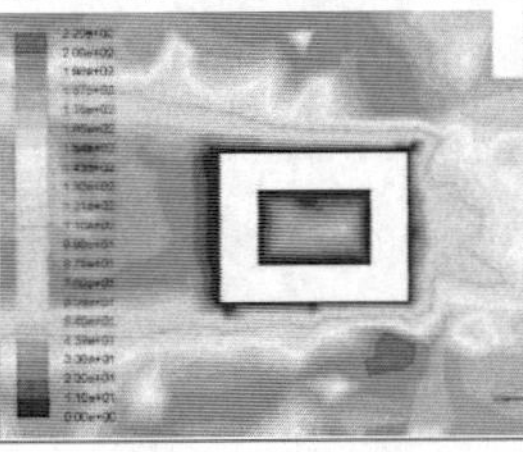	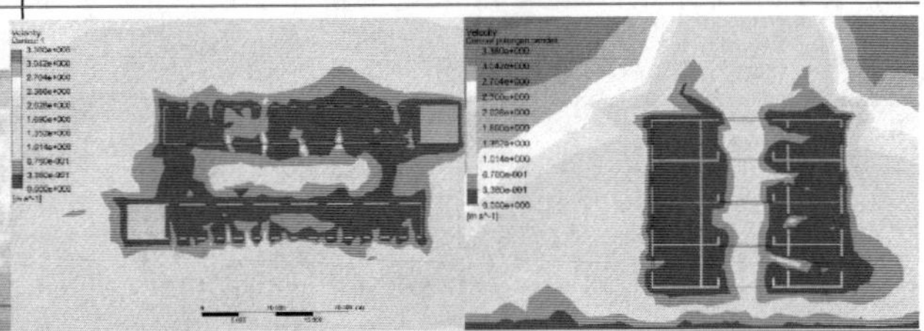
V = 1.487681 m/s	V = 1.5416522 m/s	V = 1.5423875 m/s

The second variable tested is building shape. Rectangular shape is better because wind can be distributed evenly along the building surface. From the simulation with the same capacity atrium model, we can conclude that between rectangular (1.a) and square (1.b) plan model, there isn't significant difference on wind behaviour. But 1.a model is still better because in square model, there are some units that face east – west which can increase heat radiation inside the building. Larasati (2000) on Seftyarizki [8] also stated that lifting building mass (creating a pilotis) will create better airflow inside the building which has been proven in simulation model 2.

The next simulation tested the effect of orientation to wind distribution around the building. Burnett [5] proposed to test three different orientation 0° (3.a), 45° (3.b), 90° (3.c). From the simulation we can see that 3.b has the best result where there is less shadowed area (more light-blue colour). But 3.b model need a lot of space and doesn't fit with the building regulation. Model 3.a and 3.b both have wide dark-blue areas, but model 3.c are all facing east – west so 3.a is chosen.

Table 3 Model 3 in simulation.

(3.a) 0°	**(3.b) 45°**	**(3.c) 90°**
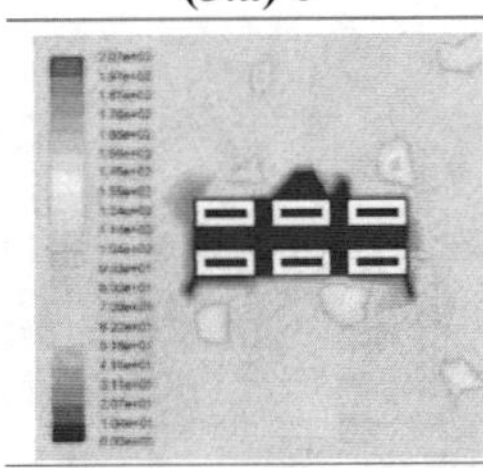	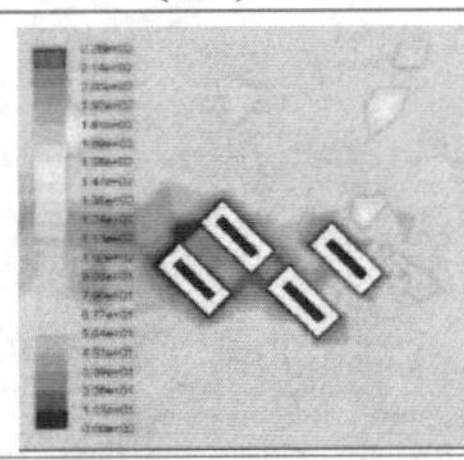	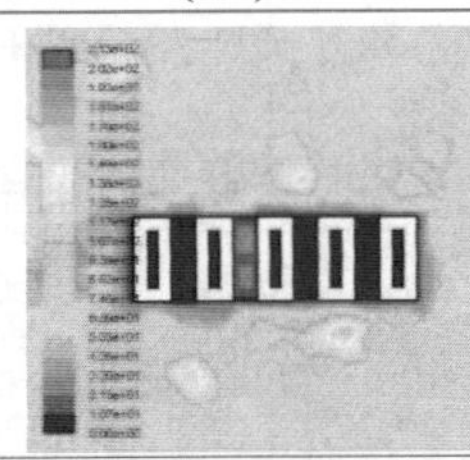

Table 4 Model 4 in simulation.

(4.a)	**(4.b)**	**(4.c)**	**(4.d)**
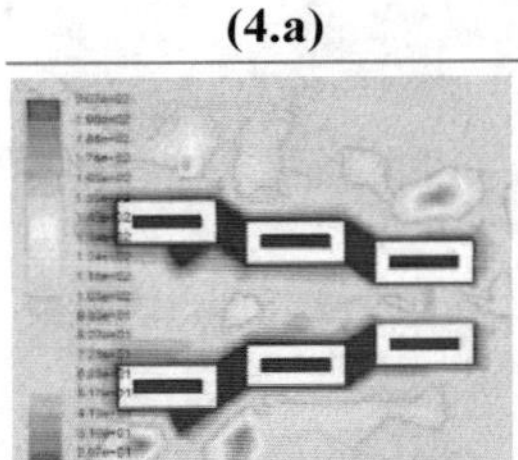	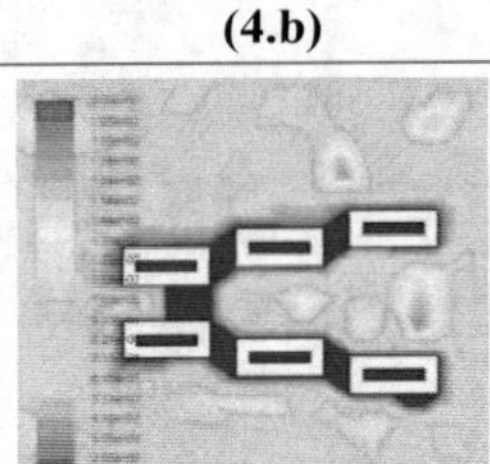	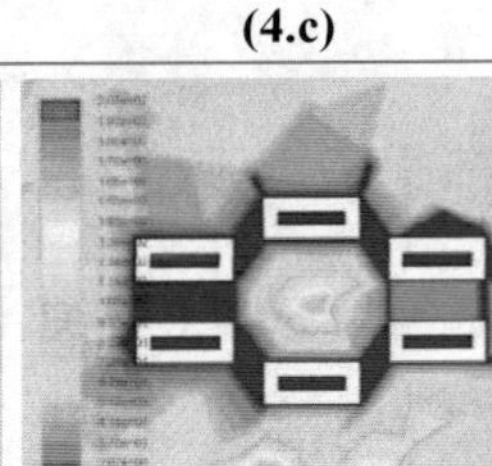	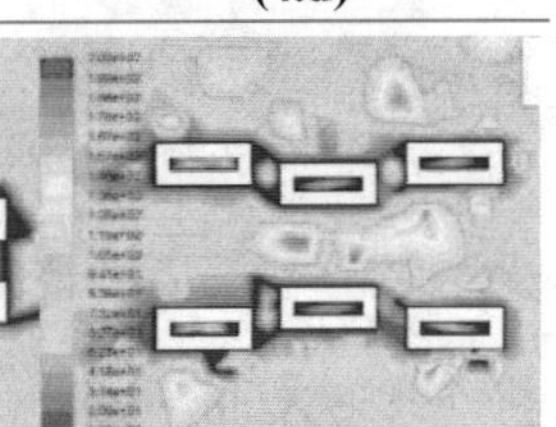

Fourth variable to be tested is building layout. For tropical region, unlinear building layout is better because it will create courtyard where airflow can pass around the buildings [9]. Each alternative has different value of distance between buildings. Combination of distance between buildings in model 4.d gives the best result (indicated by light-blue colour).

Model 4.d gives a good airflow around the buildings, but there are still some dark-blue areas in the gap between the buildings. Windcather is added in model 5, formed by the extension of building cores. This feature directs and splits airflow to evenly move inside and around the buildings.

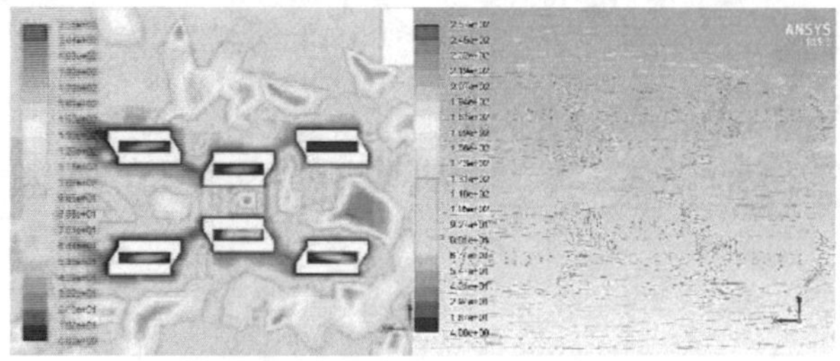

Figure 3 Model 5 in simulation.

3.2. Indoor Simulation

From the last iteration (model 5) we take data of velocity in the longer-side facade of the building, 1.4 m/s, which will be the new input for the following iterations. Due to the limitation of Ansys Modeler, in this simulation, the room model is simplified so the result is only valid for undivided room such us kitchen, dining, and living room.

The sixth variable is opening dimension. Model 6.a with dimension 0.6x1 metre, is the basic opening size taken from the existing rusunawa. This model gives enough velocity inside the room but concentrated only in inlet and outlet area. Changes made in model 6.b by expanding the outlet area as big as 0.6x0.9 metre. This changes gives much better result because airflow is distributed more evenly.

Baker (1987) in Seftyarizki [8] stated that window-to-wall ratio (WWR) of two opposing windows should be 20% each. So in model 6.c the opening size is expanded to 0.6x1.5 metre, giving a slightly slower velocity but with worse airflow distribution. In model 6.d, smaller opening dimension (0.6x0.8 metre) is tested which gives almost the same airflow result as 6.b but with higher air velocity.

Table 5 Model 6 in simulation.

(6.a)	**(6.b)**	**(6.c)**	**(6.d)**
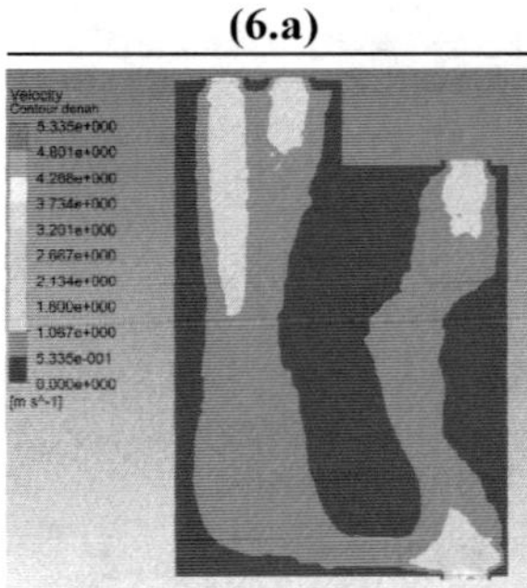	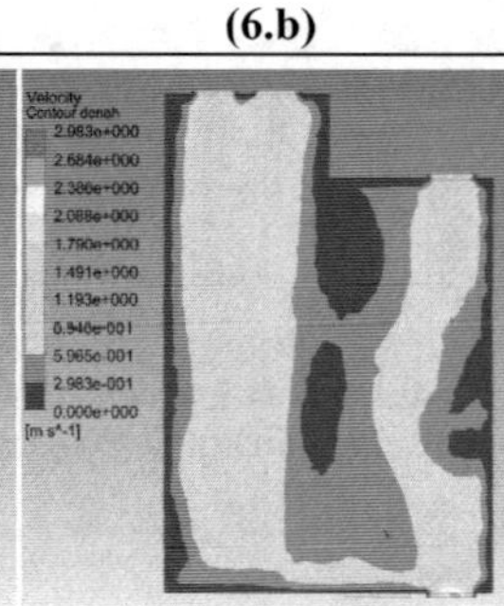	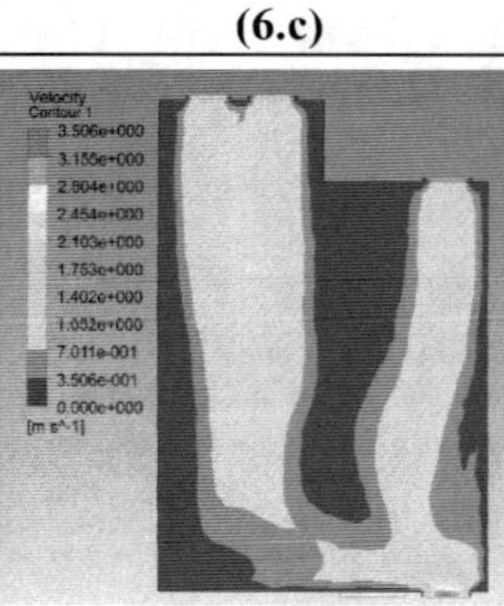	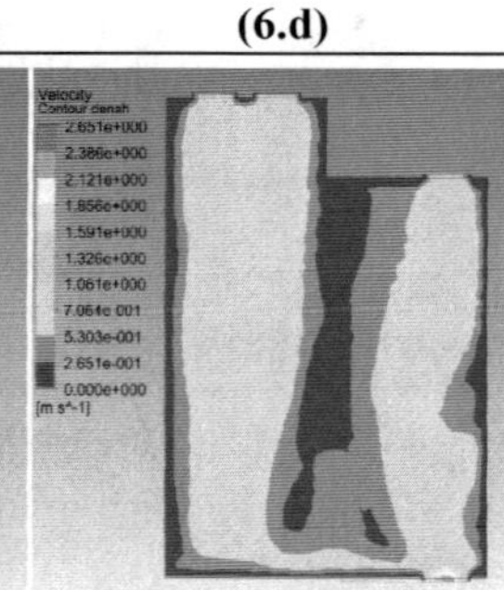
V = 0.6848 m/s	V = 0.7000 m/s	V = 0.6992 m/s	V = 0.712 m/s

From the last iteration, model 6.b is considered as the best alternative but the velocity is still higher than SNI standard.

The next variabel to be explored is opening location. Theoritically, inlet and outlet should be located in different side and different level to maximize cross-ventilation. In model 7.a, ouput is moved to higher level so hot air near the ceiling can be directly removed from the room. But this model has a problem because fresh air moves straightforward to the outlet. The air velocity is also still high and inhabitant can feel uncomfortable because air enters directly to their face or chest. So in model 7.b, inlet is separated into two location so fresh air can be distributed better and moves vertically, sweeping the room from below and exits in the upper outlet. This alternative gives air velocity of 0.36 m/s inside the room with temperature of 27°C, so thermal comfort by SNI has finally been achieved.

Table 6 Model 7 in simulation.

(7.a)	**(7.b)**
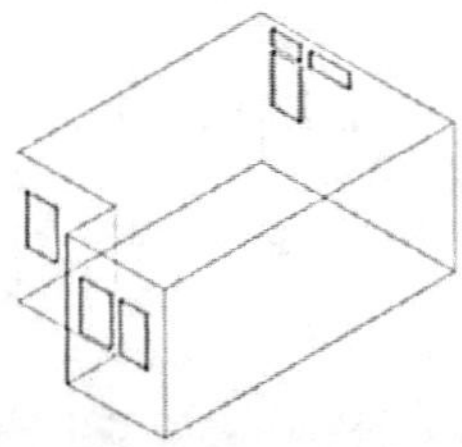	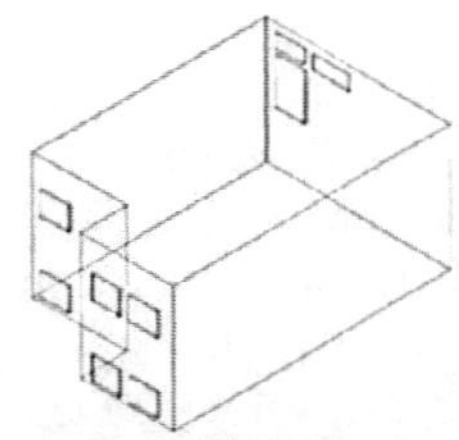
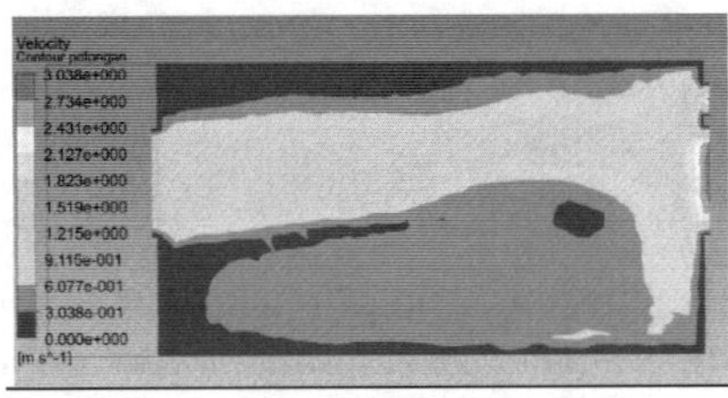	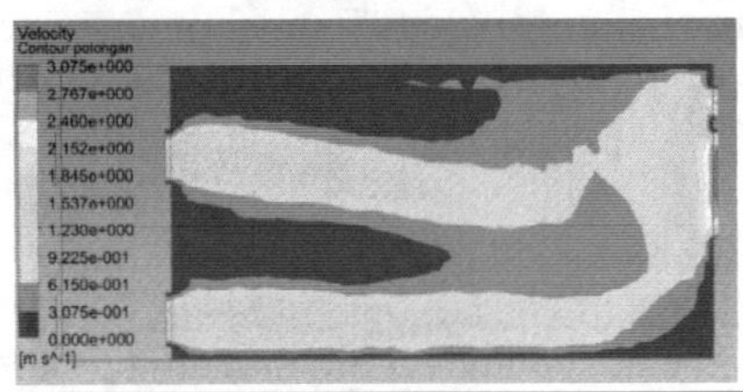
V = 0.6806 m/s	V= 0.3654 m/s

4. Discussion and Conclusion

From all the simulation that has been performed, we can conclude that building layout can give big impacts to airflow movement around the buildings. In model 4.d, the combination of building distances (wide – narrow – wide) create kind-of trap or tunnel that strengthen the wind in the middle of the site, so we get a constant wind velocity along the site. Unlinear (zig-zag) layout also allows wind to move in the small gaps between buildings producing better airflow. Windcather such us slanted or soften building corner is an effective method to direct wind, which can also be an interesting architectural element.

With smaller WWR than 20% we can still produce a good quality of airflow inside the building. In cross-ventilation, inlet and outlet area should be balance because a big difference of inlet and outlet WWR can cause uneven distribution of airflow inside the room. Opening location is also an important variable in controlling airflow. Lower inlet location and higher outlet location directs airflow to sweep along the room, maximizing cross-ventilation, and creates a more comfortable condition for inhabitant.

With passive design interventions, thermal comfort can be achieved especially by wind-driven natural ventilation. We can also reduce the usage of mechanical utilities to obtain thermal comfort. Along with the increase of housing supply, we hope this research can be a solution to reduce housing energy consumption.

In this paper, only energy for ventilation that has been researched and simulated quantitatively. With further research in other aspects of energy efficiency, this research can be developed as a design guidline for an energy-efficient vertical housing.

5. References

[1] Agency International Energy 2015 *Southeast Asia Energy Outlook 2015* (Paris: IEA)
[2] Statistik Badan Pusat 2015 *Statistik Indonesia 2015* (Jakarta: Badan Pusat Statistik)
[3] 2005. *Housing Types.* (Minnesota: College of Architecture and Landscape Architecture, University of Minnesota)

[4] Larasati Dewi 2000 *Sunshading Design Method on Preliminary Design Stage for Multi-storey Building* (Bandung: Thesis of Master Program, Institute Technology of Bandung)
[5] Baker Nick 1987 *Passive and Low Energy Building Design for Tropical Island Climates* (London: The Commonwealth Secretariat)
[6] Szokolay S 2001 The building envelope In A. Krishan, N. Baker, S. Yannas and S. Szokolay (Ed.) *Climate Responsive Architecture 1st ed.* (New Delhi: Tata McGraw-Hill Publishing Company Ltd.)
[7] Thomas R, Garnham T 2007 *Environments of Architecture: Environmental Designing Context* (New York: Taylor & Francis)
[8] Caesariadi T W, Kalsum E 2011 Climatic responsive in Melayu Pontianak House *International Conference Local Wisdom in Global Era* UKDW Yogyakarta
[9] Seftyarizki D, Yasin M P E, Wonorahardjo S 2016 Optimizing Natural Ventilation in Designing Budget Hotel for Thermal Comfort in Bandung *Arte-Polis 6 International Conference* 309-318
[10] Burnett J, Boji´c M, Yik F 2005 Wind-induced pressure at external surfaces of a high-rise residential building in Hong Kong, *Building and Environment* 40 (6) 765–777
[11] Nielsen K 2007 A Design Guide for The Built Environment in Hot Climates (London: The Cromwell Press)

6. Acknowledgments

This research is part of final year thesis project in Architectural Design Master Program, School of Architecture, Planning, and Policy Development, ITB. The local climate data assigned in the simulation is based on the data from Mr. Hastuardi from Badan Meteorologi, Klimatologi, dan Geofisika Tanjung Priok Weather Station.

IOP Conf. Series: Earth and Environmental Science **152** (2018) 012018 doi:10.1088/1755-1315/152/1/012018

Designing Tiban island as tourist destination and sustainable coastal in Bleder village

Sri Yuwanti [1], **Agung Setiawan** [2]

[1,2]STIEPARI Semarang, Pasca Sarjana, Lamongan Raya Street, No 11, Semarang, Indonesia

[1]sriyuwanti@gmail.com

Abstract. This study is on tourism development process,which involves the function of local potential tourist attraction confronted to the needs for area natural conservation. The study location is a village called Bleder, or Kartika Jaya, in Patebon sub district, Kendal regency CentralJjava. The research unit is the management of a sandy island called "pulau tiban"or 'suddenly appears island' , and village's surrounding area. Qualitative methods approach is utilized in order to understand the village situation, community social economy activities toward the sandy island, and the meaning of thatt sandy island for the village sustainability. The study result shows that some problems have evident in the exploitation of natural potential for tourist attraction as confronted to environment conservation. Conflict of interests have also emerged among groups of people and group to group regarding the management of turism attraction. To deal with the problems, comprehensive program is proposed especially to increase community's awareness on sustainable tourism, marine and coastal conservation, and the benefit and risk of ecotourism. For Bleder tourism development, the community need to concern to their village environmental safety more than on economic benefit purpose only. Introduction

1. Background

Kendal coastal area is very important area since it encompasses 23 villages stretching from west to east around 41 km long.the coastal area of Kendal is covered almost entirely by mangrove plants that serve to prevent the land from sea water erosion or abrasion. This coastal also servesas the habitat for northern Java coastal's flora and fauna. Among all sub districts of Kendal, there is one called Patebon sub-district, a coastal area which is considered as always faces and had experienced the disaster and risk of tidal wave and sea water abrasion, located at the far north east of Kendal. The area had lost almost 4 hectares of its brackish water ponds caused by sea water abrasion. The mangrove forest in Patebon becomes area's landscape distinction.there is a large mangrove nursery located in Bleder or Kartika Jaya village which supplies other locals for mangrove plant demand, such as Semarang. There are also mangrove nursery in households scale in the village. Bleder or Kartika Jaya is an unusual settlement, since it was dedicated asa local transmigration site for a group of retired Indonesian military members and their family. The settlement then is named as Kartika Jaya villlage. Recent population of Kartika Jaya is heterogeneous with occupation consists of farmers, fishermen, factory workers, private employees, and civil servants. Eventhough located at the beach, only less than

IOP Conf. Series: Earth and Environmental Science **152** (2018) 012018 doi:10.1088/1755-1315/152/1/012018

10 percentage of Kartika Jaya population work as fishermen. The main product of Kartika Jaya is agriculture.the village is also known as Bleder beach as it is the place of an estuary of two rivers. Bleder beach is close to one natural object, the sandy land arising from the sea, an accretion, called pulau Tiban. The meaning of "pulau" is island, and "Tiban" in Javanese means 'appeared suddenly'. The island has triggered outsider curiosity to visit and they enter the location through Kartika Jaya village access. Today, the sandy island grows as one marine tourism attraction in Kendal. Physically, Tiban island is a coastal sediment in the form of sand bar, with average 50 meters width and about 1 km length. The distance from kartika jaya coastline is about 200 meters and the sea depth between the beach to the Tiban island location is not more than 1 to 2 meters. After the sea pines and mangrove trees are planted, the island grows stable with the witdh and length increasing along with sandy land accumulation.

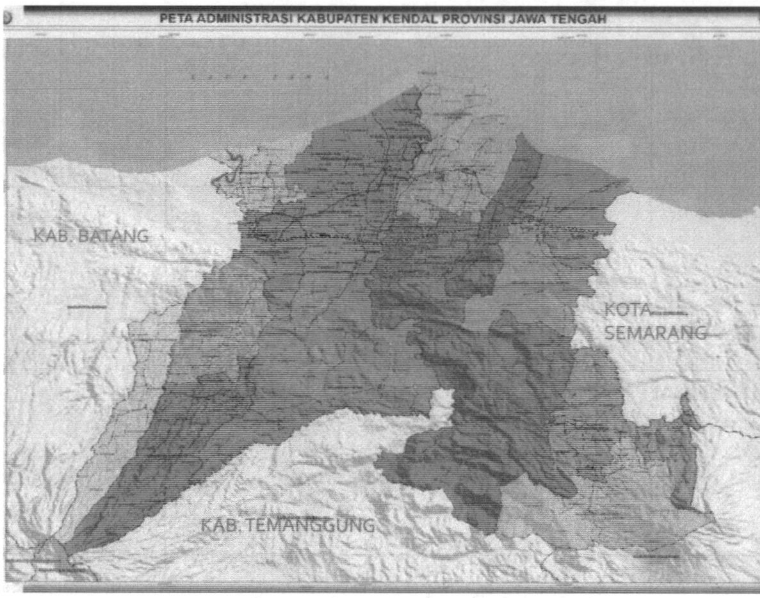

Figure 1. Map of Kendal regency

2. Problems

Pulau Tiban position is in between the Java Sea and the Bleder beach, in the form of tiny sandy island heading a lagoon surrounded by mangrove forest. A village tourism in Kartika Jaya has grown since the *gosong*, or sandy land arise on the coastal, is known locally. The tourism service is managed by the village community, that is the tourism awareness group or *'kelompok sadarwisata' (pokdarwis)*.The Bleder *pokdarw is* has been established in 2006 and named as Pokdarwis Kartika Beach. It has two divisions. One division is to handle tourist boat as transportation to go back and fortto the Tiban Island, while the other division is to do the conservation task to plant mangrove to save the coastal environment on the island, at the inland shoreline and along the river bank. There are about 200 visitors per day during weekend and holidays, and the ticket price for the boat is Rp.20.000 for modern boat and Rp.14. 000 for traditional boat. This scheme has indicated income generation from boat transportation. To increase the visitor safety and convenience, the village's tourism awareness group has been taught the basic save and rescue technique (SAR) and given guidance or training in the basic principle of tourism including institutional organization, home stay management, English for tour guide, etc.

There are some development that have been made by the local people themselves, such as building modest public infrastructure such as: simple hut or gazebo as a place to rest, humble public toilet, little mosque, and food and beverage stalls on Tiban island. The other division, has gain income from the mangrove nursery, but the amount is very small. Each plant, a seed or 'young mangrove' price is only 300 rupiahs and the group members must work hard to put each young plant into plastic bags to transport to other place when it is sold. It can be done after the young mangrove is strong enough to be moved from the nursery. It takes more than a month, and the market is scarce. In the mean time, there is no clear relation between the money accrued from tourist boat and from mangrove nursery, since nobody mention the deviding or shared income from both division organised in the *kelompok* or group.. Each group has its system on income distribution.One member of conservation division had mentioned that he did not against the growing of visitors to the island but it has caused more hardship in maintaining the environment sustainability of the island, such effort without enough budget.

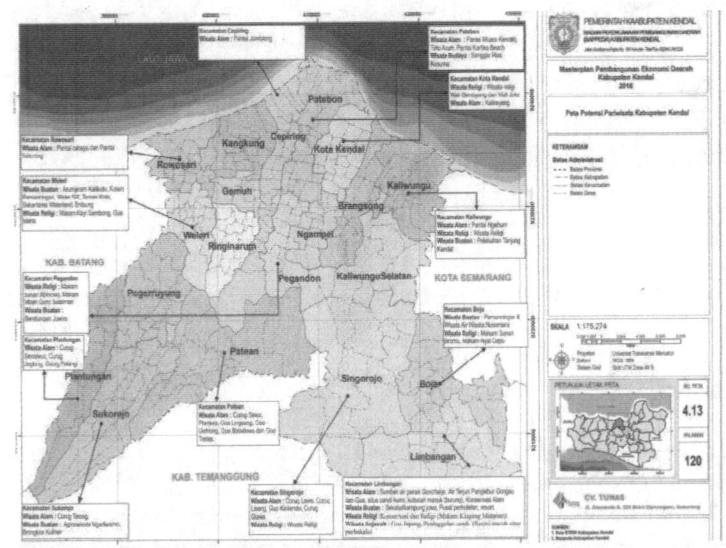

Figure 2. Map of and the location of Patebon subdistrict

3. Objective

The purpose of this research is to evaluate the process of the development of the disaster-prone settlement to be a tourist destination based on its naturalal attraction and assess its development direction. Evaluation of the resent development is done to get information on the risk and benefit of community based tourism utilizing natural potential. Assessment towards the potential for the tourism business sustainability is done by learning its current opportunities such as the growing numberof visitor and activities supporting the village tourism in relation to its basic environment. On the other hand, threats are also identified through activities which against environment sustainability and its impact. In Tiban island case, the impact of tourism activities on coastal preservation and village settlement sustainability. Since the problems associated with coastal in general may occur in all coastal villages along the Java coastline, the results obtained from the field can be exampled.

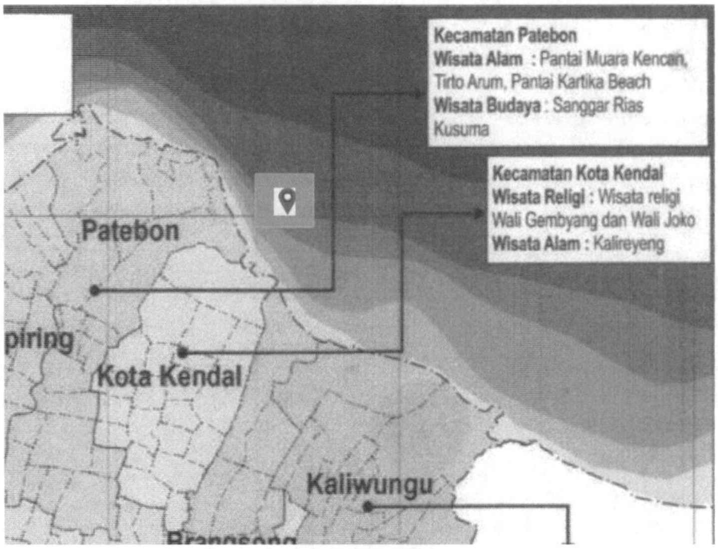

Figure 3. Map of and the location of Tiban island at Kartika Beach

4. Theory and Concept

4.1. Principles and concepts of tourism development

There are some concepts on planning and design especially related to tourism development that used to serve this study, i.e: a) Design, tourism planning, and feasibility ownership, business permit, ect. b).Spatial/landuse technical aspect, this aspect will do the inventory on layout and land use. Further is mapping the potentials and infrastructure. c).Socio economics aspect, that is related to the economic structure and its capacity to accomodate the needs of the area development and its business multiplier effect. d).Environmental impact aspect, which relates to the needs to analyse the impact of tourism activities to natural resources and it influences to the people living condition and their physical environment. Gunn's model on tourism development which is based on regional factor, exposes the area geography and spatial aspect a lots. One of the four major phases of the planning work is a research of geographic position, related to its content and its landscape expression, by which the domination of the landscape will be the location distinction [1]

4.2. Holistic approach in designing tourism destination

Fagense (1991) in Gunn [2] offers a new tool in his time, on the spatial aspect of tourism which help him to identify the suitability of tourism destination zone design. The geographical framework he provided was able to express the spatial aspects of location. The most relevant to the environment issuehe worked on is that the tool help to more efficiently identify location of zone for tourism development, to maximize the utilization of indigenous spatial, economic and environmental resources and to pursue specialization and balance to achieve optimization of the realistic opportunities. Further, Fagense's insight tool can help in formulating integrated strategy to accommodatethe local to national initiative in entrepreneurial at local level so as to provide a context of investment and development confidence, and so as to coordinate transportation, communication, utility infrastructure, and public capital work programme [2]. According to Gunn, the strength of Fagense's approach is:".... its practical applicability which emphasizes tourism management tool that are not only spatially sound but also are related to local integration, business success, and environmental protection..." [2].

5. Methods

A qualitative approach is used to have a better understanding on the growing situation in the field. The main observational unit is the physical area of Kartika Jaya or Bleder village and the pattern of economic activity of the community especially in village based tourism. Data used are primary and secondary. The primary data is obtained by conducting interviews and observing the activities of the informen consisted of the tourism awareness group, village apparatus, and the community which is divided into groups: local people, the residents involved in tourism activities (food stall, boat tours, homestays), and visitors whom exclusively comes to see the island. The analysis mainly is to evaluating developments and assessing the impact of the current situation. Theories and concepts used to assist the analysis consist of the concepts of tourism planning, environmental sustainability, and conservation-based tourism programm and activities design. Tourism development link to the design program concerned with the real situation of the village observed. This research aim is to find the way in designing tourism development activities in the form of a sustainable community-based tourism, in a coastal village in Central Java, as well as the need to conserve it physical environment. The process of thinking pattern is illustrated as shown inthe mindset flow chart figure.

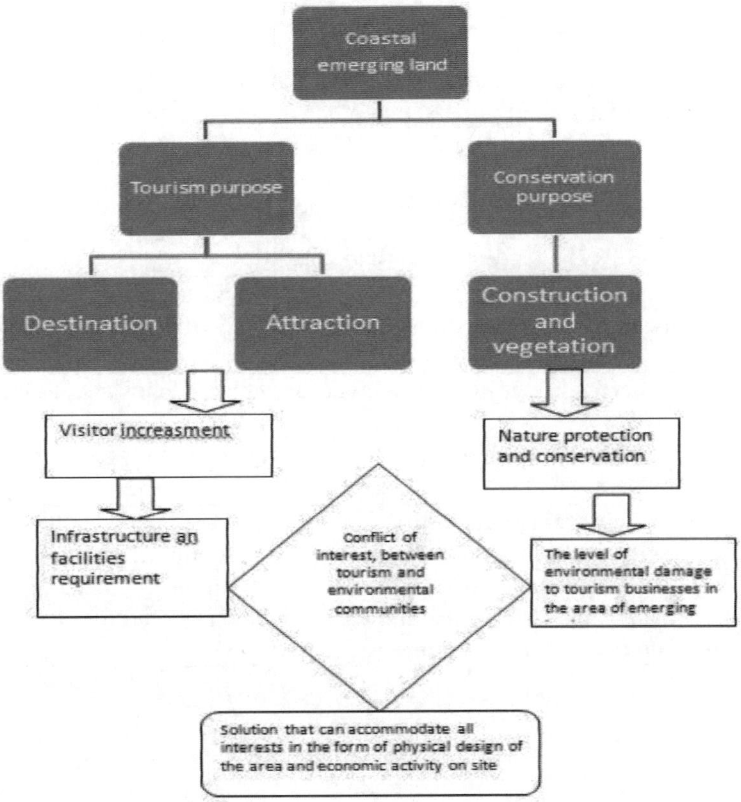

Figure 4. Thinking pattern

6. Results and discussion

6.1. Results

6.1.1. Village social economic potentials. Kartika Jaya village has various potentials of natural and especially of cultural asetsThere are some of theuntapped existing potentialsthat still need to be dig furtherto be utilized as the atraction and way to conserve the Tiban Island. (see table 1).

Tabel 1. Kartika Jaya's village natural potential and its utilization level

Current	Condition	Utilization rate	General Trend	Future Prospect
Tiban Islan	Good	High	Most visited object	Without control will capable to damage the island
Forest and Mangrove cluster	Good	Low	Not yet dig and utilize	Potential to be developed as new attraction
River as attraction	Poor	Low	Not yet utilized	Potential to be developed as new attraction
Bledder beach lagoon	Good	Low	Not yet utilized	Potential to be developed as new attraction

6.1.2. Problems related to village tourism development. Problems found in the real condition consist of village infrastructure, conflict among groups, and other related to public facilities and environment knowledge of the people.(see Table 2)

Tabel 2. Problems found at the site

Problems	**Real conditions**
Poor infrastructure	Small and damage road, only limited traditional rural transport available up to a certain point outside the village, The access pass through the conflicting village.
Internal conflict of interest	Different perception between conservation groups and tourism group.
Eksternal conflict	Conflict Between Kartika Jaya and Wonosari village regarding the possession and tourism benefit from Tiban Island
No Source of clean water	Clean water are taken from the village to the island for the public toilet and the food and beverage stall's requirement
Conservation and visitor management knowledge	Low knowledge on sustainable tourism on the visitors and management sides, the knowledge which required for environment conservation and visitor management.

6.2. Discussion

6.2.1. Regulation based the development of tourism on the coastal. The practice of coastal management and sustainability in Indonesia is ruled by the country regulation especially the Law number 27 year 2007 on Coastal and Small Island Management which must be consulted in Tiban island case. For the coastal tourism development case, the rule of HP3 or Hak Pengusahaan Perairan Pesisir (the right for utilizing and managing coastal) and the need on coastal area conservation, in order to protect the environment resource while there also many efforts to use it. In this law, it is defined that the coastal line is a shoreland or coastal physical border in a proporsional width, minimum 100 meters away to the inland measured from the highest sea water level surface. This position then is used as the main theme or focus for environment isue in the evaluation of existing Tiban island tourism activity since the community have exploited the coastal right in the 0 meter borderline seawater surface level. In the consepts, the protection of shoreline with mangrove plants and all regulations about it is intended to prevent the inland from any coastal disaster by either human activity or natural causes. The protection through disaster mitigation might be done physically or structurally, and solved with construction or vegetation program.The protection and rehabilitation is done in order to lift the coastal carrying capacity to give more space for human dan other living creatures to life. In the community level, this regulation is not only does offer an understanding and knowledge about a better environment, but forces the community to involve in guarding and protecting the coastal, and also limits them from destructive actions.When the coastal is supposed to give them the highest benefit from any related business activity, they tend to ignore the regulation when doing their business there. Basically, they must know and understand the risk if they broke the regulation. The usual activity unconform with the regulation on coastal sustainabilty done in the field of tourism are: - developing tourism activity in the prohibited area, include building permanently physical infrastructure without permission-littering on the beach, -disrupting the coral reef complex, -cutting mangrove plants or disrupting the mangrove ecosystem, -fishing using disappropriate tool like fish-bom, and the likes as prohibited or against the coastal ecologycal function [3]. Beside of this law, there are regulations that support and give detail explanation on the main law, to guide and direct further regulation aplication in the field. Based on the cases, many more regulations must be consulted by the community before they run tourism industrial or business activities. Similarly, many regulations and programs must be looked over and learned before designing tourism activity development, wheter it is in the coastal, on the beach, or inland.

6.2.2. Programs potential at Tiban island. Ecologicaly,Tiban island functions as a coastal belt from seawater abrasion for the Kartika Jaya. When Tiban island is treated as tourist attraction or destination then it has to follow the tourism business regulation and has to choose the standard for ecotourism. There are at least three reasons to further develop tourism in Tiban island: economic calculation on benefit and cost, infrastructure for serving visitors, and appropriate activities for conservation site and its proper management. First, related to the economic concept of tourism business which rely on the greater number of nature uniqueness.The benefits depend on the cost and income that comes from visitors. Costs incurred will be used to increase access and transportation, and it will require greater number of the motor boat. Unfortunately, the more motor boats operated, the more vulnerable condition of the the river wall embankment due to heavily exposed to the rip current of the modern motor boat. It also means more cost required to reinforce the riverside bank from the village to the lagoon of Tiban island, which of course can not be covered by income accumulated from boat ticket prices. Secondly is related to the infrastructure provision to serve the market segment. Aside of the touring type of visit, meant shorter time, the stall will not much required by visitor who usually are characterized as allocentricist. Third, related to the design of tourism activities in environmentally conserved area. Knowledge on the land utilization regulation, community land ownership, village power on the arise land on the sea or a gosong, an accretion, the environment conservation knowledge of community, the impact of increasing tourist arrivals on the economic activities of the village and on the environmental conditions. The question is then on what kind of design is most appropriate for Tiban island alike condition. Protected areas are surely not the recommended places for tourism, especially for the mass tourism type activities. Nevertheless there is still space that can be utilized for tourism business, in the form of special interest tourism activities. Tourism activities in the protected areas or restricted areas are included in ecotourism . In the case of tourism business managed by the community, as the Kartika Jaya case, the most important is the sustainability of business and forever existent of tourism asets, whether it is cultural assets, or natural assets. Public commitment must be drawn and agreed by Kartika Jaya's villagers, to save Tiban island ecosystem and its main function,to prevent the village and its surrounding from sea water abrasion disaster. Therefore what needs to be done by Kartika Jaya and the surrounding village, is to free the island from the burden of mass visits but still make it as the site for coastal study and marine research. The final question might be on how to increase the benefit from environment to lift income and welfare of the whole local community through limited tourism activities without disturbing the function of the nature. In Tiban island case, the answer will need more creative thinking using local potentials and done by the locals themselves. The first thing proposed is on creating ecotourism and agrotourism attractions, then trying to manage the attraction for the village community welfare. Based on the village geographycal situation, some of the village economic potentials are worthed tobe utilized as means to develop Kartika Jaya as a village for special interest tourism destination. In example, as the village has two rivers, its can be used forriver tracking activity program while heading to Tiban island using traditional boat and many other aquaculture ideas. In one hand, it will widen the village economic growth and people income through opening employment opportunity for mangrove group activists, home industry producers, village culinary business, ponds and other fishing culture owners, etc. In the other hand, the activities will trigger the village toward development as a tourism destination by involving the whole member of community and pose the village as the service zone of the Tiban island. In addition, Bleder beach mangrove forest can also be used as a place to learn in mangrove planting and nursing maintenace. Many home industries that have already operated related to mangrove product such as batik processing, mangrove crackers (tancang) production process, culinary based on brackishwater fish or aquaculture product, and some curiousity goods made of sea shell, mangrove, or sea pine seed. All of those are included in attraction definition, the first ingredient in the concept of tourist development. The second ingredients is on access which is served by the traditional boat and other village ecofriendly vehicle. This is not easy, since it needs to change the mindset of motorcycle and motorboat users.The third ingredient of principle of the tourism development concept, is the amenities in the form of accomodation, such as homestays, food and beverage, rest areas, etc.

7. Conclusion

7.1. The conluding remarks

Based on the field findings and discussion, it can be conluded, that: First, the tourism activities of Tiban Island as a marine tourism object on the coast of Bleder, tend to influence the environment with the negative impact such as, erosion along the riverside up to Tiban island pier, and the sea pine trees logging to establish the food stalls on the Tiban island to serve visitors. The cost of rehabilitation will be greater than the income yield obtained today. In addition, the coastal disaster may happen because of the damage caused by human error and nature will. Second, claim by Wonosari village people that Tiban island was and is part of the Wonosari village due to its past location was within the boundary of their lost ponds as abrasion impact, need to be solved and explained legally and formally by government institution related, to eliminate the posible growth of grassroot conflict seed. Basically, the new land that arises on the sea or *accretion*, is to be owned by the central government, must be reported to the ministry of landuse and spatial planning, and managed under the authority of the marine fisheries ministry and the forest and environment ministry. By this limitation, the Tiban island formally is an orphan, not owned by either Bleder Kartika Jaya nor Wonosari community or each village administration. Third, the village settlement can be developed to be tourism destination, as replacement or substitution of Tiban island, by creating various tourist attractions inland. When Tiban island is treated as the magnet, the village settlement can be designed as tourist services zone. This will need for feasibility studies on village readiness, the availability of infrastructure, either the existing, the need for reconstruction or provision for the new one, and improving the skills of the community's tourism awareness group members on ecotourism and visitor management.

7.2. Further actions recommended

There are alternatives for developing Tiban island'secotracking activity and Kartika Jaya village ecotourism,such as: First, tourism in Kartika Jaya or Bleder village can still be developed by utilizing the safe zone in the village. Basically, all Kendal coastal area is defined as conservation area for avoiding seawater abrasion disaster, therefore Tiban island cannot be developed for mass tourism.

IOP Conf. Series: Earth and Environmental Science **152** (2018) 012018 doi:10.1088/1755-1315/152/1/012018

However, it is still possible to be designed for limited visitor tourism object, considering the ecological function of the island as a natural abrasion inhibitor for the surrounding environment. Second, the need to continuously monitor and to measure the conservation level conducted by mangrove and environmental groups, as an effort to maintain the assets of nature tourism, while protecting the island from further damage. Third, the need to empower village community by providing assistance and coaching, guidance and training especially to the existing tourism awareness group, concerning environment friendly tourist attractions, in order to establish skill on tourism and visitor management inline with environment management. Fourth, the need to built a stronger cooperation with the neighboring coastal villages along the coast of Kendal since those areas considered as a conservation zone that needs to be protected together. Fifth, utilization of other geographic potentials such as rivers, forests, etc. must be carefully planned to avoid another new environment disruption. Further detailed research is required to determine the potential and zones of village setlement, or between villages, that can be used for mass tourism or agrotourism activities while utilizing the Tiban Island as a learning spot on marine ecology and coastal conservation or other eco tourism activities.

References

[1] Gunn C A 1994. *Tourism Planning Basics, Concepts, Case*, vol 3 (Washington DC: Taylor and Francis) p 120-127
[2] Gunn C A 1994. *Tourism Planning Basics, Concepts, Case*, vol 3 (Washington DC: Taylor and Francis) p 119-120
[3] Regulation of Indonesian Republic, Act Number 27 Year 2007, Concerning Coastal Area and Small Island Mangement.

Acknowlegdment

STIEPARI research team convey the special gratitude for informants and Kartika Jaya village apparatus, those are: The village head of Kartika Jaya, Pokdarwis and all members, the members of Kartika Jaya PKK, the members of Karang Taruna Kartika Jaya,the stall sellers at Tiban Island, the stall sellers in the Kartika Jaya village, the owners of homestay, and for MP3 students of STIEPARI.

The Development of Historical and Eco-Tourism District of Setu Babakan in South Jakarta, Indonesia: Ecodistrict Planning Approach

Samsirina[1], Wiwik Dwi Pratiwi[2], and Reiska M. Putri[3]

[1,2] Lecturer of Architecture Study Program, ITB, Indonesia
[3] Research Assistant of Architecture Study Program, ITB, Indonesia

E-mail: samsirina@ar.itb.ac.id

Abstract. On the development of tourism within a settlement, its sustainability depends on how to make the lowest possible environment impact and provide long-term economic and social benefits to local communities. This paper describes the results of analysis on a settlement where the tourism activities developed in it. The principal of eco-district was the main theory used in the analysis in order to accelerate district-scale sustainability. Eco-district is the principle of urban planning that aims to integrate the objectives of sustainable development and reduce the ecological impact of the development. The case study was located in Setu Babakan, a lakeside and a historical tourism district in South Jakarta, Indonesia. The focus of the study was on the physical condition of the open spaces and activities inside them. In some tourism destination areas, many of the attractions took place in open spaces. The results of the evaluation were several recommendations for the development concept of open spaces in the historical and ecotourism district of Setu Babakan which hopefully will guarantee it's sustainability in the future.

Keywords: eco-district planning, ecotourism, evaluation, historical district

1. Introduction

The area of the lakeside has its own appeal, so as this area is often used as a tourist spot or as a place for recreation and socialization. The appeal is even greater if the location of the lake is located in the inner city area that has been very densely built, due to two major driving factors. The first factor is its appealing natural condition, contrary to the crowded city condition. The second factor is accessibility where the location relatively easy to reach by the people who live in the city.

Tourism activities that develop around the lake will certainly have an impact on the lake and the surrounding area. One such impact is the transformation of the physical, social and cultural environments around the lake due to tourism activities. The transformation can have positive or negative impacts. When the transformation leads to the type of change that has a positive value for tourism activities, then it will be able to support the sustainability of the area as a tourist area. Positive transformation can occur in the form of the improvement in physical quality through the provision of infrastructure facilities that can support tourism activities, as well as the improvement of the social and economic quality of the community with the opportunity to open small business in their homes or the employment opportunities in tourism business. Nevertheless, the impacts are often in the form of negative transformations such as the degradation of physical environment due to waste pollution and the increased of traffic load that also lead to infrastructure damage, the marginalization of local communities, acculturation of local culture and other negative impacts.

One of the lake tourism areas in Jakarta is located in Setu Babakan. Setu Babakan is a cultural heritage area located in Srengseng Sawah, Jagakarsa Subdistrict, South Jakarta. The plan to establish Setu Babakan as a Betawi Cultural Heritage area began in 1996, however, an official decision only then appeared in 2000 with the issuance of Governor's Decree no. 9 in 2000. The Betawi Cultural Heritage of Setu Babakan was inaugurated in 2004 after going through various preparation processes. The presence of the lake also become one of the attractions sold by the manager Setu Babakan in addition to other main tourist attractions directly related to Betawi culture such as the local tradition, art activities, traditional houses etc.

Tourism activities are closely related to open space. Open space can attract tourists and become the main place for recreational activities. Based on the observations, the majority of recreational activities in tourism areas are taking place in open spaces than in enclosed spaces. Therefore, the increase of tourism activities in one area will lead to an increase in the need for the availability of open space as a place to accommodate those tourism activities. Tourism activities related to the lake and cultural tourism activities in Setu Babakan take more places in the open spaces of this area.

This paper will describe the results of analysis on the settlement of Setu Babakan. The principal of eco-district was the main theory used in the analysis in order to accelerate district-scale sustainability. Eco-district is the principle of urban planning that aims to integrate the objectives of sustainable development and reduce the ecological impact of the development. The focus of the study was on the physical condition of the open spaces and activities inside them. The results of the evaluation were several recommendations for the development concept of open spaces in the historical and ecotourism district of Setu Babakan which hopefully will guarantee it's sustainability in the future.

2. Eco-district Principles

Eco-district is the principle of urban planning that aims to integrate the objectives of sustainable development and reduce the ecological impact of the development. According to PoSI as stated by Seltzer [1], Eco-districts are characterized by values of diversity and participation, equity in decision making and investment, health and well-being of the community, positive environmental impacts, and conservation and stewardship. There are 7 (seven) sustainability performance goals of eco-districts:

1. Community Vitality – Healthy, equitable, and vital communities with active and diverse participation
2. Air Quality and Carbon – Beyond carbon neutrality and healthy air quality
3. Energy – Net-zero energy usage annually
4. Access and Mobility – Healthy, clean, and affordable transportation options
5. Water – Water, in all its forms, meets both natural and human needs
6. Habitat and Ecosystem Function – Integrate built and natural environments for healthy urban ecosystems
7. Materials Management – Zero waste and optimized material

The definitions put forward by some researchers and planners on Eco-district itself are quite diverse. In this research, the patterns of open space in the settlement becomes the main variable to evaluate the current condition based on the eco-district principles and then proposed a number of improvement recommendations in order to meet the principles. Therefore, the definition of eco-district to be used is the definition of Bennet mentioned by Brown [2]. In Brown [2], it is mentioned that this definition was produced by Bennet who developed the model of an ecologically sustainable district. This definition builds the framework for a comprehensive strategy to build communities with the lowest possible environmental impact and highest long-term economic and community returns [2].

According to Bennet in Brown [2], An Eco-District is a neighborhood that generates all its energy from on-site renewables, collects and recycles rainwater and waste, and prioritizes pedestrian, bike, and transit access. It combines mixed use, mixed-income development; neighborhood scale parks, schools, community centers; and services, and enhanced IT infrastructure. The objective of the program is to test, accelerate and eventually codify the next generation of best practices in green development and civic infrastructure that can be scaled to create neighborhoods with the lowest environmental impact and highest economic and social resiliency [2].

By drawing a number of keywords from its definition, there are 6 (six) elements of a settlement pattern that can meet the criteria for ecologically sustainable districts:

1. On-site Renewable Resources
2. Rainwater & Waste Collection & Recycling
3. Pedestrian, Bicycle, & Transit Access
4. Mixed-Use & Mixed-Income Development
5. Neighborhood-scale Parks
6. Schools, Community Centers, & Services

To analyze the pattern of a district whether it meets the principle of eco-district or not, Brown [2] by improving the district pattern from Bennet has produced an eco-pattern district evaluation table as can be seen in the following table.

Table 1. Eco-pattern language

Ecodistrict Patterns	1	2	3	4	5	6
Walkable Community			X	X	X	X
Pedestrian Density			X	X	X	X
Web of Public Transportation			X	X		
Activity Nodes			X		X	
Green Streets		X	X		X	
Corner Grocery				X		
Parallel Roads			X			
Tangent Auto Roads			X	X		
Self-Sufficiency	X	X		X		X
Scattered Work				X		X
Zero Waste		X				
Energy Independence	X					
Water Neutral		X				
Urban Agriculture	X				X	X
Neighborhood Identity			X	X	X	X
Diverse Neighborhood				X	X	X
Manageable Scale			X	X	X	
Life Cycle		X		X	X	X
Building Fronts			X	X		
Individually Owned Shops				X		
Accessible Green		X			X	
Small Public Squares			X		X	
Neighborhood Boundary		X			X	
Main Gateway				X		X

Source: [2]

This table is adapted in the Setu Babakan study to evaluate the ecological level of the existing district pattern in the Setu Babakan area so that some recommendation can be proposed in the development of Setu Babakan as a neighborhoods with the lowest environmental impacts and highest economic and social resiliency.

3. Setu Babakan, South Jakarta

3.1. The Development Zones of Setu Babakan

This tourist destination of Kampung Budaya Betawi stands on a land area of 32 HA. Setu Babakan has great potential as a tourist attraction due to the presence of the lake within and it has a vast land that can be utilized for a variety of tourist activities.

Setu Babakan was divided into 4 (four) development zones. The four zones are zone A, B, C, and embryo zone. Of the 4 zones, zones A, B, and C have not been completed due to land ownership and development constraints. While embryo zones already can be visited by the public and usually crowded on weekends.

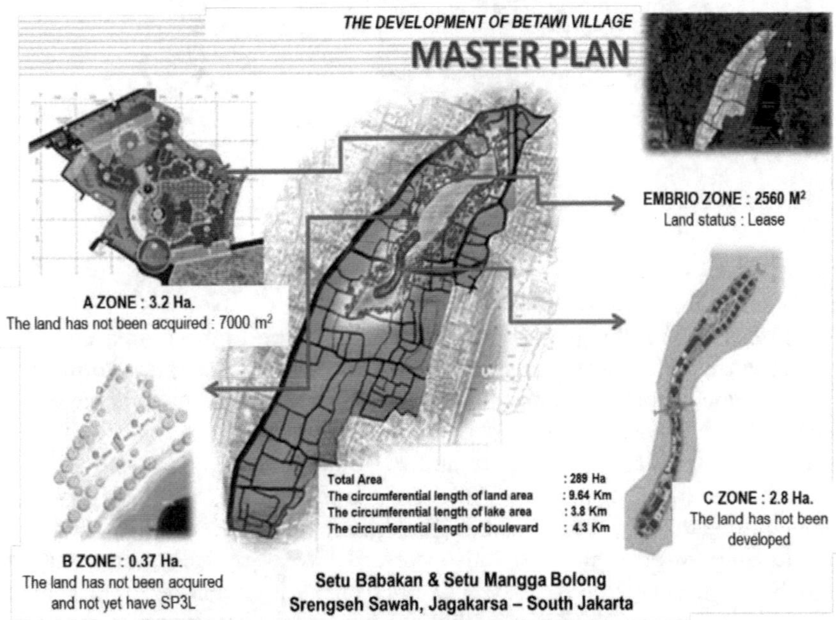

Figure 1. The Masterplan of Setu Babakan [3]

3.2. The Physical Condition of Open Spaces in Setu Babakan

From a village with an area of 289 hectares, 65 hectares of them are owned the government in which it is managed is only 32 hectares [4]. This settlement is inhabited by at least 3,000 families [4]. Most of the inhabitants are native Betawi people who have been hereditary living in the area. The rest are migrants, such as migrants from West Java, Central Java, Kalimantan and others who have lived more than 30 years in this area [4].

In the study area, there are some green open spaces with fairly clean and well-maintained condition as these open spaces also become the main destination for tourists visiting Setu Babakan. These open spaces have some functions that can be commonly used by tourists. However, in terms of quantities there are some decreases in the areas of open spaces around Setu Babakan. This transformation was driven more by an increase in the need for a place to live rather than to accommodate tourism facilities.

Figure 2. The Transformation of the Areas of Open Spaces in one of The Zone
Source: modification from Pratiwi [5]

Based on its function there are two types of open space: active open space and passive open space. The active open spaces within the area are mostly located in the Embryo area and used as the venue for some art performances. Meanwhile, passive open spaces mostly are in the form of parks and green paths. Open spaces in the Embryo area are also often used by tourists who visit Setu Babakan just for picnics or family gathering because the atmosphere in this place tends to be cooler than other areas in Setu Babakan and the area is quite extensive and there are many traders selling foods in this area. Besides being used by tourists, open spaces in the Embryo area are also used for daily activities by local people, such as to gather, hold meetings and as a place for children to play.

Aside from the Embryo area, there are also other active open spaces in Setu Babakan namely Warung Riung & Soban '22 which is a fairly wide open area with a number of nearby merchant kiosks and also there are some children playgrounds that can be visited by the public. Similar to the Embryo area, the presence of lush trees have made the atmosphere of the area of Warung Riung shady and cool, so that visitors pretty much stop to this place.

The other active open space is in the form of fish pond or called empang. Empangs are generally located behind the houses or stalls. Empang is one of the income sources of the Setu Babakan community. The embankment of the empang mainly planted with plants such as banana trees.

Green open spaces alongside of the lake have been equipped with trash cans, garden lights, and pedestrian pathway. The pedestrian pathway has a width of approximately 1.5 meters and paved with conblock material. On the side along the pedestrian pathway there is a vehicle lane that also paved with the same material as pedestrian pathway. This vehicle lane is also extensively used by pedestrians to circulate. Meanwhile, the pedestrian pathway is more widely used as a place to sit and eat, and equipped with benches and tables.

In the nearest area there is Setu Mangga Bolong which is still in one management unit with Setu Babakan. In contrast to the Setu Babakan, Setu Mangga Bolong area is not as much visited as in Setu Babakan and mostly used as a fishing spot.

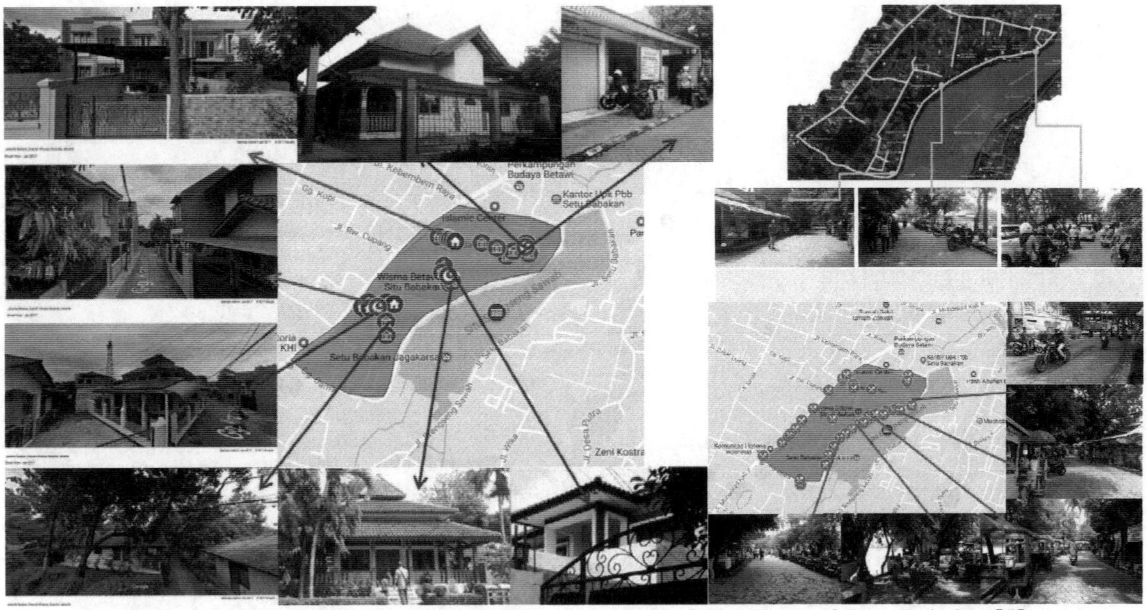

Figure 3. The Condition of Open Spaces in One of The Zone of Setu Babakan[5]

4. The Evaluation of Open Spaces in Setu Babakan based on the Principles of Eco-district

The evaluation of open spaces in Setu Babakan was conducted based on eco-district principles put forward by Bennet (2009) and by using the table to evaluate district patterns generated by Bennet (2009) and Brown (2009). The following table shows the evaluation results.

Table 2. Evaluation of Open Spaces in Setu Babakan

The Principles of Ecodistrict	The Evaluation of Setu Babakan				
	Y		N		Description
	1	2	3	4	
Walkable Community Encourage walking as a primary mode of transportation by creating an environment that is not only safe, but also easy an interesting for pedestrians to use. (Brown and Dykema, 2009)				X	The local streets of settlement around the lake are often used by visitor vehicles to enter and/or off the lake area which led to an increase of the vehicle load on those narrow local streets. Meanwhile, the Irregular circulation patterns as well as several dead-end streets in the settlement also make it difficult for visitors to be oriented in the district.
Pedestrian Density For public squares, courts, pedestrian streets, any places where crowds are drawn together, estimate the mean number of people in the place at any given moment (P), and make the area of the place between 150P and 300P square feet. (Alexander in Brown and Dykema, 2009)			X		There is a very significant difference in the density of pedestrians between weekdays and weekends/holidays. Pedestrian density also concentrated in embryo zones.
Web of Public Transportation The system of public transportation can only work if all the parts are well connected. (Alexander in Brown and Dykema, 2009)			X		The nearest road that is served by public transportation is Jl. Mohamad Kahfi II. Near one of the main entrance gates, the gate of Si Pitung, there is a public transport stop: Setu Babakan's Halte. The lake of Babakan itself is located inside the local settlement and has a distance of about 400 meters from the gate of Si Pitung.
Activity Nodes Create nodes of activity throughout the community, spread about 300 yards apart. (Alexander in Brown and Dykema, 2009)				X	The lake, the promenade along the lakeside, the performance stage, are some potential activity nodes that still need to be improved especially in zones other than embryo zones. At present, the activity nodes are more concentrated in the embryo zone. In addition,

The Principles of Ecodistrict	The Evaluation of Setu Babakan				
	Y		N		Description
	1	2	3	4	
					although Setu Mangga Bolong is still included in the same management area as Setu Babakan, this area seems to be cut off and there is no continuity of development activities with Setu Babakan area.
Green Streets Avoid the use of hot hard asphalt, especially for local roads. On local roads, put in more paving stones, gravel or grass blocks to form a surface for the wheels of the cars. (Alexander in Brown and Dykema, 2009)		X			Alongside of the lake has been provided promenade with paving block material that can still absorb rain water into the soil. In addition, shade trees also planted on along this side. Other than that, local streets in the settlement are still using asphalt.
Corner Grocery Give every neighborhood at least one corner grocery, somewhere near its heart. Place them on corners, where large numbers of people are going past. (Alexander in Brown and Dykema, 2009)	X				Minimarkets can be easily found on the side of the main road, while in the settlement around the lake, there are quite a lot of houses that open small shops so that residents can easily buy daily necessities. Tourism activities also increase the number of residents who open food and souvenir stalls in front of their houses especially those facing the promenade along the lakeside.
Parallel Roads Keep parallel roads at least 100 yards apart (to make room for neighborhoods between them) and no more than 300 or 400 yards apart. (Alexander in Brown and Dykema, 2009)	X				The main entrance road to the Setu Babakan area which are parallel between one another are Jl. Mohamad Kahfi II and Jl. Srengseng Sawah.
Self-Sufficiency A self-sufficient community is a sustainable one. Encourage a balanced mix of land uses that can accommodate industry, commerce, farming and housing by providing a diverse and integrated network of zoning laws. (Brown and Dykema, 2009)				X	Relies heavily on the presence of outside visitors. The main performance stage on the region still stands on private land (not yet released by local government). There is a mix of residential, commercial and retail, but still need to be intensified.
Scattered Work Prohibit large concentrations of work, without family life around them. Prohibit large concentrations of family life, without workplaces around them. (Alexander in Brown and Dykema, 2009)		X			There is a mix of residential, commercial and retail, but still need to be intensified.
Zero Waste Treat all waste onsite with biological systems. Reclaim as much water as possible for non-potable uses and harvest biomass for use as fertilizer on local farms. Collect sewage sludge, kitchen waste and yard waste and convert it into gas for energy. (Brown and Dykema, 2009)				X	There is no strategy related to the recycling of waste in the area. The handling of waste done only on the sorting between organic, an-organic and B3 kinds of waste.
Energy Independence Minimize dependence on a greater municipal grid for the energy needs of the community. Take advantage of onsite renewable resources such as solar, hydro, geothermal and wind power to generate the energy required to make the district operate. (Brown and Dykema, 2009)				X	Energy needs are still highly dependent on PLN (state-owned electricity company)
Water Neutral Minimize dependence on a greater municipal grid for the water needs of the community. Collect rainwater and treat blackwater onsite using biological systems. (Brown and Dykema, 2009)				X	There is no water-related strategy within the region yet. Formerly, lake water can be used by local residents, but because the water condition of the lake is now polluted then it is no longer used by the residents for their daily needs. The majority of residents still take water from the ground (pump wells) and there is no

The Principles of Ecodistrict	The Evaluation of Setu Babakan				
	Y		N		Description
	1	2	3	4	
					waste water recycling efforts.
Urban Agriculture Encourage the community to support local farmers and grocers by providing them with tax credits so that their prices are competitive. (Brown and Dykema, 2009)		X			Residents are still conducting the old tradition where on the yard of their houses they plant productive trees. The local government also planted productive trees in the development program of this Betawi cultural heritage tourism area.
Neighborhood Identity People need an identifiable spatial unit to belong to. Help people to define the neighborhoods they live in. (Alexander in Brown and Dykema, 2009)		X			Strong historical and cultural identity of the neighborhood close to the heart area of tourism. But only several houses located deep inside the settlement area are applying Betawi architecture elements especially for the façade.
Diverse Neighborhood The homogeneous and undifferentiated character of modern cities kills all variety of life styles and arrests the growth of individual character. Do everything possible to enrich the cultures and subcultures of the city. (Alexander in Brown and Dykema, 2009)		X			Diverse community in terms of their level of economic ability, and ethnicity / religion. There have been many non-Betawi residents who have been residing in the area.
Manageable Scale Give each community the power to initiate, decide, and execute the affairs that concern it closely: land use, housing, maintenance, streets, parks, police, schooling, welfare, neighborhood services. (Alexander in Brown and Dykema, 2009)	X				The preservation of Betawi culture in the Setu Babakan district involves the local community. Festivals and performances are performed alternately by residents from various RT and RW.
Life Cycle Make certain that the full cycle of life is represented and balanced in each community. (Alexander in Brown and Dykema, 2009)				X	Facilities for resident's life cycle that already exist are: school (kindergarten, elementary, junior high, high school), university, business, and cemetery. Need more upgraded with facilities that do not yet exist: nursing home, barrier free environment etc. There is a plan for the development of a Child-Friendly Integrated Public Space with an area of 6000 m2.
Building Fronts Building set-backs from the street, have actually helped greatly to destroy the street as a social space. Build right up to the paths. (Alexander in Brown and Dykema, 2009)		X			Most of the buildings are facing toward the lake. However, there is a potential for the occurrence of the street without facades on the side of the lake due to several lands fortify their peripheries adjacent to the promenade alongside of the lake. Originally, the local streets in the area were still in the form of soil paths that were later widened (the residents alongside the streets donated some of their lands for street expansion). Some of the houses whose land has been cut off for the expansion of the street then change their orientation towards this local street.
Individually Owned Shops When shops are too large, or controlled by absentee owners, they become plastic, bland, and abstract. Encourage the development of individually owned shops. (Alexander in Brown and Dykema, 2009)	X				Stalls and small retail stores are the foundation of the livelihood / economy of the district.
Accessible Green People need green open places to go to. Build one open public green within three minutes' walk—about 750 feet—of every house and workplace. (Alexander in Brown and Dykema, 2009)	X				There is no tariff to enter the area (free of charge). Visitors who bring a vehicle are only required to pay for parking tickets for the vehicles.
Small Public Squares Make a public square much smaller than you	X				Promenade along the side of the lake is a public social space which can be accessed by anyone.

The Principles of Ecodistrict	The Evaluation of Setu Babakan				
	Y		N		Description
	1	2	3	4	
would at first imagine; usually no more than 45 to 60 feet across, never more than 70 feet across. (Alexander in Brown and Dykema, 2009)					
Neighborhood Boundary The strength of the boundary is essential to a neighborhood. If the boundary is too weak the neighborhood will not be able to maintain its own identifiable character. (Alexander in Brown and Dykema, 2009)		X			The main gates and houses with strong Betawian identity near the entrance area create a strong perceptual boundary of this Betawi Cultural Heritage area.
Main Gateway Mark every boundary in the city which has important human meaning— the boundary of a building cluster, a neighborhood, a precinct— by great gateways where the major entering paths cross the boundary. (Alexander in Brown and Dykema, 2009)	X				There are three entrance gates scattered in the arterial road of Setu Babakan area.

Notes:
1 = appropriate
2 = slightly appropriate
3 = slightly un-appropriate
4 = un-appropriate

Based on the table above, it can be concluded that the district patterns that have met the ecological criterias but still need some improvements are: (1) green street, (2) corner grocery, (3) parallel roads, (4) scattered works, (5) urban agriculture , (6) neighborhood identity, (7) diverse neighborhood, (8) manageable scale, (9) building fronts, (10) individually owned shops, (11) accessible green, (12) small public squares, (13) neighborhood boundary, and (14) main gateway. Setu Babakan area already equipped with several infrastructures and facilities to support the activities of tourism and living. Local community participation in the activities of tourism also becomes another added value for its social resiliency as a settlement and as a tourism area.

Meanwhile, the ecological patterns that have not existed in Setu Babakan area and need to be a priority are: (1) walkable community, (2) pedestrian density, (3) web of public transportation, (4) activity nodes, (5) self-sufficiency, (6) zero waste, (7) energy independence, (8) water neutral, and (9) life cycle. Connectivity and integration of existing infrastructure become main aspects that have been missing in Setu Babakan area. Inclusivity becomes another main issue due to the lack of facilities especially for disable and elderly. The last issue is related to the sustainability of the area: economic and environmental sustainability.

5. The Recommendation for The Development of Setu Babakan Eco-district

Potential development of Setu Babakan Ecotourism can be formulated based on the evaluation result above. More than a half of eco-district pattern criteria from Bennet (2009) and Brown (2009) have been met in the current Setu Babakan area. Thus, it can be concluded that the Setu Babakan area has the potential to be further developed as an eco-district tourism area. Recommendations are divided into two types: recommendations for the improvement of existing eco-district patterns as well as the recommendations to develop the eco-district patterns that do not yet exist.

On the existing eco-district patterns, improvements can be made primarily by improving the quality of the tourist route from the main gateway to the lake area as well as the green pathways connecting all open/green spaces and public open spaces in the Setu Babakan area. The improvements of tourist routes are not only by improving the quality of the pavement but also by improving the comfort and environmental attraction around those routes. The Increased comfort of tourist route is realized through the provision of shade trees, seating and resting area at a number of points and the separation between the vehicle circulation and the pedestrian circulation. Pedestrian circulation should be a top priority within the district. Meanwhile, the increased attractiveness of the environment around the tourist route is realized through the increased maintenance level of buildings and open spaces especially those passed by the visitors along the tourist route.

Meanwhile, the followings are recommendations in a number of district patterns from Setu Babakan that have not yet ecological and should be a priority for the development.

5.1. Walkable Community

The concept of walkable community in Setu Babakan area can be presented by the functioning of pedestrian and promenade path alongside of the lake purely as the area for pedestrian to circulate. In some parts of the area, especially near the entrance area, it is necessary to provide adequate parking spaces so that cross-circulation does not occur between pedestrians and vehicles as happened in the current conditions. In addition, to make pedestrian circulation as the main mode of circulation in the district, there should be a provision of safe, easy and attractive pedestrian environments. Houses and other buildings should be directed toward pedestrian paths, the quality of pavements for the pedestrian path as well as for the vehicle street should be improved and from these paths, there should be a possibility to view towards attractive sights.

5.2. Pedestrian density

Spread some points of activity to attract visitors throughout the development zone in Setu Babakan region so that the densities of visitors are not concentrated in one zone only. This concept is reinforced by the provision of a convenient pedestrian link network that connects the entire zones.

5.3. Web of public transportation

Provide a convenient place for public transportation transit. Provide additional facilities to increase comfort such as adequate seating with good conditions and durable materials, a place that appeals to children so they can play while waiting for public transport such as providing swing seats for children.

5.4. Activity nodes

The development and management of all Zones in Setu Babakan district: Embryo zone, A, B, and C zones need to be done equally and not focusing on one zone only. The development of Setu Mangga Bolong area needs to be improved and provide a continuous path of tourism activities between Setu Babakan with Setu Mangga Bolong. For circulation, the circulation of visitor's vehicles should be limited to the main routes of the area designated by the district's manager. This restriction is aimed to make the circulation of visitor's vehicles not penetrate the local residential streets. Thus the tourist activities will not disturb the comfort of the local residents who lived around Setu Babakan.

5.5. Self sufficiency

Increase diverse land uses that can accommodate several functions such as industrial, commercial, agricultural/ plantation and residential areas. Household small industries that do not pollute, especially to the lake water, can be integrated into the land use of Setu Babakan area. The types of small industries developed are those that can support tourism activities within the Setu Babakan district such as the Betawi food production and handicraft industries.

5.6. Zero waste

The handling of waste in Setu Babakan area should not stop only at the separation between dry waste, wet waste, and B3 waste. But it needs to continue until the recycling process so as to minimize the waste generated out of this area. There have been several examples on the application of zero waste concepts in some settlements in Indonesia. For example the settlement of Banjarsari, Cilandak Barat, in South of Jakarta. The same concept can be applied to Setu Babakan area in order to achieve environmental sustainability, and the concept can also generate additional sources of income for local residents.

5.7. Energy independence
Energy independence in Setu Babakan region can be achieved through the utilization of solar energy panels. In addition, solar energy lights can be installed in the promenade area along the lakeside and along other circulation pathways.

5.8. Water neutral
Water conservation within Setu Babakan area can be achieved through the efficient use of water as well as rainwater storage and the recycling of grey water for bathroom flushing and garden watering. In open spaces that are either public open spaces or on the private yard of the houses, absorbing wells need to be applied to increase water absorption into the soil. In addition, to maintain ground water quality, it is necessary to control groundwater retrieval. Groundwater taken from deep wells must be prevented. Utilization of government water supply network (PDAM) needs to be improved and with water quality that needs to be maintained so that people are interested to use it.

5.9. Lifecycle
Residential environments that are responsive to the occupant's life cycle are those whose physical condition and facilities available in it support the development of the place as a safe and comfortable living place for its inhabitants from birth to death. In Setu Babakan, this concept can be realized through the implementation of barrier-free design of open spaces for disabled and elderly.

Through all of the above recommendations, whether for improving existing conditions or meeting the criteria of ecological district patterns, Setu Babakan is expected to become a sustainable cultural tourism area with the lowest environmental impact and highest economic and social resiliency.

6. Acknowledgement
Our gratitude to Kementerian Riset Teknologi dan Pendidikan Tinggi, for funding to ongoing research of ‘The Transformation of Public Spaces in Settlements with Tourism Activities in the Urban Historical Area’ in 2017, carried out by Housing and Settlement Research Group in the scope of Architecture Study Program, SAPPK ITB. (http://www.ar.itb.ac.id/pp/)

References
[1] Selzter, Ethan, et al. 2013. Making EcoDistricts: Concepts and Methods for Advancing Sustainability in Neighborhoods. Portland, Oregon.
[2] Brown, Gabriel, and Dykema, Pete. 2009. EcoPattern District. A Paper for The Portland Urban Architecture Research Laboratory. University of Oregon: Portland.
[3] Disparbud, Pemprov DKI Jakarta. 2015. Perkampungan Budaya Betawi (PBB) di Setu Babakan Kecamatan Jagakarsa Jakarta Selatan.
[4] Maryetti, et al. 2016. Pengembangan Berkelanjutan Kampung Budaya Setu Babakan sebagai Daya Tarik Wisata. Jurnal Destinasi Kepariwisataan Indonesia Vol. 1 No. 1 Juni 2016, page 27-44.
[5] Pratiwi, Wiwik Dwi, et al. 2017. Transformasi Ruang Luar Publik Permukiman dengan Kegiatan Wisata di Kawasan Bersejarah Perkotaan. Laporan Akhir Penelitian P3MI KK Perumahan Permukiman, SAPPK ITB. Institut Teknologi Bandung: Bandung.

IOP Conf. Series: Earth and Environmental Science **152** (2018) 012020 doi:10.1088/1755-1315/152/1/012020

Habitat for Innovative Milieu: A Place-Making Study of University and Start-up Enterprises Relationship

Agus Suharjono Ekomadyo[1], Tyas Santri[2], and Andhika Riyadi[3]

[1] Departemen of Architecture School of Architecture, Planning and Policy Development, Institute of Technology Bandung, INDONESIA
[2] Departemen of Architecture, Langlangbuana University, INDONESIA
[3] Departemen of Architecture School of Architecture, Planning and Policy Development, Institute of Technology Bandung, INDONESIA

Abstract. Milieu is ecological verb to explain how innovation sustain and development through innovative mileu. Responding the innovation park issues as place for innovation, this paper studies how places contribute as habitat for transforming knowledge to business through innovation. By Actor-Network Theory (ANT) approach, this research investigates the relationship between start-up enterprises, university, their place-making, traces how knowledge is produced and consumed. It is found that there are three kind of relationship: linear, eventual mutual, and continuing mutual. Linear relationship is usual relationship between university and alumni. Eventual mutual relationship where mutual relationship between start-up enterprises and university construted in certain events. Continuing mutual relationship is happen when start-up enterprises create permanent interaction with university. It is also found that laboratory is delegated artifacts as mediator for transforming research to innovation. Several start-up enterprises are generated by enterpreneurship events, but most of them do not have mutual relationship with the university anymore. It is recomended that succesful innovation park is the capability to create intensive interaction among start-up office, laboratory, and classroom as places of innovative milieu. This research is hope to contribute real-practice innovation description, in order to support innovation park development program to build habitat that create innovative milieu.

1. Introduction

Innovation discourse has its ecological verb to explain how innovation can sustain and develop: innovative milieu. In ecological term, milieu is defined as the environmental setting to support the species to breed and develop. So, innovative milieu is understood as how environment -physical and social- support innovation to survive, sustain, and develop.

Innovation is the source for entrepreneurs to empower their competitiveness to exploit change as an opportunity for a different products or services. Entrepreneurship itself emerged in society to create more adaptive economic activities while the industrial economy is going to be stagnant [1]. Innovation occurs where entrepreneurial act consists to produce a new method of a commodity [2]. Schumpeter argued that innovation –based on knowledge and technology- has been main factor of national economic development and criticized the neo-classical focus on capital and labour as productive factor accumulation. Etzkowitz [3] formulated the innovation economic as a Triple-helix: synergize of industry who generate economic activities, universities who produce knowledge, and government who regulate the development.

Ezkowitz's Triple-helix stimulated the emergence of special place to incubate innovation through industry, universities, and government collaboration, commonly called innovation park. There were many success stories, also failure ones, of innovation park. Etzkowitz identified that the success or failure of innovation park was depend on creation of consensus space, where actors is attracted and committed to run innovation together.

This paper is based on research to uncover the place-making for innovative milieu by tracing university and start-up enterprises relationship. Institut Teknologi Bandung (ITB) and several start-up enterprises founded by its alumni -which are chosen by faculty proportion and recommended by "innovative" lecturers- are chosen as research case. It is assumed that innovative milieu is constructed depend on the intensity of mutual relationship between university as knowledge production agent and its entrepreneur alumni as knowledge consumers for their business purposes.

Research is conducted by Actor-Network Theory (ANT) method to describe how place is constructed by the relation of actors and artefacts. This method is exercised to explain social assemblage of informal place-making [4], to uncover collaborative atmosphere of coworking space [5], and to develop design strategies among many actors [6]. It is hope ANT method is useful to describe place that is constructed by actor-artefacts relation, in order to answer where consensus space is happen in real practice of innovation.

2. Innovative Milieu, Innovation Park, and Consensus Space

Innovative milleu is an idea that drives many researches about innovation related to spaces and environment [7]. Ideas to build innovation parks, which is later known as science or technology park, is an effort to create a place where innovation activity is being handled carefully, as a place where intensified interaction between researchers, industrial decision makers, and government happened to build innovative milleu. In certain places, these innovation parks built in city-scale, which involving local government, local universities or research bureau, and local industry. Later on, these efforts known as techno-polis, which means technological parks in city-scale [8]. These innovative milleu, in few particular cases, not only contribute to local economic growth, but also taken a significant role in local economic security when critical condition occurs [9].

Innovation parks cannot be understood separately from innovation concept itself. In the last three decades, understanding about innovation as a subject has been greatly improved. The most recent understanding that has been widely used is Triple Helix Theory by [3]. His books became widely used models by various institutions related to innovations. Government, academics and industry became the tip of the spear for the innovation that is yet to come. In the end, this could be seen as renewal of certain values in broader society.

Collaboration between three institutions became the start and the primary key of innovation. Etzkowitz argues that innovation needs a neutral place, where every involved actors that have different background, can be involved deeply in development of new ideas that drives economic growth Etzkowitz [3]. Etzkowitz called these neutral places as Consensus Space. Various discussions about consensus space is being carried by many institutions, there are those who succeed, many who fails. Many different factors when brought about success and failure of this consensus space initiatives, but all comes down to one single reason: strength of relations between actors in innovation processes.

One of many examples of successful consensus space is the infamous Silicon Valley in Standford. An experimental company carried out in this place eventually became huge business and creates many world-class companies that operate globally. From the founding of Stanford because lack of manpower in the United States, until the founding of Joint-Venture Silicon Valley (JVSV), is not a momentary achievement. Many relations are forged and many are disconnected in the process of building JVSV. In this regard, analysis of JVSV cannot be performed under specific point of time, disconnected from the bigger picture.

There are bad stories of efforts to copy JVSV. Such as discussion in New York Academy of Sciences, it is performed to accommodate certain action plan in technology application. This effort to create a consensus space in high level has been meeting a dead end. The failure mainly related to

IOP Conf. Series: Earth and Environmental Science **152** (2018) 012020 doi:10.1088/1755-1315/152/1/012020

creates a consensus space to foster innovation, one needs interest of the opposite parties, other than credibility and ability in decision making, to fill the consensus space and not only planned it [3].

Triple Helix argues that creating consensus space in regional innovation, there is an "office" that specifically handles technology transfer, this office also known as Technology Transfer Office [3]. Later on, this office became an institution where convergence of technology, needs of society and industrial necessity meets and seek mutual benefit, which is the end is utilization of knowledge. The only down point of this theory is lack of reasoning on how relevance or hybrids of academics that put science as their primary goals, government that gives public services, and industry that seeks profit became related and durable during their synergy.

The concept about innovation park that suggested by Etzkowitz ended in something later known as "Technopolis". Whatever the label, the meaning is to form consensus space that foster innovation. If techno is attached in innovation process of a region, then what kind of innovation that has to be done? How can this technorelated innovation can solve local society needs? How can we validate these things?

The same criticism answered by the concept of Pro-Innovation Situation [10]. An effort to expand network that support the needs of academics, society needs and industrial enrichment also became important factor, besides adapting the market demand. This effort has to be done with proper methods, with regard to where consensus space is created. In this method, many things need to be considered when creating a consensus space. Besides people that involved in creating consensus space, there is also local non-governmental organization, local institution, to global development agendas that are related become very important when calculating consensus space.

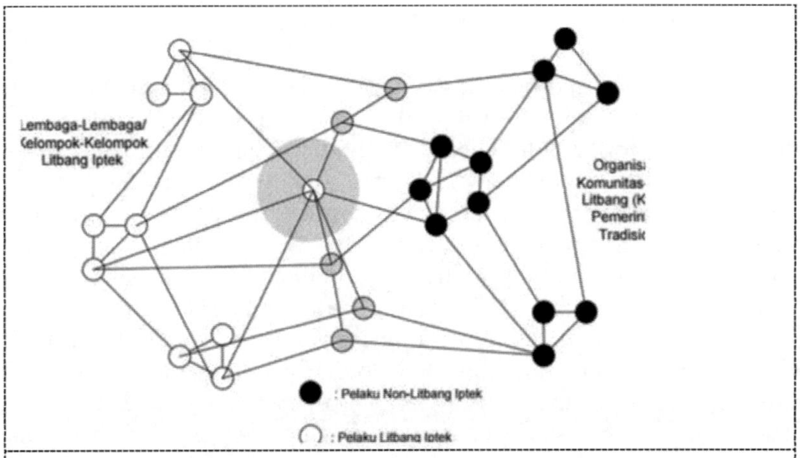

Figure 1. Pro-Innovation Situation [10].

Broader social analysis, than only observing human actor, is needed in pro-innovation situation analysis. Diversity of defining elements, such as choosing a location, is rarely defined by humans. Defining a location for pro-innovation situation mainly based on earth's surface. Institutional relatedness, generally represented by specific landmark, such as industrial area or campus, Technology intensive-activity that is done by available resources, such as WiFi, electricity, distance to center of crowd, All of these is forming pro-innovation situation, and not only depend on human.

The involvement of many technical objects is what makes ANT relevant to be used in this research. ANT relied on concept of translation [11] to explain how relations between entities established. Latour argues that translation is transform action into a format that supported by another actor in achieving goals. This kind of transformation needs a medium that also active in transform its goals, this medium can be non-human. Because of this medium is active in transforming the goals of actors, then this transformation that needs a deeper analysis. The conformity of this transformation between actors with their specific goals, until the process of fulfilment of the needs itself become accommodated with ANT's concept on translation. To differentiate those who analyze and analyzer without disregarding transformation that being done by non-human actor, on this case, ANT give a name for actor that transform with the name of Mediator [12]. In short, ANT is an analysis of transformation by mediators that build relations among them.

3. University, Entrepreneurship, and Start-Up Enterprises

The idea of entrepreneurship development at Bandung Institute of Technology (ITB) has been going on for a long time. It is realized that science, technology, and art (ipteks) will be more effectively useful in society if it can be developed through business activities or entrepreneurship that will provide economic value added. In the field, this activity has been going on lecturers personally or in groups in a particular study or laboratory group. Officially, a few years ago ITB established Business Incubator Center (PIB) to bridge ideas of science-based innovation into business, and then transformed into ITB Innovation and Entrepreneurship Development Institute (LPIK) ITB.

ITB alumni are many who become businessmen by setting up start-up enterprises this happens because of the very real role modeling of their senior who managed to build big companies that contribute to the national economy. Each faculty in ITB actively produces to alumni who established the start-up company. ITB Alumni who founded a start-uping company that started from LPIK-ITB support and some who founded his company independently after graduating ITB. The data from ITB tracer study in 2012, 2013, 2014, 2015 show some alumni recorded to set up a start-uping company.

This research took the case of 24 alumni pioneer company Institut Teknologi Bandung (ITB). The selection of cases is based on the representation of the faculties that produce alumni who choose a career as an entrepreneur based on the results of the tracer study. The case of the start-up company studied was traced through lecturers of the program study who were considered pro-innovation (consistently conducting research and business activities), in the hope that the start-up company was still linked to the almamater. Data retrieval is done through structured interviews, with question structures adopting the Network-Actor Theory, with the founder of the start-up company as a respondent.

In general, start-up company profiles are viewed from a long range of standing, number of permanent employees, and product type selected. The data obtained is expected to provide an overview of the stability of the business from the start-up studied. The ranges of long standing startups vary from 1 to over 15 years. The number of permanent employees also varies, from less than 5 people to more than 25 people, depending on the type of business selected. The type of product selected also shows variations, from services, software, food and clothing products, and hardware. Judging from the age of the company, it appears dominantly under 5 years old, and this can be interpreted that the strengthening of entrepreneurship issues in universities that strongly influence the student is still ongoing not long enough, and has not been able to produce stable start-up companies in running their business. However, judging by the number of permanent employees and the type of product chosen, it appears to have the courage to take risks, and this shows a strong entrepreneurial ethos from the founders of the start-up company studied.

4. University and Start-up Enterprises Relation: Sociotechnogram Analysis of Place-Making

Relationship between university and start-up enterprises is important to trace the innovation in practice, by assumption that start-up founders use knowledge produced in university to improve and develop their business. In university, knowledge is produced through several ways: textbooks interpretation taught in the classroom, reflection of practical knowledge, and scientific knowledge extracted from research activities. By Yuliar's pro-innovation situation model, it is assumed that innovation milieu strength is depend on the intensity of mutual relation between start-up enterprises as knowledge consumers and university as knowledge producer. By case of observed start-up enterprises, it is mapped 3 type of relationship based on interaction intensity: linear relationship, eventual mutual relationship, and continuing mutual relationship.

IOP Conf. Series: Earth and Environmental Science **152** (2018) 012020 doi:10.1088/1755-1315/152/1/012020

Linear relationship is usual relationship between university and its alumni, whatever alumni's choice of work. Students got basic knowledge taught by lecturers in the classroom (figure.2). After graduation, they became alumni and some chose to be entrepreneurs by founding start-up enterprises. Their business knowledge is developed through practical experience by solving consumers or products/ services problems. There are no significant feed-back connection to the university. Most of observed start-up enterprises (42%) is shown have linear relationship with their alma mater.

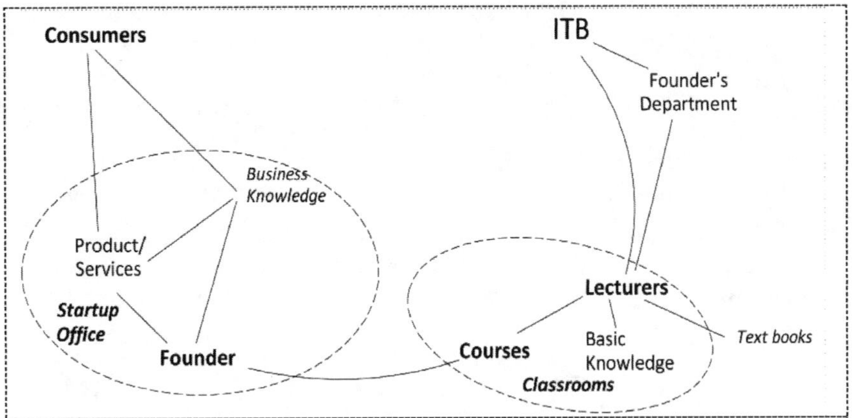

Figure 2. Linear relationship between start-up enterprises and university.

Eventual mutual relationship where mutual relationship between start-up enterprises and university construted in certain events (figure.3,4 and 5). Several observed start-up enterprises (33%) shows this relationship. There are three variants here: relationship through informal consultation with the lecturer, as visiting lecturer in student courses, and business partnership with lecturers.

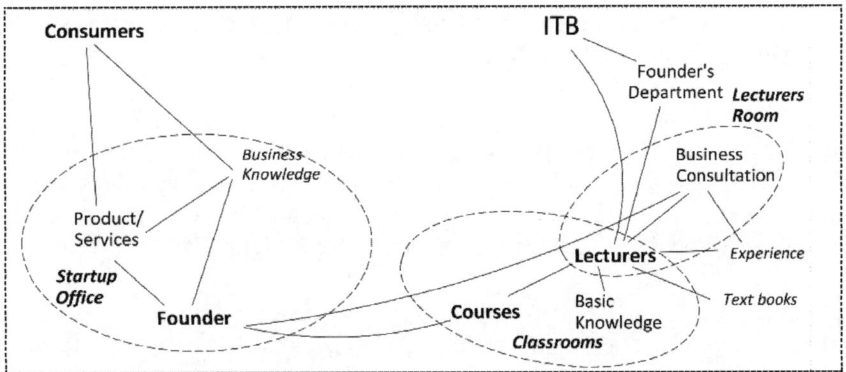

Figure 3. Eventual mutual relationship between start-up enterprises and university: informal consultation.

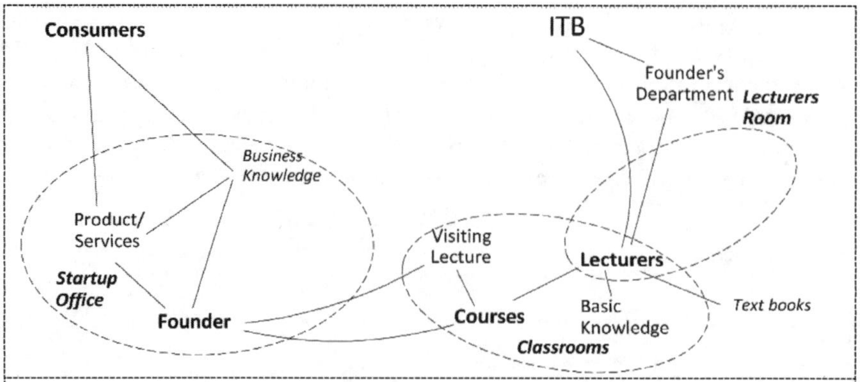

Figure 4. Eventual mutual relationship between start-up enterprises and university: visiting lecturers.

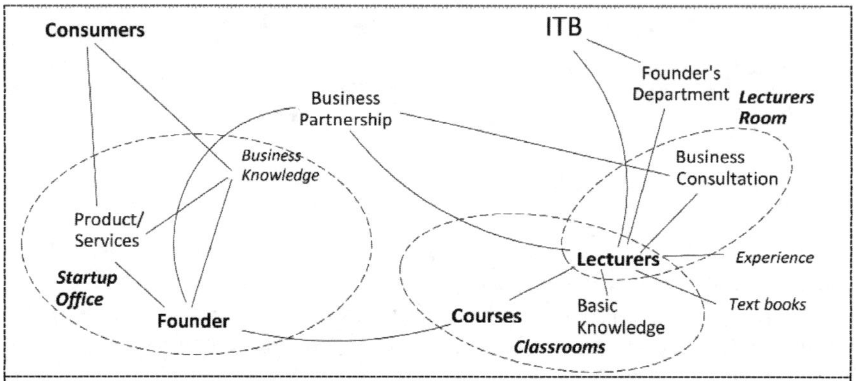

Figure 5. Eventual mutual relationship between start-up enterprises and university: business partnership.

Continuing mutual relationship is happen when start-up enterprises create permanent interaction with university. There are 21% observed start-up enterprises have this kind of relationship. Two variants is shown: lecturer as co-founder, and relation with the laboratory or research center. In first varian, knowledge production refers to lecturer personal practical experience, while in second varian knowledge is produced through research activities (figure.6 and figure.7).

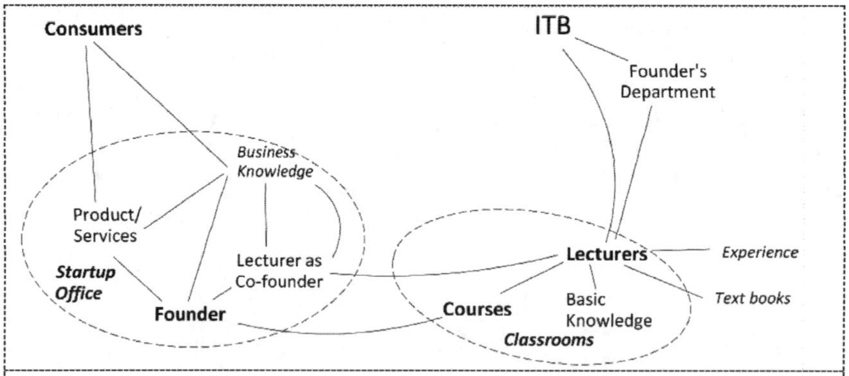

Figure 6. Continuing mutual relationship between start-up enterprises and university: lecturer as co-founder.

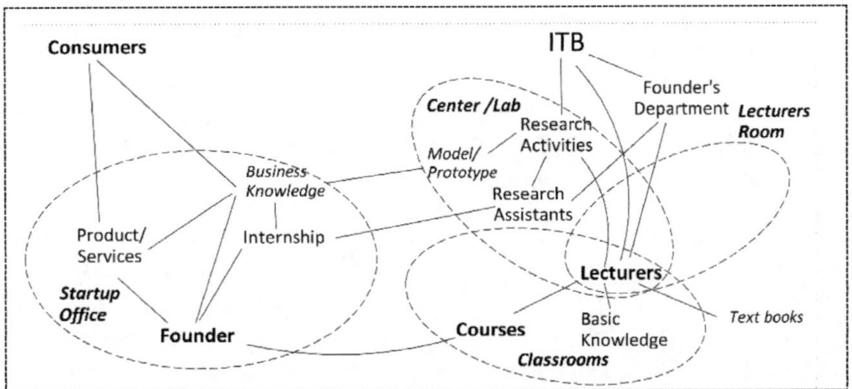

Figure 7. Continuing mutual relationship between start-up enterprises and university: relation with laboratory or research center.

In continuing mutual relationship, interesting finding is shown in Dycode enterprise. Although the founder do not have any relation with lecturers yet, Dycode office is used as internship place for the laboratory's research assistants. Indirectly, university research based knowledge is used to develop Dycode business through discussion among interns, founder, and community. Here, laboratory become the lecturer delegated artefacts as mediator for transforming research to innovation.

This finding shown that entrepreneurship events, by case of ITB, have not strongly relate yet with knowledge production activities through lecturers or laboratories/ research centre. These events are taken place but have not successfully create consensus space between business and research actors in order to transform university's knowledge into start-up enterprises' innovation. It is worried that will be the other kind of knowledge economy bubble [13]. Any kind of entrepreneurship events are able to attract public attention broadly, but just gives a little impact to create innovative milieu.

By ANT's sociotechnogram analysis, this research delivers important findings about real practice of innovation place-making. Start-up offices, laboratories/ research centre, classrooms, and lecturer rooms are important places to mediate the knowledge production and consumption. It is recommended to govern these places relationship as consensus spaces as the key factor to create innovative milieu of innovation park.

5. Conclusion

This research argues that relation among start-up enterprise offices, laboratories/ research centres, lecturer rooms, and classrooms become significant places to create innovative milieu for knowledge production and consumption mutual relation between university and start-up enterprises. Laboratories and research centres can be delegated as mediators to transform university's knowledge into start-up enterprises' innovation. Several entrepreunership events are able to generate start-up enterprises, but have not to be able yet to create mutual relation with university's knowledge production especially based on research activities

The problem of entrepreneurial university programs in Indonesia emerges from disconnection betwen institutional policy and real-practice of innovation. Drucker said that innovation and entrepreneurship appeared as adaptive economic effort to respond industrial stagnancy. Innovation park should be designed as a place with intensive activities, as habitat for innovation to survive, sustain, and develop

References

[1] Drucker, P.F. (1985). *Innovation and Entrepreneurship: Practice and Principles.* Harper and Row, New York.
[2] Schumpeter, J.A. (1939). *Business Cycles: A Theoretical, Historical and Statistical Analysis of*

the Capitalist Process. McGraw-Hill Book Company, New York
[3] Etzkowitz (2008). *The Triple Helix: University–Industry–Government Innovation in Action.* New York: Routledge
[4] Ekomadyo, A.S., Prasetyo, F., and Yuliar, S. (2013). Place Construction and Urban Social Transformation: An Actor Network Theory Analysis for Creative-Kampung Phenomena In Bandung. *HABITechno International Seminar,* November 11, 2013. Bandung, Institut Teknologi Bandung
[5] Agrirachman, F.A., and Ekomadyo, A.S. (2017) Analisis Teori Jaringan Aktor Pada Co-Working Space Dan Komunitas Startup Di Bandung. National Seminar *Kearifan Lokal dalam Perspektif Global*, Universitas Sumatera Utara Medan, 25-26 January 2017
[6] Ekomadyo, A.S., and Yuliar, S. (2014) Social Reassembling as Design Strategies. 5th Arte-Polis International Conference *Reflections on Creativity: Public Engagement and The Making of Place*". Procedia - Social and Behavioral Sciences 184 (2015) 152 – 160, 1877-0428 © 2015. doi: 10.1016/j.sbspro.2015.05.075
[7] Simmie, J. (2005). Innovation and Space: A Critical Review of the Literature. *Regional Studies,* Vol. 39.6, pp. 791–806, August 2005
[8] Syamwil, I.B. (2010) Enhancing knowledge-based industry: The Case of ITB's New Industrial Research Campus Plan in Bekasi, West Java. Arte-Polis 3 International Conference on *Creative Collaboration and the Making of Place.* Bandung: Institut Teknologi Bandung.
[9] Simmie, J., and Martin, R. (2010). The economic resilience of regions: towards an evolutionary approach. *Cambridge Journal of Regions, Economy and Society,* 2009, 1–17. doi:10.1093/cjres/rsp029
[10] Yuliar, S. (2011). Research Transformation to Innovation. National Research Council, Jakarta
[11] Latour, B. (1978) *Science in Action: How to Follow Scientist and Engineers Through Society.* Harvard: Harvard University Press.
[12] Latour, B. (2005). *Reassembling the Social: an Introduction to Actor-Network Theory*. New York: Oxford University Press.
[13] Gandini, A. (2015). The rise of coworking spaces: A literature review. *Ephemera: theory & politics in organization,* volume 15(1): 193-205

IOP Conf. Series: Earth and Environmental Science **152** (2018) 012021 doi:10.1088/1755-1315/152/1/012021

Building Facade Transformation of Shophouses Viewed From Sustainability Principles Case Study: Shophouses in Bandung City

Tika Novis Putri[1], Boedi Darma Sidi[2], Marlisa Rahmi[3], Samsirina[4]

[1,2,3,4]Architecture Study Program, Institut Teknologi Bandung, Indonesia

tikanovisputri@gmail.com[1],boedsi@yahoo.com[2], marlisa.rahmi@gmail.com[3], samsirina@ar.itb.ac.id[4]

Abstract. This study is based on phenomenon of shophouse's facade transformation in cities which is made after shophouses are built for business identity. Viewed from sustainability perspective where 'sustainability' is defined as able to maintain continuation with minimal long-term effect on the environment—the changes on shophouses building façade materials is indicated as energy wasting problems.

The study was conducted in Bandung, Indonesia. The first part of this study is intended to draw typology out of the quantity changes of building facade and to trace the driving factors for its typology. The Conclusion of this study shows a high requirement for facade transformation which is categorized as Minor, Medium and Major changes that forced by inadequacy of the original design in representing the accommodated function, activities or products. The next part of the study aims to provide several recommendations and innovative solution to prevent excessive use of building materials as well as energy waste that goes against sustainability principles.

Keywords: bandung, building facade, shophouse, sustainability principles, transformation

1. Introduction

Shop-house is designed to accommodate both dwelling and business unit needs [1]. This type of architecture could be found almost at every emerging city in Indonesia. These buildings are typically characterized with rectangular form with its narrow front size and has two or three floor levels. In Bandung, shop-house buildings are commonly found in districts along the city's main downtown roads

Shop-houses' facade transformation could be viewed as a response to specific needs related to the shop's functions and formed a "new quality" of urban streetscape [2]. Viewed from sustainability perspective, the transformation of building facade can potentially contribute to building materials wastage as well as specific energy wastage in term of quantity and quality [3].

The first stage of this study is concentrated on the amount/quantity of material being used in each type of changes that range from minor to major changes which is indicated as total dismantling and replacing old facade. Another form of transformation is related to blocking or allowing natural light into the building which is increasing energy consumption either for cooling system or lighting.

This study attempts to analyze the transformation of shop-houses' facade in Bandung City through the perspective of sustainability principles which specifically first based on wasted materials quantity. The result of this study would serve as a foundation for the formulation of concept and recommendation that can contributes to more efficient and sustainable design for shop-houses building facade. Literature Study

2. Literature Study

2.1. Facade Review

According to Taywade [4] the word "facade" originates from French *façade*, which means "front" or "face". Thus, building facade can be simply defined as the front part of a building.

Facade is the visual element an observer would first notice of a building. Through facade, one can form a general idea of the building's interior function [5]. As well meaning and interior space organization [6].

The appearance and image of a building are strongly influenced by various elements that form the building's character [7]. The facade configuration elements that make up a building's image are:

1. Openings elements such as doors, windows and aesthetical openings.
2. Geometries/ shapes which can either be solid (massive) or transparent (void).
3. Dominant materials such as glass, timber, etc.
4. Finishing type and method i.e., exposed concrete or natural stone that creates different impression compare with painting method.
5. Color combination.

The facade of shophouse in this case—is usually constructed by several single elements that together form a particular unity to express a particular characteristic. These single elements, among others, are material, color, texture, door, window, wall, roof and ornament. These elements are arranged in such a way to create an image or dramatic effect according to the function of the building.

2.2. Transformation Review

Widyastomo in Antoniades [8] said architectural transformation is a change process that occurs gradually until an ultimate phase is reached whereby the change is conducted as response to external and internal influence. Transformation of form in architecture can be subtractive, additive or substitutive to a building form or volume. Debaigts [9] identifies five main reasons of shophouse's facade transformation as follow: (1) The type of business conducted and product sold; (2) Commercial policy; (3) The nature, appearance, and quality of the product sold; (4) The competition; and (5) Architectural constraints.

2.3. Sustainability Principle

Architecture sustainability of building encompasses all of the building's cycle; from construction to operation, conservation and demolition phase. In this context, facade as a part of a building possesses an important role in ensuring sustainability and energy-efficiency.

Building and sustainability are inseparable in the context of management of materials and energy efficiency [10]. Building material changes in finish-constructed architecture product is generally could be counted in two categories such as materials inefficiency and energy inefficiency [11]. The amount of materials inefficiency is not always parallel with high energy inefficiency. Dependent to each material used the energy inefficiency is counted from its exploration at the environment until seen as part of constructed architecture elements. This includes the energy used for transportation from (mining) site to production, production process, transportation to construction site, site preparation, its construction and maintenance. Both of those basic thinking also direct or indirectly have an impact to environmental condition.

In this research, general variables related to the studied sustainability principles are generated from *greenship rating tools* assessment criteria (Verse 1.2) which was formulated by *Green Building Council Indonesia* (GBCI) in 2013 [12]. With regard to the topic of shophouse's facade transformation, this study gives emphasis on this following GBCI criteria: (1) Energy efficiency and conservation that includes natural lighting and ventilation; (2) Indoor health and comfort that deals also with view to the outside of the building, visual comfort and thermal comfort; (3) Air quality and comfort.

The assessment of the aforementioned criteria is then identified with the transformation that affects shophouse's facade elements such as exterior color, signage/banner installation, door modification, wall demolishment, etc

3. Research Methodology

This research applies qualitative method and employs direct observation technique to study the physical condition of the objects also equipped by interview with several informants with relevant knowledge on the studied issues.

Based on the result of the observation and interview could be identified type of transformation. Which is then classified into several categories based on the significance of the changes. Impacts on sustainability are further studied through each type of transformation.

3.1. Sample Location and Determination

One-hundred shophouse's samples and their owners/tenants are chosen as the subject of this research's observation and interview which are located in 3 sub-regions (SWK): Cibeunying, Gedebage, and Bojonagara. Geographically. These three Sub-Regions are situated on the eastern, central and western part of Bandung City.

The three Sub Regions are valued as good representations of the city's general characteristic, as:

a) Cibeunying is characterized through its higher education institution, culinary and trade services as well as heritage area (including chinatown shophouses).

b) Gedebage could be identified form its industrial and residential area and is prepared as primary service area of Bandung.

c) Bojonagara which is located in the western part of Bandung serves as one of the city's main gates. It is also characterized through its major transportation hubs such as airport and train station beside its commercial functions.

Shophouses taken as samples are of row type—each row consists of several shophouses with at least one shophouse has undergone facade transformation.

3.2. Data Analysis Technique

Data from observation and interview are used to determine the changes on the facade elements of the shop-house buildings from their original form. A scoring method is then applied to classify those changes into three categories: Minor, Medium and Major changes. Those changes could fundamentally be an addition or subtraction to the facade elements. Addition of facade element that includes installation of signage, addition of ornament, installation of secondary skin and other exterior structures such as canopy and stairs. Meanwhile, subtraction of facade element could be manifested in changes from opening to massive wall or demolition of facade wall, etc.

Result of the transformation—which takes into account business type, road classification and other influenced factors on the building's facade—is subsequently reviewed in accordance to sustainability principles, with emphasis on material usage and energy wastage aspects

4. Analysis And Result

Based on the data gathered from 100 shophouse samples in three Urban Sub-Regions (SWK)—Cibeunying, Bojonagara and Gedebage—researchers assess each type of facade modification using a scoring method. For the purpose of this assessment, the types of modification are distinguished as

IOP Conf. Series: Earth and Environmental Science **152** (2018) 012021 doi:10.1088/1755-1315/152/1/012021

follow: (1) addition of building floors, (2) modification of building roof, (3) modification of doors, (4) modification of windows, (5) installation of signage components on the facade wall displaying business brand, advertisement, etc. (6) addition of secondary skin/wall panel/sunshading /double skin, (7) addition of building material; and (8) addition of exterior elements (such as stair and terrace canopy).

4.1. Analysis of Facade Transformation Typology

Based on the assessment and the total score accumulated the transformation is determined in three classifications as follow:

1. Minor Facade Transformation Typology
   This typology encompasses changes mostly in the form of ornament addition that does not require deconstruction of existing facade elements. Common practices are addition of name board, terrace canopy etc. In this scene replacement of the building's second skin material that does not cover the whole facade.
2. Medium Facade Transformation Typology
   Observation shows several cases where changes are made through the addition of giant signboard that blocks the building openings, the addition of secondary skin, the installation of stairs/steps, as well as the modification of doors, windows and even roofs. These changes has visually transformed part of the facade into a completely different form.
3. Major Facade Transformation Typology
   Several shophouse buildings exhibit transformation on almost all facade elements, including wall, openings, roof, and the number of the building's floor.

These changes often trigger the dismantling of existing walls as well as openings, which in turn transforms the building facade into a completely different form. This type of transformation is classified as major facade transformation.

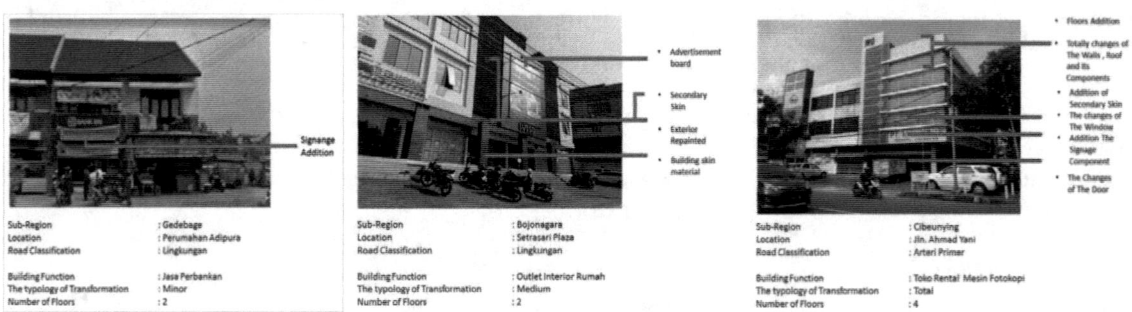

a b c

Figure 1. Example of Minor-Medium and Major Facade Transformation Typology
Source: researcher's documentation (2017)

4.2. Analysis of Shophouse's Facade Transformation in Three Urban Sub-Regions (SWKs) of Bandung

From a total of 100 shophouse samples made up by:

- 40 samples from Bojonagara Urban Sub-Region,
- 30 samples from Cibeunying Urban Sub-Region, and
- 30 samples from Gedebage Urban Sub-Regions;

As many as 37% have undergone medium transformation, 34% minor transformation, and the rest 29% total transformation. The cases of shophouse's facade transformation can be found almost evenly on all types of road, from arterial to collector and local road but are more prominent on secondary collector road. The majority of the cases also occurred to shophouses with two floors or more. In general, the most common types of business accommodated by shophouses with minor, partial or major facade transformation are retail, service and culinary.

Table 1. Sample of Transformation Classification Based On Material Components

Pictures	Sub-Region	Location	Building Func-tion	Specific Building Func-tion	Number of Floors	Road Classification	GSB (m)	Building Facade Components													
								Minor Typology					Tipologi Medium						Tipologi Mayor		
								Exte-rior Re-paint-ed	Signange Compon ent	Building Skin Material	Terrace Canopy	A-nother Ele-ments Addi-tion	Adver-tise-ment board	Secon-dary Skin	Stairs	Doors	Win-dows	Roof	Rail-ing/ Fence	The Walls and Its Componen ts	Floors Addi-tion
	Cibeun ying	Djuanda	kuliner dan jasa	lt.1 rumah makan, lt.2 studio	2	Kolektor Primer	5	V	V		V										
	Cibeun ying	ABC	pertok oan	toko elektronik	3.5	lingkungan							V								
	Cibeun ying	Ahmad Yani	pertok oan	Toko rental mesin fotokopi	4	Arteri Primer	6		V	V				V		V	V	V		V	V
	Gedeb age	Rancabola ng	jasa	bank	3	Kolektor Primer	5		V	V											
	Gedeb age	Sekejati	kuliner	rumah makan	2	Arteri Primer	2						V						V		
	Gedeb age	Derwanti	jasa	bank	2	Kolektor Primer	5		V							V	V	V		V	V
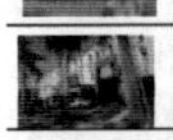	Bojona gara	Setrasari	jasa	Lembaga Pelatihan	3	Kolektor Sekunder		V	V	V											
	Bojona gara	Setrasari Plaza	pertok oan & jasa	Outlet interior rumah	3	lingkungan	6	V					V	V							
	Bojona gara	Djunjunan	jasa	Bank	4	Arteri Primer	6		V							V	V	V		V	V

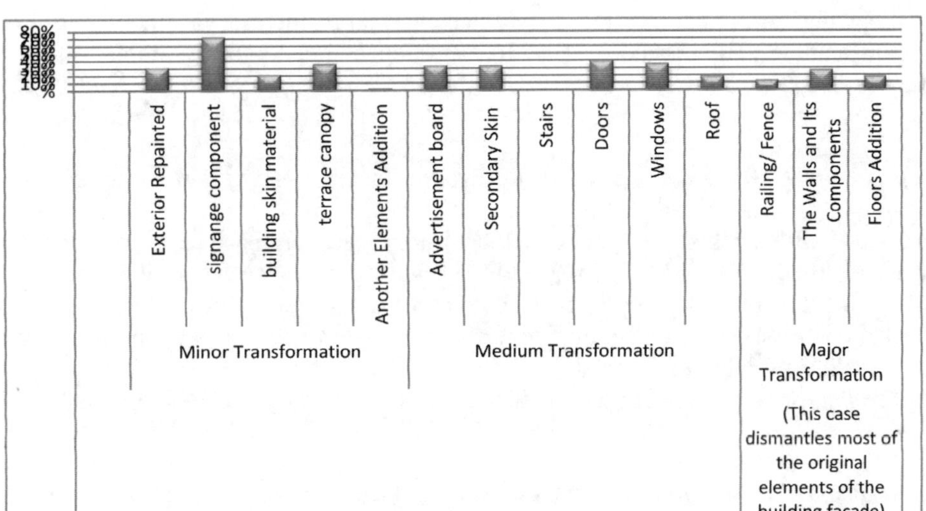

Figure 2. Transformation Classification Result Based On Material Component

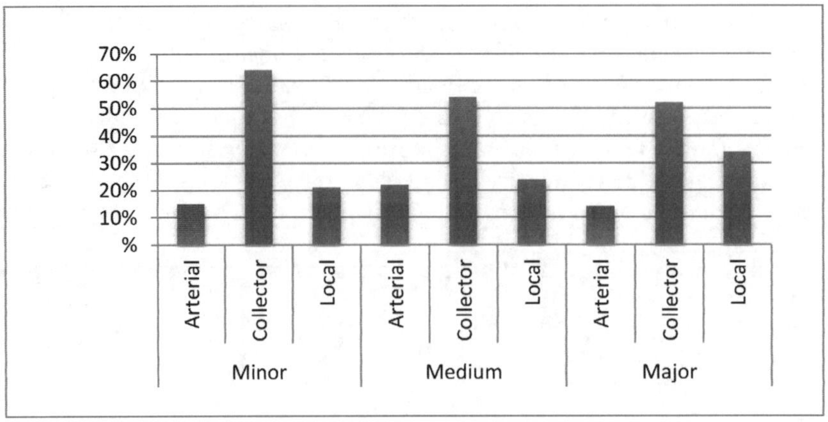

Figure 3. Percentage of Shophouses Facade Transformations on Different Classification of Roads
Source: researcher's analysis (2017)

4.2.1. Minor Facade Transformation. Cases of minor facade transformation are most prevalent on collector roads and found in almost equal number in all three Urban Sub-Regions. In Bojonagara Urban Sub-Region, minor facade transformation amounts to 35% from a total of 40 samples; in Cibeunying it makes up 29% of total 30 samples; while in Gedebage, 38% of total 30 samples (Figure 3a and Figure 4a).

The most frequent change found in this typology includes the installation of name board (72%, and more prevalent in shophouses with service business); the addition of terrace canopy (36%, typically in shophouses with culinary business); and the modification of exterior paint (31%, in shophouses with culinary and retail business).

The most common types of business catered by shophouses within this minor category are service (26.47%) and retail (38.23%). The service businesses are generally locally-scoped; such as beauty shop, bank, travel agent and printing service. Similarly, the retail businesses within this classification are primarily shops that aim to cater to the daily needs of local residents, with their scope of service limited to the surrounding neighborhood. These locally-scoped shops include, among other, clothing store, stationary store and food store.

4.2.2. Medium Facade Transformation. Cases of medium facade transformation are also commonly found in the area alongside collector roads.

Transformations within this category include modification/replacement of doors (40%), modification of windows (36%), installation of advertisement board (34%), installation of secondary skin (33%), modification of roofs (18%) and addition of stairs/steps (1%). In 29% of the total samples, the additions of secondary skin and advertisement board have put a stop to the function of existing windows (Figure 3b and Figure 4b).

The most prevalent types of business within this category are service (24.33%) and retail (35.14%).

4.2.3. Major Facade Transformation. In the case of major facade transformation, 45% of samples are located in Bojonegara Urban Sub-Region and 52% are on collector roads.

About 27% of the total shophouses sample has dismantled their original facade walls and openings, replacing them with a completely new facade design. Adding new floors onto existing building structure is also another common practice that prompts major facade transformation, as found in 17%

of total sample. Lastly, the addition of railing, which requires dismantling of some of facade elements, amounts to 12% (Figure 3c and Figure 4c).

The types of business catered by shophouses in this category are culinary (34.5%), retail (17.23%), service (20.69%) and others (27.58%).

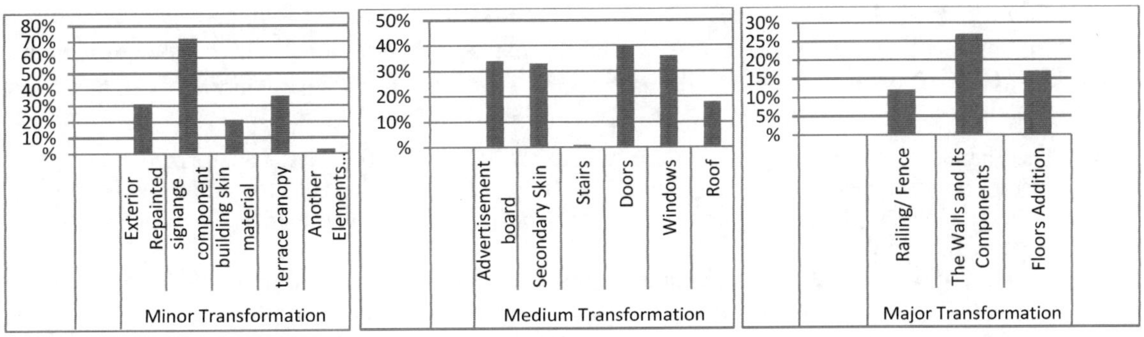

a b c

Figure 4. Percentages of Minor, Medium and Major Facade Transformation by Elements

a b c

Figure 5. Shophouses with Minor, Medium and Major Facade Transformation

4.3. Analysis of Shophouses Facade Transformation Based On Sustainability Principles

Viewed from sustainability principle, the medium and major typology are two classifications of facade transformation that encourage most significant material wastage. In medium transformation, change on doors and windows causes wastage of existing material, while simultaneously encourages considerable energy consumption through the use of substitute material such as aluminum. Other changes such as the installation of secondary skin also potentially increase energy consumption rate by way of the materials used; especially when the secondary skin is constructed from metals such as iron, aluminum, etc.

In major transformation, material wastage becomes a byproduct of changes on almost every aspect of facade element; considering that almost all existing material are discarded and replaced by the new ones. Old materials—which supposedly could be used for longer period—are dismantled and replaced by newer ones which could probably be produced through a process that requires higher energy consumption.

4.4. Factors Causing Shophouses Facade Transformation

Through observation and interview, researchers are able to identify several factors causing shophouses facade transformation as follow:

- The needs of signage to represent the products, brand and business; and the lack of space on the original facade that could cater to such needs. Owners/tenants often do not have other options but to install signage in front of the existing facade, although this practice would subsequently obstruct several original elements such as walls and window openings from view.
- The change of the building's function; where the 2nd and 3rd floor are transformed from living space into business/commercial space such as cafe, restaurant and shop. This prompts owners/tenants to dismantle parts or the whole outer walls on the upper floors to accommodate the new function.
- Shophouse's design as a whole cannot represent the building's function as well as the products and brand, which prompts owners/tenants to modify the facade according to their needs and preferences.
- Some tenants already had a certain standard operational procedure (SOP) for their business design, especially in types of business such as bank, health clinic and fastfood restaurant.
- The needs to provide additional protection shield to the building (i.e. secondary skin and canopy) with various purposes such as reducing heat intensity from direct exposure to sunlight and preventing rainwater from entering the building.
- The absence of regulation concerning the function, business type and building facade of shophouses means that owners/tenants have a complete freedom to make change as well as exploration on the facade.
- The needs to replace poor quality construction material and facade structure used by developer with better ones.

Table 2. Tabulation of Analysis on Shophouse's Transformation in Each Business Type

The Typology of Transformation	Leased (un function yet)	Hotel	Education	Services	Office	Culinary	Shops	Mix-used			
								Shops and Office	Shops and Culinary	Services and Culinary	Shops anda Services
Minor	0	0	3	9	2	5	13	2			
	0%	0%	8.82%	26.47%	5.89%	14.70%	38.23%	5.89%			
Medium	0	0	1	9	7	5	13	2			
	0%	0%	2.70%	24.33%	18.92%	13.51%	35.14%	5.40%			
Major	1	2	1	6	0	10	5	4			
	3.45%	6.59%	3.45%	20.69%	0%	34.5%	17.23%	13.79%			
TOTAL	1	2	5	24	9	20	31	8			
	1%	2%	5%	24%	9%	20%	31%	8%			

5. Conclusions

The result of this study shows a high requirement for facade transformation in shophouses as a result of the inadequacy of the original design in representing the shophouses business' function, activity and product. The transformation contributes to energy wastage especially through the use of construction material against sustainability principles.

In conclusion, the above phenomenon should be treated as an evaluation whereby an all-inclusive synergy is needed, encompassing all parties from developer, consumer (owner and tenant), designer and builder, to government, to ensure that facade be designed according to the shophouses function so as to minimize changes and energy wastage.

6. Recommendation

This research will be conducted through several stages. This paper represents the first stage, where analysis on facade transformation with regard to sustainability principles is only based on preliminary observation and literature study.

The next stage will deal with further study, where the transformation will be researched more specifically by using GBCI criteria with emphasis on source and cycle of material being used in the transformation.

Recommendations drawn through this study are elaborated as follow:

- In the area of spatial planning, a categorization system of shophouses needs to be provided; developed based on the identity of each activity to ensure a more distinguishable facade design.
- In the aspect of architectural design, shophouses facade should be able to address the need for signage elements such as name board, advertisement board and so on.
- The use of recyclable materials should be encouraged in shophouses with high risk of transformation in order to reduce environmental load.
- The prospect of developers selling shouphouses with just in-construction-frame facade should be taken into consideration to allow would-be owners to design their own facade.

Figure 6. Example of design recommendation for shophouse's facade with space for advertisement
Source: researcher's documentation (2017)

References

[1] Sudrajat Rifqi 2014 *Transformasi Fasade Ruko (Studi Kasus di Koridor Mayjend Sutoyo)* Departement Arsitektur Fakultas Teknik Universitas Sumatera Utara

[2] Syahrir 2010 *Kajian Perubahan Pemanfaatan Lahan Perumahan Menjadi Perdagangan dan Jasa Komersial di Perumahan Tumbuh I dan Perumahan Tumbuh II Kota Kendari* Program Pasca Sarjana Magister Teknik Pembangunan Wilayah dan Kota Universitas Diponegoro Semarang

[3] Akadiri O. Peter, dkk 2012 Design of A Sustainable Building: A Conceptual Framework for Implementing Sustainability in the Building Sector *Journal Building* Vol2 Page 126-152

[4] Taywade Pallavi, Shejwal Santosh 2015 Structural Design of a Glass Façade International *Journal of Scientific and Research Publication* Vol5 Issue3

[5] Setiawan, Dedi 2016 Tipologi Perubahan Elemen Fasad Bangunan Ruko Pada Penggal jalan Puri Indah, Jakarta Barat *Jurnal Arsitektur Bangunan dan Lingkungan* Vol6 No1

[6] Zubaidi Nada Cholid, dkk 2015 Karakteristik Fasade Bangunan untuk Pelastarian Koridor Jalan Panggung Surabaya *Jurnal Arsitektur Universitas Brawijaya*

[7] Pattileamonia Vensia Ria Aprilia 2016 *Landasan Konseptual Perencanaan dan Perancangan*

Pusat Kebudayaan Maluku di Yogyakarta Program Studi Arsitektur Fakultas Teknik UniversitasAtmajaya Yogyakarta
[8] Widyastomo Deasy, dkk 2015 The Transformation of Value and Meaning of Sentani Residential Tribe in Papua Indonesia The *International Journal Of Engineering And Science (IJES)* Vol4 Issue12 Pages 21-32
[9] Debaigts, Jacques 1974 *Shop Fronts Architectural Book* Publishing Co., New York
[10] Ruuska Antti, Hakkinen Tarja 2014 Material Efficiency of Building Construction *Journal Building* Vol4 Pages 266-294
[11] Sukawi 2008 Ekologi Arsitektur: Menuju Perancangan Arsitektur Hemat Energi dan Berkelanjutan *Simposium Nasional RAPI VII*
[12] Greenship Rating Tools; Greenship Untuk Bangunan Baru Versi 1.2 *Green Building Council Indonesia*, Divisi Rating dan Teknologi, April 2013

Acknowledgments
We are grateful to the Research and Community Service Department (LPPM) of Bandung Institute of Technology (ITB) for their funding and support of this research. This research is a collaborative research conducted in 2017 under the Housing and Human Settlement Research Group, an expertise group in Architecture Department; School of Architecture, Planning and Policy Development; ITB.

IOP Conf. Series: Earth and Environmental Science **152** (2018) 012022 doi:10.1088/1755-1315/152/1/012022

A Review on The Application of Phase Change Material for Indoor Temperature Management in Tropical Area

A Jurizat[1], S Wonorahardjo[2]

[1,2]School of Architecture, Planning and Policy Development, Institut Teknologi Bandung

aldissainjrz@gmail.com[1], titus@itb.ac.id[2]

Abstract. Phase change material (PCM) is widely use on any products such as electronic product, airplane industry, agricultural and medical container, etc. However in building sectors the usage of PCM is quite uncommon. Since the potential of PCM as thermal energy storage (TES) is very promising, the usage of PCM in cooling system of buildings should be pursued. The cooling strategy of PCM is based on thermal balance concept that it absorb heat energy at charging period and release back at discharging period. In this study, the strenght and challenge on the usage of PCM in building is discussed in several factors. The results of this research shown the potential advantages of PCM in retrofitting the thermal performance of buildings in tropical area.

1. Introduction

The research of Phase Change Material (PCM) has been increased in recent decades. Published papers in the Web of Science database related to PCM has been significantly increased from 154 papers in 2000 to 10752 papers in 2014 [1]. These number are spread for lots of field study. Furthermore, PCMs are also beneficial for lots of industry such as electronic manufacturing, solar power plants, agricultural, photovoltaic electricity, pharmaceutical products, domestic hot water and space industry [1]. The trends of PCMs will continue to increase in the further.

For building performance, Phase Change Material (PCM) could be useful for passive cooling strategy through thermal energy storage. This strategy could manage energy use in the building and also capable for reducing dependency on the fossil fuels which is the major cause of carbon dioxide (CO_2) production [2]. Few temperate countries have been conducted extensively studies about PCM including China, European countries and the US [3]. However, the study of PCM for building application in tropical climate is quite uncommon, especially for cooling strategy [1, 3].

This study will review the past studies about PCM application for the indoor environment in the tropical area. This climate is facing the sunshine for an entire year, thus make the people adapt with the high temperature. Moreover, the aim of this study is to create a map of recent studies about PCM application in tropical climate.

2. Previous Study of PCM Application For Indoor Temperature In Tropical Climate

As mention above, lots of study about PCMs in various field, especially for the building performance. The main benefit of the study is to balance the temperature between indoor temperature and the human physiological temperature. Few studies about this topic have been found from digital journals (See Table 1.). Moreover, the findings have various approaches in particular part of the study. However, these materials will be comprehensively discussed in this review.

Relevant studies about PCM for tropical climate were collected (See Table 1.). Table 1 figured out the information of the studies such as the location, the PCM type, the melting temperature, the method, the PCM location on the building and the factors. The first study has been conducted in 2008 and the latter study is in 2016. However, the PCM studies in tropical area are still uncommon if compared to another climate.

Table 1. The study of PCM on the tropical area

No.	Author(s)	Location	PCM Name and Type	Melting Point (°C)	Method of study	PCM Location
1	Pasupathy & Velraj, 2008 [11]	Chennai, India	Eutectic PCM (48% $CaCl_2$ + 4.3% NaCl + 0.4% KCl + 47.3% H_2O)	26-28	Full Scale Experiment	Roof
2	Mettawee & Ead, 2013 [10]	Cairo, Egypt	Coconut Oil	22-24	Full scale experiment	Ceiling
3	Biswas & Abhari, 2014 [3]	Charlerston, South Carolina, US	Fatty acids/glycerides	21	Full scale experiment and numerical study	Wall
4	Guichard, Miranville, Bigot, & Boyer, 2014 [7]	Reunion Island	DupontTM Energain® (parrafin wax)	23.4	Full scale experiment	Wall
5	Biswas et al., 2014 [4]	Charlerston, South Carolina, US	Paraffin	21.1	Full scale experiment and numerical simulation	Wallboard
6	(Chaiyat, 2015 [5]	Thailand	Paraffin	22	Full scale experiment	Wall
7	Lei et al., 2016 [9]	Singapore	-The idealized PCMs (temp. range 0.1°C) -The PCMs with temp. range of 4°C	24 (outdoor)	Experimental and numerical	Wall

3. Recent Condition of Tropical Climate

Climatic condition has a huge influence for the building performance in the building. The climates were divided by the latitute. One of the most popular climate classification systems is Köppen-Geiger. The climate is splitted to five zones namely, A: equatorial, B: arid, C: warm temperate, D: snow and E: polar. Furthermore, it determines the level of precipitation W: desert, S: steppe, f: fully humid, s:

summer dry, w: winter dry, m: monsoonal. Lastly, it gives details about temperature as h: hot arid, k: cold arid, a: hot summer, b: warm summer, c: cool summer, d: extremely continental, F: polar frost.

Mostly, the tropical climate is defined as 'A' (equatorial). This area is inclined to have less variance in temperature throughout the year. Equatorial region has significantly to large amount of incoming solar radiation. This solar radiation becomes the main factor for microclimate in a tropical area, which potentially induces the heat.

The differences of the climate create an adaptation process for the living. People on the northern and southern part deal with the cold condition and people living in the tropical climate must be adapted with the sunshine for an entire year. The use of PCM for the temperate climate is to reduce the peak of coldest temperature with some heat energy that saved on the materials. However, it is similar while PCM would be used for cooling strategy likely for the tropical climate. The PCM will absorb the heat along the day and release the heat on the night when the temperature is lower down.

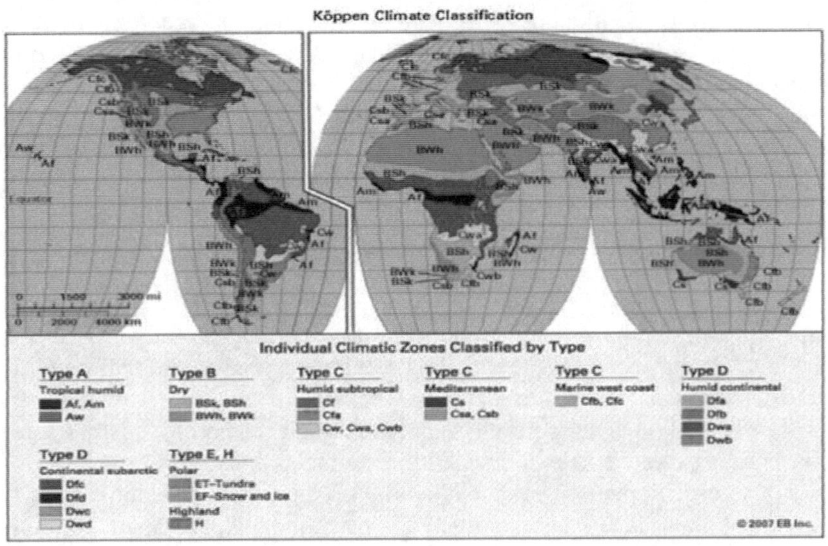

Figure 1. Köppen–Geiger's climate classification [4]

Geographic location determines the building design and building adaptation. The annual hours between 9 am and 5 pm is about 95% and make the temperature is quite high. This condition affects the building performance as well depending on the climate area. For tropical climate, it is usually utilized the air flow through the indoor environment to reduce heat. This strategy could be obtained with active cooling and passive cooling system. For active cooling system, occupant use air conditioner to lower the temperature inside the house. Passive cooling strategy could manage the indoor air temperature with utilize the building envelope.

Research for PCMs are mainly conducted the study in subtropical and humid continental. However, very few studies reported the applications of PCMs to building envelopes in tropical climates [9]. The study shows that the main problem of PCM application in tropical area is the effectiveness of PCM melting-solidifying cycle cannot reach the maximum performance because of variation temperature.

4. Results and Discussion

In this section, the findings of each study will be separated by its motive or background, the methods and the results.

4.1. Study Background

The prone background of all research listed in the Table 1 is building energy consumption. This is due to the fact that building consume 30-40% of total energy use. Moreover, 60% energy for the bulding is used by the heating, ventilating and air conditioning system (HVAC).

Unsuitable condition of the indoor thermal comfort force the occupants to use a strategy such as air conditioner or heater to create a balance condition between occupants' body and environment. Common condition in tropical area is the high temperature and induce people to utilize air conditioner to reduce its temperature. This is causing the highly used for the energy consumption with approximately 48% [5]. The topic for reducing energy consumption in the building, therefore, becomes a main concern owing to diminish fossil fuel reserves and the impact of greenhouse emission [3]. However, the passive cooling strategy in the tropical climate is a good choice due to the minimum impact to the environment. This strategy is mainly affected by the building envelope which is play a role as an adjacent between indoor and outdoor environment. Moreover, building envelope is a key factor for indoor comfort that influences the heating and cooling loads and therefore presents an opportunity to reduce the energy demands [3]. The performance of these enclosure must take an act to control and manage the indoor environment.

Phase change material (PCM) for the building application was studied in many literatures by applying for cooling and heating strategies. For the cooling process, PCM could replace the air conditioner system with electrical power to high performance of building envelope and its material [6]. To conclude, the PCM has many opportunity to generate the passive cooling strategy for the building and lessen the energy consumption produced by the building.

4.2. Study Methods

The listed studies (Table 1) have been conducted to explore the PCM performance by using experimental method. This method sets few factors to be measured in the real condition. Apart from this, there is also numerical method to describe the set up condition by mathematical models. However, experimental method is more popular than numerical method.

Few PCM have been selected to combine with the object study. In tropical area, all the related research (Table 1) use organic and eutectic PCM. The organic PCM that have been used in tropical area are coconut oil [7], fatty acids/glycerides [8] and parrafin [6, 8, 9]. Apart from this, Pasupathy & Velraj used eutectic PCM (48% $CaCl_2$ + 4.3% NaCl + 0.4% KCl + 47.3% H_2O) on their study [11].

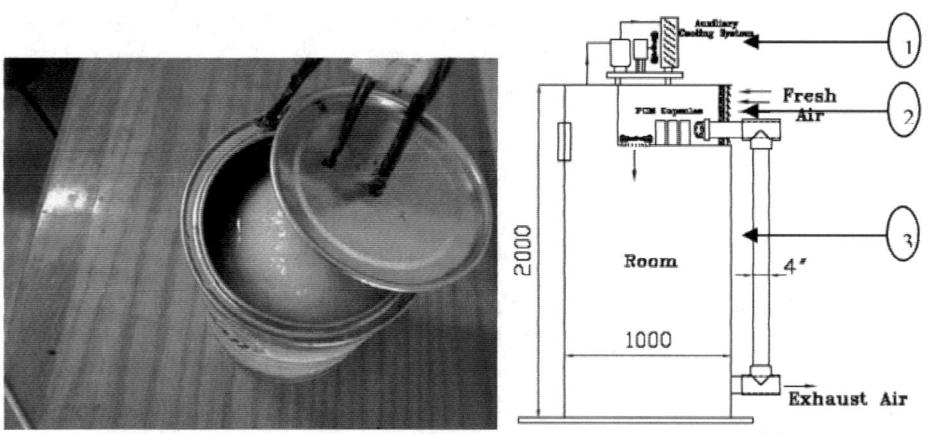

Figure 2 The use of coconut oil for PCM (left) and schematic prototype (right) with (1) auxiliary cooling system, (2) PCM capsule and (3) chamber [10]

The study from Metawee [7] used organic coconut fatty acid as PCM because the suitable performance for tropical climate with its melting and freezing temperature and also has low price.

PCM is encapsulated in few cans as capsules and placed on the ceiling of experimental chamber. During the night, as the first process, the ambient air enters the ceiling chamber for cooling the capsules. Next, the air continue to spead over the indoor chamber before it exhausts to be cycled again through the first process.

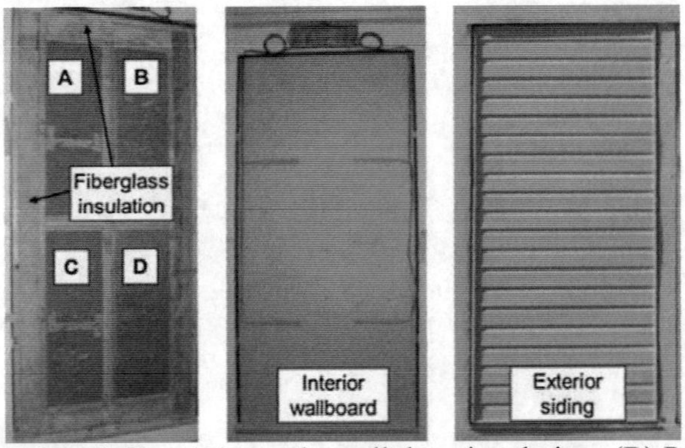

Figure 3 Test wall construction (left) – (A) regular cellulose insulation, (B) PCM– cellulose insulation, (C) cellulose HDPE mix, and (D) cellulose–PCM–cellulose sandwich structure. Finished interior (center). Finished exterior with vinyl siding (right) [8]

Biswas and Abhari [8] conducted a research by using both experimental study and numerical study. Some parts of building envelope combined with the PCM and exposed to the natural wathering. Figure 3 shows the detailed information about the set up. Next, few sensors and instrument were placed in the test wall and the building (Figure 4). The sensors and other instruments measure the condition of temperature, weather condition, humidity, solar irradiance and wind velocity.

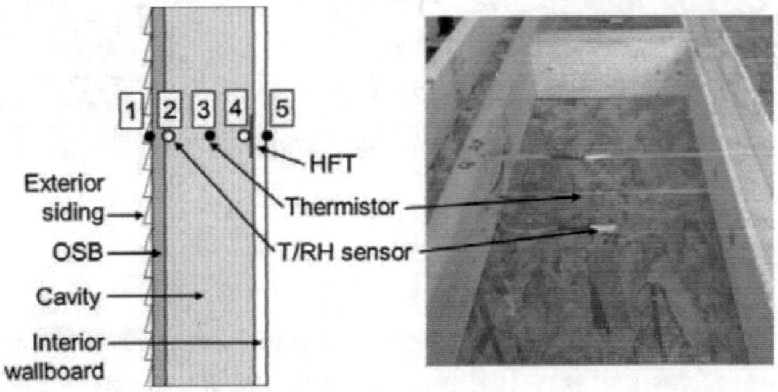

Figure 4 Sensors placement in the test wall [8]

Pasupathy & Velraj use two constructed rooms to study the effect of having PCM panel on the roof (see Figure 6)[7]. The first room is used PCM panel located between the bottom of concrete slab and the rooftop slab, another has no placed PCM panel. Next, the condition of each room is measured to find the thermal performance for the both rooms. Figure 5 illustrates the section of the roof inserted with PCM and without PCM. Plywood is used for the inner walls except ceiling with 6 mm thickness on all the sides to study the effect of PCM panel placed on the roof. The material of PCM panel is made up of stainless streel with 2 m x 2 m and the thickness is 2.5 cm which contains inorganic salt hydrate ($CaCl_2$ + 4.3% NaCl + 0.4% KCl + 47.3% H_2O) as PCM.

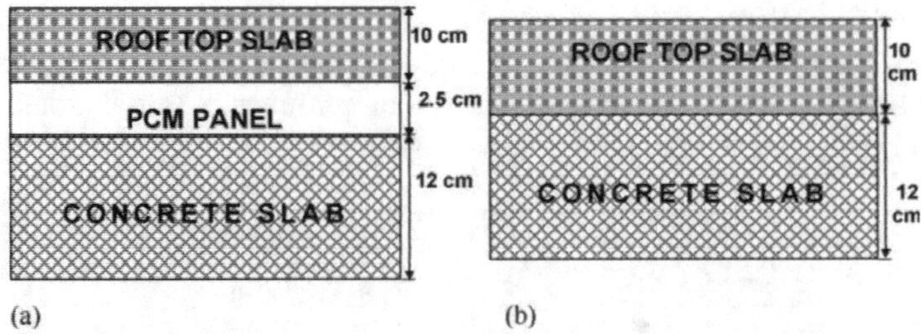

Figure 5 Cross-sectional of the roof with PCM panel (a) and without PCM panel (b) [7]

Table 1 also figured out the influenced factors for the study. There are many factors immersed on the study of PCM. The parameters involved to create sharply analyst. There are three main factors on the research of the PCM, i.e. the environmental temperature, the developed model and the PCM properties. The environmental temperature has many parameters, such as weather condition [3], ambient temperature [7, 8], heat transfer coefficients [7], sky temperature [7], temperature distribution [7], heat flux [8, 11] and relative humidity heat [11].

Figure 6. Two test rooms for experimental study [10]

Developed PCM models are the physically equipment that its function to influence the environmental condition. Pasupathy & Velraj described the parameters are radiation properties of the surface, geometrical parameters and physical properties of the roof material (Roof top slab, PCM and concrete slab).

The PCM properties are the set factor to reveal the explanation of PCM performance. The properties are has been conducted a study of PCM with three parameters such as the temperature range of phase change [9], the shape of enthalpy–temperature curve [3] and the amount of PCMs [3, 7].

4.3. Study Results

The results of the studies are varies depend on the goal and influenced factors. Study from Pasupathy & Velraj [12] shows that the environment has little effect on the inner surface of the concrete slab as all the heat energy is absorbed by the PCM placed in the roof.

In the research from Metawee and Ead conducted the study in Cairo, Egypt. The study used PCM capsules in a ceiling space to manage peak load of heating energy in a chamber. From these results, it can be concluded that the PCM ceiling system becomes effective strategy to reduce the peak load of heating. Large amounts of latent heat can be stored in the PCM, which enables to make more compatible, clean, durable, and cheaper systems for air-conditioning.

Interesting finding was that PCM with inner portion of the wall cavity has no difference results for cooling electricity as adding PCM to the entire cavity [8]. Exterior wall with PCM inside has a better performance at lower cost than the wall with PCM inserted in all cavities. Moreover, the application of PCM has a good energy-savings potential depends on the wall orientation. This factor has a highly influenced for passive cooling strategy rather than other factors.

Guichard et al. have been conducted a research to study most influent parameters for PCM performance [9]. According the study, the parameters are convective exchange coefficients and a specific parameter of PCM model. It has been proved that this coefficient plays an important role in the predictions and should be chosen precisely.

5. Conclusion

The previous studies about PCM for building application in tropical climate is still uncommon. Most of these studies have a research background on energy consumption. In tropical climate, passive cooling system becomes suitable to manage peak load of heating with minimum impact for the environment.

It is shows that convective exchange coefficient has an important role to reach the maximum performance of PCM. The specific parameters for PCM application should be chosen and measured precisely. This case is related to the fact that PCM cannot reach the maximum performance in tropical climate because the diurnal of ambient temperature is narrow. Further research may develop the PCM performance with described parameters to find the best strategy to manage the peak energy load period.

6. References

[1] Akeiber H, Nejat P, Zaimi M, Majid A, Wahid M A, Jomehzadeh F and Ahmad S 2016 A review on phase change material (PCM) for sustainable passive cooling in building envelopes *Renewable and Sustainable Energy Reviews* **60** pp 1470–1497

[2] Tyagi V V, Pandey A K, Buddhi D and Kothari R 2016 Thermal performance assessment of encapsulated PCM based thermal management system to reduce peak energy demand in buildings *Energy and Buildings* **117** pp 44–52

[3] Lei J, Yang J and Yang E 2016 Energy performance of building envelopes integrated with phase change materials for cooling load reduction in tropical Singapore *Applied Energy* **162** pp 207–217

[4] Kottek M, Grieser J, Beck C, Rudolf B, Rubel F 2006 World Map of the Köppen-Geiger climate classification updated *Meteorol* **15** 259-263

[5] United States Energy Information Administration (EIA) 2009 Residential Energy Consumption Survey (RECS) (www.eia.gov)

[6] Chaiyat N 2015 Energy and economic analysis of a building air-conditioner with a phase change material (PCM) *Energy Conversion and Management* **94** 150–158

[7] Mettawee E S and Ead a I 2013 Energy Saving in Building with Latent Heat Storage *Int. J. of Thermal & Environmental Engineering* **5(1)** pp 21–30.

[8] Biswas K and Abhari R 2014 Low-cost phase change material as an energy storage medium in building envelopes: Experimental and numerical analyses. Energy Conversion and Management **88** pp 1020–1031

[9] Guichard S, Miranville F, Bigot D and Boyer H 2014 A thermal model for phase change materials in a building roof for a tropical and humid climate: Model description and elements of validation *Energy and Buildings*

[10] Behzadi S and Farid M M 2014 Long term thermal stability of organic PCMs *Applied Energy* **122** pp 11–16

[11] Biswas K, Lu J, Soroushian P and Shrestha S 2014 Combined experimental and numerical evaluation of a prototype nano-PCM enhanced wallboard *Applied Energy* **131** pp 517–529

[12] Pasupathy A and Velraj R 2008 Effect of double layer phase change material in building roof for year round thermal management. *Energy and Buildings* **40(3)** pp 193–203

7. Acknowledgements

This publication and research supported by Indonesia Endowment Fund for Education (Beasiswa LPDP) and Research Schema of Desentralisasi Ristekdikti Indonesia 2017 contract No.009/SP2H/LT/DRPM/IV/2017

IOP Conf. Series: Earth and Environmental Science **152** (2018) 012023 doi:10.1088/1755-1315/152/1/012023

'An Interactive Web of Water, Space and Life': A Paradox of Urban Dwelling in Setu Rawa Besar, Depok

K R Kurniawan[1,2], **M N Fadhil**[1], **S Abdilah**[1]

[1]4th Floor Department of Architecture Bldg, Faculty of Engineering, Universitas Indonesia Depok Campus, Depok, Jawa Barat, 16424

[2]r_kemas@yahoo.co.uk

Abstract. This paper discusses an interactive web of water, space and life amidst the paradox of formal and informal urban dwelling in *Setu* (Lake) Rawa Besar in Depok, Indonesia. Its high-density population lives in a such critical environmental condition with the amount of garbage that has polluted the *Setu*. The meshwork of human activity in terms of architecture and space strengthens the linkages between the natural and manmade environments. This interactive web highlights the paradox of urbanization and modernization of the city amidst environmental and socio-cultural issues. This research was initially based on findings of the International Joint Studio and Workshop at Universitas Indonesia, in collaboration with University of Florida and Cardiff University in 2016 and 2017. The methods of the study are through qualitative analysis with the advanced methods of in-depth interviews, observation and site analysis, and socio-cultural mapping of the informal and formal community. In addition, this research also reviews the existing urban master plan that has not been fully implemented yet. This research identifies the ways in which municipal authorities can collaborate with the informal community to help manage the waste within the *setu*, to propose further plans to create a 'healthier' settlement for the informal community.

1. Introduction

Depok is a medium-density city that is situated in the southern part of Jakarta and the northern part of Bogor. Depok plays an important role to function as a water reservoir/water buffer zone (with flows from Bogor to DKI Jakarta). In order to prevent flooding downstream, these water reservoir areas in Depok consist of many lakes, known locally as 'Setu'. One of these lakes, namely Setu Rawa Besar became the main research topic for this paper.

IOP Conf. Series: Earth and Environmental Science **152** (2018) 012023 doi:10.1088/1755-1315/152/1/012023

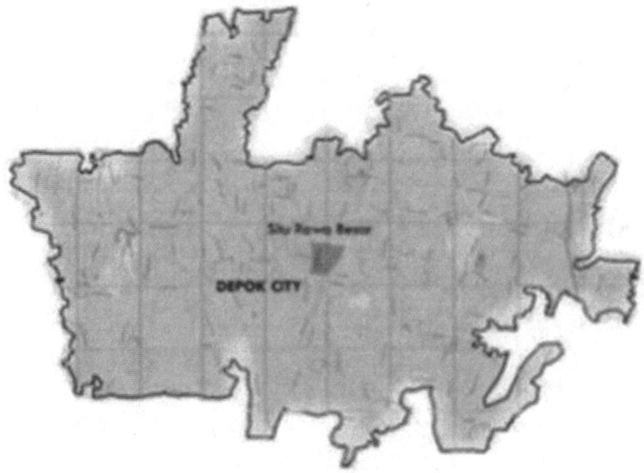

Figure 1. Appointing Setu Rawa Besar location (highlighted with red color) on Depok City map, illustrated by Reza Yuniar Sandi.

Depok with its high-density population is now facing an environmental challenge in which the amount of waste (mostly from upstream outside the Setu) pollutes the water and reduces the size of the Setu. This condition creates some health and social problems.

According to the local government, the condition of Setu is very poor. Since there is an informal settlement located around the Setu, the Setu's size has been narrowed down to 15 Ha (from the original size of 25 Ha). Besides that, the Setu also has suffered due to the amount of waste disposal that comes from nearby settlements, local industries and markets around the Setu. [1]

Figure 2. Informal settlements and working space that becomes dumping ground located surrounding the Setu.
Source: Author's Documentation

On the other hand, Setu Rawa Besar is one of most contaminated lakes in Depok with a very dense population. Nonethless, there are certain advantages and a relative degree of potential for development. For instance, its location is in the center of Depok, near the Office of Mayor of Depok (local government headquarters). It is flanked by three main roads and is located close to the center of transportation interchanges (trains, buses/minibus and 'ojek'/rental motorbikes). With the lack of public space in Depok, therefore, Setu Rawa Besar has the potential to be developed as a new public space as well as one of strategic water reservoirs, which can also serve as an eco-tourist destination. There are several existing small workshops and industries, such as local craftsmen's enterprises, food preparation venues, freshwater fish cultivation and some other household scale industries that may increase the site's economic value.

Setu Rawa Besar is considered to be the biggest Setu in Depok. It plays an important role within people's everyday lives, especially to those who reside in the informal settlements around the Setu. In order to approach these issues and the problems relating to an interactive web of Water, Space and Life, we have conducted literature searches and made some direct observations on site. We also applied a method of being objective, third party observers, in the form of photographs and video recordings. We used a combination of research techniques, varying between qualitative and quantitative methods, in order to open the possibility of findings from different socio-cultural perspectives. The research process is monitored through in-depth interviews with formal and informal residents on their perceptions of the issues and their use of the Setu. Qualitative data collection is often used when trying to gain an understanding of other people.

From our observations and interviews, we found out about many problems that Setu Rawa Besar faces on a daily basis. We identified these problems as four main issues (water, waste, living space and regulations) that are inter-related with one another.

2. Research

The initial observation found several facts in Setu Rawa Besar. First, the water appears to be unclean; second, the densely-packed population is composed of formal and informal settlers; third, self-made homes are built from improper, salvaged materials; four, the informal community is located too close to the water's edge; and five, the Setu has multiple uses- fishing, waste dumping, sewage disposal.

Based on the research conducted by the local and international students and the academic team during the joint workshops in 2016 and 2017, the research team found three interactive networks on site. These are: Water and Waste, Water and Living Space, and Regulations and Living Space [2].

2.1. Water and Waste

> *'Over the last three decades, Indonesia has enjoyed substantial achievements in poverty reduction, human development and improvements in service delivery. However, enormous challenges remain. Improving access to safe water and sanitation is a key priority, due to the severe consequences of poor sanitation infrastructure on public health, the economy, and the environment. Of the four most important causes of under-5 mortality in Indonesia, two—diarrhea and typhoid—are fecal-borne illnesses directly linked to inadequate water supply, sanitation and hygiene issues. Difficult access to improved water supply also means that poor households, particularly women and children, spend too much time fetching water.'* [5]

In the Setu Rawa Besar neighbourhood, the settlements that are located in the eastern part of the lake have a decent waste and disposal system. They have to pay for the trash management regularly and very few of them combust the trach. In addition, all of those formal settlement have their own septic tank as the results of a planned residential area.

However, unlikely that eastern part, the northern and western part of the lake are fulfilled with informal settlements. The settlers live there inexpensively as the consequence of their occupation in small service sectors, such as *becak* drivers, carpenters, laborers, and small retail vendors. As the results of their status as informal settler and exacerbated with the unplanned residential area, those informal residents have no access to a fixed or formal system to manage their waste and disposal.

Figure 3. Blue, Green, and Grey Water Diagram.
Illustrated by Fachry Ardani

The diagram above illustrate the interactive connection between water and waste in the Setu Rawa Besar. Blue water represents the clear water for daily consumption. The conditions of lake water that is not reliable and worsened by the polluted private well have forced the settler to buy the water for consumption purpose from the local seller. It is distributed directly from mobile cart seller to formal and informal settlements. They also do the rain water harvesting which is marked with green water in the diagram. The recycled or reused rain water is collected in the buckets at the houses. The residents use it for bathing, watering the plants, and washing the dishes. The last category is grey water which is waste water. This end of the cycle of water usage between Setu's residents are directly transferred to the lake. This water is one of the main pollutant in the Setu Rawa Besar because of the absence of water purification and process before disposes into the lake. Home industries and market in the lake surroundings are also at the bottom of the lake's pollutions.

2.2. Water and Living Space

Some residents come from a Perumnas neighborhood (formal houses provided by the national housing board) have lived by the lake since 1970s. They are from various professions, such civil servants, teachers, and entrepreneurs.

'Water is a living thing.' [4] Based on interviews in 2017 with residents from the formal settlement near the Setu, they said that in the 1980's people could still see the lake view clearly through the Perumnas housing. This observation shows that more than 30 years later, the Setu has been degraded with additional dwellings along the lake.

The paradox is that with its strategic location in the center of Depok and behind Mayor of Depok Office which should have potential as the main living space of Depok, the Setu seems to be the backyard of Depok, where waste is thrown into the lake.

Moreover, the people from formal settlement also have built many "rental houses" and are renting them out to settlers, outsiders and newcomers. With this business, they earn about Rp. 200,000-300,000 per month. Some of the neighbours in the formal area also "own" and claim part of the informal land, and they are also renting it out.

It is a paradox when it comes to people's dwellings and livelihoods. Although some residents from the formal settlements gain financial benefit (from renting the claimed land) to informal residents. Yet, other formal residents would prefer the informal residents to be evicted.

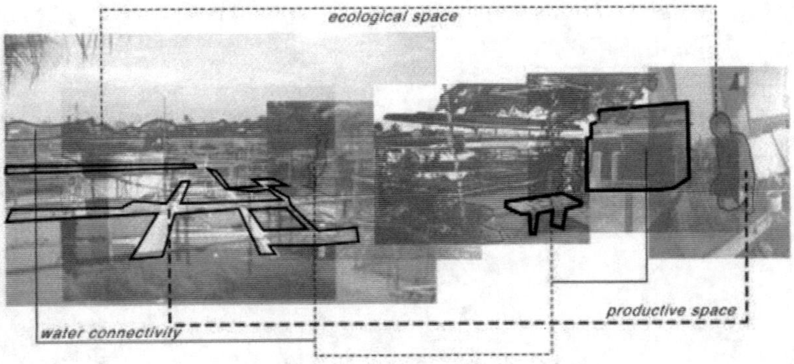

Figure 4. Integrated Water and Living Space Connection Diagram.
Illustrated by: Sutanrai Abdilah

They would like to see the lake cleaned up. Informal dwellers are blamed as the main cause of the unhealthy conditions around the Setu. This situation has created some tensions between the communities. Formal and informal residents say:

"We get our drinking and washing water from a 10-meter deep well," (Formal). We only rent the house for working nearby," (Informal).

2.3. Regulations and Living Space
The dwellers from informal settlement acknowledge that they throw trash and waste (both solid and liquid) directly into the lake. Residents' grey and black water does flow into the lake. Everyday they collect the trash from the lake, and from their homes. Then the trash is burnt in the settlements's backyard. They burn trash rather than collect it because it costs Rp. 20,000 for trash collection, and that is too expensive for them. Sometimes they do *'kerja bakti'* or *'gotong royong'*. But it is more a community initiative rather than a government intervention.

'The main problem that the government faces for living space is land authority' [2]. People in the informal settlement keep to built illegal setlements as government lack of resources to control the area. It is a phenomena of the existence of informal actors that are socially very influential to produce unplanned urban squattres.

Another problem is government capability to do law enforcement for private business which caused environmental problems on the area. There are market and tofu factory that are located outside the Setu which are identified as the main sources of waste in the Setu. It is a paradox that both factories are still in full operation, and the factory owners are seemingly unaware about ecological problems created by their activities.

3. Findings

3.1. Zoning Diagram
The diagram below was the result of zoning conducted by the Research Team to determine the allotment and function of the space contained around Setu Rawa Besar. The human activities and movements generally occur around Setu, especially in the parts of the informal settlement.

IOP Conf. Series: Earth and Environmental Science **152** (2018) 012023 doi:10.1088/1755-1315/152/1/012023

Figure 5. The division of zoning area.
Illustrated by: Sutanrai Abdilah

This situation results are then followed by the location of a high accumulation of waste that occurs in Setu, where the location of waste accumulation is adjacent to the residents' activities. Beside that, the waste sedimentation is also in the flow close to the Setu's upstream source. The waste is actually from outside Setu. The residents dammed the inlet so the trash would not enter the Setu water catchment area. The waste comes from markets upstream that are close to Setu.

On the lower ground on the northern side of Setu, the Research Team found several places of waste accumulation, which resulted in flooding. The floods are not only caused by lower geographical conditions of Lio Village, but also by the waste sedimentation on the edge of Setu, which clogs the drainage ways. On the other hand, in the formal settlement there are only minimal activities that are directly related to the Setu. This is due to their daily occupations that occur far away and not related to the Setu itself. Any local activities occur on the weekend.

3.2. Mapping the Activity Diagram

The myriad types of activities that occur in the area near Setu Rawa Besar can be grouped into 3 major interactive webs, namely 'Space - Water,' 'Space - Live,' and 'Water – Live'. These three basic elements represent the 'Quality of Life' diagram of the interactive web of activities in Setu Rawa Besar.

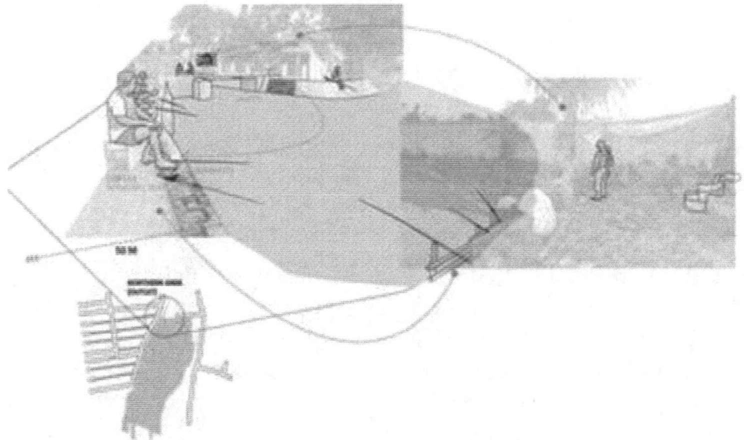

Figure 6. Mapping Activity Diagram.
[2]

These diagram shown that the activities of the informal settlements that are in-direct contact with the Setu. Water is the basic element to support the activities (washing, fishing, cultivating fish, showering, etc) which have important role in daily life of people in informal settlements of Situ Rawa Besar. These activities creates 'living aspect of water' [4]. However, for the provision of clean water, micro-economic and dietary activities in the informal settlements do not have direct contact with the water in Setu. The blue water (clean water) was bought from mobile cart seller. This connection has been shown in Figure 3 about grey, green, and blue water diagram.

3.3. Diagramming the Quality of Life

This identification process then concludes that three basic spheres form of the interactive web into an integrated connectivity between water, space and 'life' (Live) in Setu Rawa Besar to produce the 'Quality of Life'. The objectives need to be an overall concept that solves problems more holistically. In this case, the government can create regulations that not only play an important role to improve the quality of water and the environment, but also in the form of education to improve human resources in Setu Rawa Besar. Government can adopt recycling and waste management education programs from other places that have success stories, for instance in Australia [6].

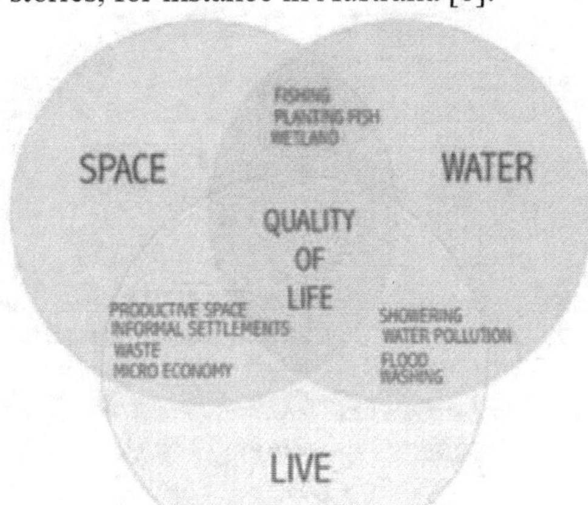

Figure 7. Quality of Life Diagram.
Illustrated by: Sutanrai Abdilah

4. Conclusions

This research raise our awareness that the Setu is not safe in terms of its health and environmental conditions. The research found the paradox of urban dwelling that is represented through the ambiguity implicit in interactive web of 'water, space and people' in Setu Rawa Besar. The residents from the formal settlement maintain benefits from the residents of the informal settlements, but also they blamed people from informal settlements as the cause of the Setu's degradation.

It is the fact that, formal and informal settlers respectively need each other in terms of doing their business. On the other hand, the lake should be restored and cleaned. Therefore, the International Joint Studio and Workshop produced four recommendations in dealing with this interactive web. During the interim period, it is suggested that the Depok City government engage with the local community with rehabilitation projects, communicate the issues with local NGO and seek possible funding from enterprises or private individuals. In the short term, it is suggested to work through education programs that involve NGOs and schools. Additionally, another program would be to apply floating treatment wetlands. In the long term, it is suggested that the government build community sanitation and community projects to remove waste from the water. Other initiatives include applying green infrastructure, relocating the informal settlers to safer place nearby in order to enlarge the size of Setu, and developing Setu Rawa Besar as a new public space offers potential for an eco-tourism concept for the City of Depok. This research also found that unoccupied land has the potential to be developed as a green area for water catchment, beautification and ecological revitalization.

Acknowledgments
We would like to thank and to mention the important contribution of students and facilitators from Universitas Indonesia, University of Florida, and Cardiff University (Eco City Studio) and who were involved in the International Joint Studio and Seminar in 2016 and in the International Joint Studio 2017 in Jakarta and Depok. The activities provided important data and information for this paper. We would also like to give special thanks to Prof. Christopher Silver, Dr. Andrea Frank, Ms. Diane Wildsmith, Mushab Abdu Asy Syahid, Meredith Anne Fowler, and Laura Callaghan who provided access to the materials from the joint studio and workshop. This Joint Research was funded under the Global Green Initiative (GGI) grant from the Institute of Education (USA) in cooperation with funding from Cardiff University and Universitas Indonesia.

References

[1] Bappeda Kota Depok 2016 *Rencana Penataan Permukiman Kumuh Kampung Lio – Kota Depok. Chap.* **3** 50-56
[2] Wildsmith D, Kurniawan K R, Frank A, and Silver, C (eds.) 2016 Global Innovation Initiative, Urban Green Infrastructure: An Interactive Web of Water, Space and Life *The International Joint Studio and Seminar, in collaboration with Universitas Indonesia, University of Florida and Cardiff University. Department of Architecture, Universitas Indonesia Depok. Chap.* **3** 183-226
[3] Silverman D 2013 *Doing Qualitative Research: A Practical Handbook 4th edition* (Thousand
[4] Moore C W 1994 *Water + Architecture* (London: Thames and Hudson) Oaks, California: Sage Publications)
[5] Indonesia: Expanding access to clean water for the poor (2013) Available at: http://www.worldbank.org/en/results/2013/04/04/indonesiaexpanding-access-to-clean-waterfor-the-poor (Accessed: 18 January 2017)
[6] Recycling education programs (2013) Available at https://www.sunshinecoast.qld.gov.au/Living-and-Community/Waste-and-

Recycling/Recycling-and- Waste-Education/Recycling-Education-Programs (Accessed: 18 January 2017)

IOP Conf. Series: Earth and Environmental Science **152** (2018) 012024 doi:10.1088/1755-1315/152/1/012024

The Assessment of Aural Disturbances Experienced by A Church Located in Housing District of Bandung Case Study: GKII Pniel Bandung

Bayu Andika Putra[1], S Wonorahardjo[2]

[1,2]School of Architecture, Planning and Policy Development, Institut Teknologi Bandung
[2]wonorahardjo@yahoo.com

Abstract. Church as a sacred building is best situated in a surrounding which has an ideal aural environmental quality to enhance its acoustic performance as well as to enrich aural experiences of the worshippers. However, due to the lack of suitable site in urban context, in many cases church has to be located in the housing district with all its demerit factors, among others are aural disturbances which emanated from different sources. This study will assess the aural disturbances experienced by a church located in housing district of Bandung. The assessment will include major sources of aural disturbances, its intensity, and level of concern of the worshippers, using direct field measurement method and interview. Results of this study will display level of aural quality attained by this church, and appropriately propose relevant strategies to overcome the aural disturbances being faced.

1. Introduction

Aural architecture is known as a science that combines architecture, acoustics, sound, perception, and anthropology in one perspective in view of human experience of sound and space. Aural architecture encourages emotional and behavioral responses in both real and virtual environments [1]. The aural environment is the space where the sense of hearing becomes the main tool in feeling the sound space. Architecture, sound, perception, and anthropology become a perspective in looking at human experience of sound and space. A room with a good aural can provide an experience of intimacy, anxiety, connectedness, alienation, warmth, and an atmosphere of spirituality [1]. This experience can be felt anywhere as long as the sound plays an important role in it.

The worship space has a complex acoustic environment that requires hearing sound clarity and aesthetic sound experience. This experience can create emotions through perception of sound [2]. The Church as a sacred building must be located in an area of ideal environmental quality to enhance its acoustic performance while enriching the aural experience of its congregations. Creating a worship space is not just about creating space inside, but also about how the environment can support the procession and atmosphere while worshiping. Quiet and away from noise sources is the ideal aural environment for the church.

However, because of the urban planning in Indonesia that places churches into residential areas, many churches must be in the residential district with all the demerit factors, including aural disturbances derived from vehicles, private activities, and neighborhood activities. Aural comfort depends on the noise level of the environment. Noise, defined as an unwanted noise, can interfere and cause attention to become distracted or unfocused.

2. Selected Church

The church is a Protestant church located in Bandung City. This church is located on Cimanuk Street with a land area of about 500 m2. This one-story church building was built in 1990 by a local congregation who wanted a place of worship with a strategic location but also quite calm.

Figure 1. GKII Pniel Bandung

Based on Regional Spatial Plan of Bandung, the location of this church stands in the residential area. The rules of the church building functions set into the zone of residential areas and arranged in accordance with the needs of the region. In addition there is also the educational function of SMP N 44 Bandung right next to the church.

Figure 2. GKII Pniel Bandung
(maps.google.com)

The development of the economy and the rapid growth of the culinary industry in Bandung has had an impact on the area around the church. Housing that was quite quiet, now turned into a row of restaurants that every day is always crowded by visitors. As a result the church atmosphere is not the same as before. If the church was once considered a building of concern to the environment and sacred, then due to increased activity around it the church began to lose its sanctity and slowly barely visible again. The intensity of the vehicle is increasing and the sounds of music from the restaurant add to the hustle and bustle of the environment.

3. Methodologies

The assessment will include major sources of aural disturbances, their intensity, and level of concern of the worshippers.

The method in this research is direct field measurement and auralization [2]. Direct field measurement is one of the methods used to measure the physical aspects of the aural conditions from the location. Direct field measurements were made at the time of non-church worship activities to measure noise levels using a Sound Level Meter around the church environment when the church was empty.

Auralization is a method to assess aspects of aural perception. Auralization can also be called by observation or feeling directly. Observations were carried out to assess the level of aural disturbances from various sources and intensity.

4. Observation and Measurements 1

This observation is to observe field conditions, noting any sounds or audible sound disturbances, and also including their intensity.

Figure 3. Church Environments

The first observations and measurements were made at 6 am before the First Sunday Service. At this morning the church environment was quite calm. There is almost no activity going on around the church. Lack of activity at this hour makes the smallest sound can be heard. Some of the voices heard include: the sound of birds, the sound of the door open, and the sound of falling leaves.

The closer to the hour of worship will begin, the congregations have begun to visit the church. The sound of the piano from the church marks the activity of worship in the neighborhood. Until the beginning of worship activity, the piano sound has become the main voice is heard. In addition, occasionally heard the sound of traffic, either crossing the road environment and the vehicle of congregation who came to the church. The noise is described in Table 1.

Table 1. Description of sound noise observation 1

No.	Location	Kinds
1	Bird	High pitch Quite Clear Non vocal Natural Decay
2	Opened door	Low pitch Quite Clear Non vocal Artificial Attack
3	Falling leaves	Low pitch Quite

		Clear Non vocal Natural Decay
4	Traffic	Low pitch Loud Clear Non vocal Artificial Attack
5	Piano	Low pitch Quite Reverberant Non Vocal Artificial Decay

Measurements were made outside the church with a specified measurement point from across the road (MP1), from inside the church fence (MP2), and from the church's closest wall to the street (MP3). This measurement will be done a few minutes before the worship starts.

Table 2. Description of sound noise observation 1

No.	Measurement Point	Minute-n (dB)											Average
		0	1	2	3	4	5	6	7	8	9	10	
1	MP1 (Street)	70.9	70.3	66.7	72.1	67.2	68.7	70.9	74.2	68.7	73.5	72.4	70.5
2	MP2	70.4	66	66.6	69.2	64.7	66.1	67.2	72.1	67.3	70.9	70.7	68.2
3	MP3 (Wall)	67.2	65	64.9	67.2	62.9	64.6	65.3	69	69.5	67.1	68.4	66.4

The first measurement is done during Sunday Service 1 at 6 to 6.30 am. Measurements were made for 10 minutes.

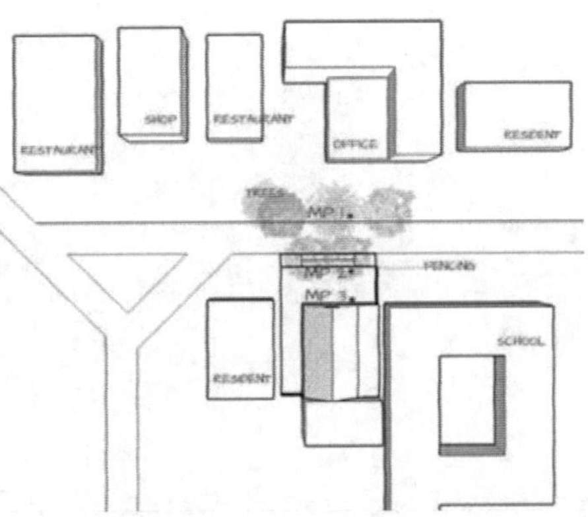

Figure 4. Measurement points

The measuring point 1 (MP1) shows a relatively higher intensity compared to the MP2 and the MP3. It is because this measuring point closer to the sound of traffic. The average value on MP1 is 70.5 dB, MP2 68.2 dB, and MP3 66.1 dB. The peak point on the 3rd and 7th minutes indicates the sound of traffic recorded through the measuring point.

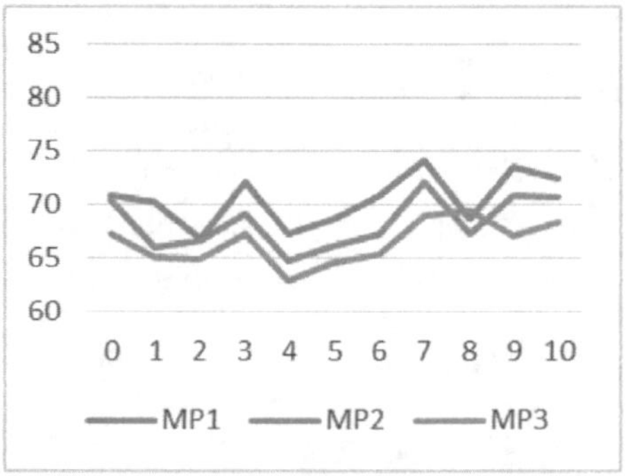

Figure 5. Measurement 1 Chart

5. Observation and Measurements 2

The second observation was made at the time of of the Second Sunday Service at 9 am. At this time activity in the church environment has started crowded. The aftermath of the first worship service at 8 am was the first observation.

Noise at this service is more varied due to activity in residential and commercial environments around. The aural disturbances observed include: the sound of birds, the sound of high traffic, children playing, music from neighbors, singing, and street vendors.

Table 3. Description of sound noise observation 1

No.	Location	Kinds
1	Bird	High pitch Quite Clear Non vocal Natural Decay
2	Traffic	Low pitch Loud Clear Non vocal Artificial Attack
3	Children	Low-High pitch Loud Clear Vocal Attack
4	Music	Low pitch Quite Clear Non Vocal Artificial Decay
5	Man singing	High pitch Quite

		Reveberant Vocal Artificial Attack
6	Street vendors	High pitch Loud Clear Non vocal Artificial Attack

Table 4. Description of sound noise observation 1

No.	Measurement Point	Minute-n (dB)											Average
		0	1	2	3	4	5	6	7	8	9	10	
1	MP1 (Street)	75	71.2	79.7	72.3	77.4	72.9	71.6	82	73	76.6	83.3	75.9
2	MP2	71.8	66.1	73.7	68.3	75.8	71.9	68.9	78.8	68	70.4	73.1	71.5
3	MP3 (Wall)	69.7	63.5	65	66.6	71.2	68.1	68.5	69	64.3	70	68.8	67.7

The second measurement is done during the Second Sunday Service at 8 to 9 am. Measurements are also done for 10 minutes.

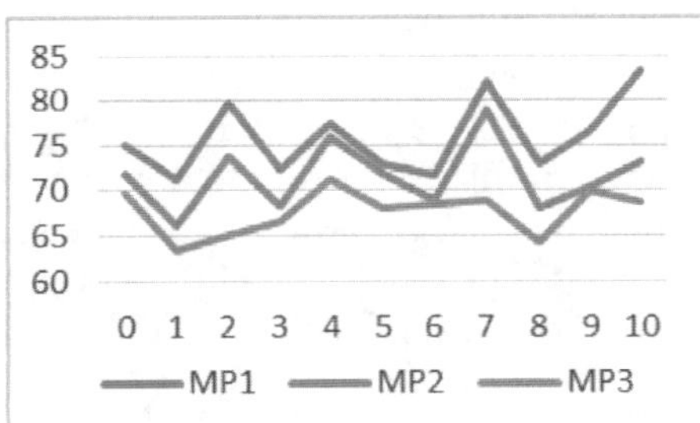

Figure 6. Measurement 2 Chart

The diagram shows the fluctuation of the sound intensity. This is enough to describe the atmosphere and sound disturbances that exist in the church environment. The lowest intensity at MP1 is at 71.2 dB (minute 2) and the average value is 75.9 dB. The average value in MP2 is 71.5 dB while the MP3 is 67.7 dB. This average value is higher than the measurements of First Sunday Service.

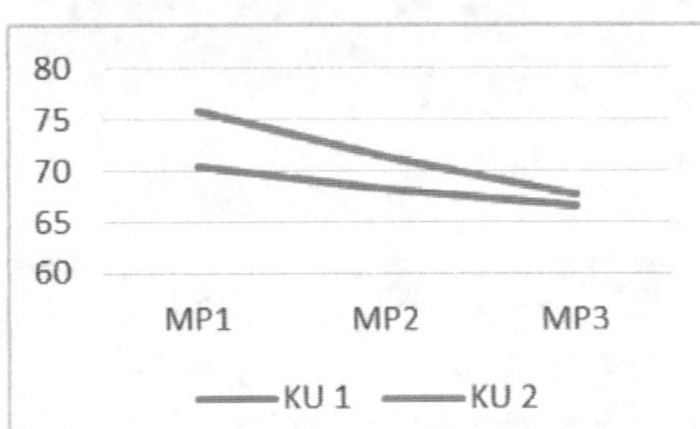

Figure 7. Decrease in the mean value of the measuring point

Based on the graph, the distribution of sound in the church environment can be understood with reasonable. The MP1 taken from the curb across the church has a higher average value than the other measurements point. Noise from the traffic and neighbors activity are the cause. While the comparison

IOP Conf. Series: Earth and Environmental Science **152** (2018) 012024 doi:10.1088/1755-1315/152/1/012024

between measuring results service at 6 am with the service at 9 am shows a clear difference about the higher environmental noise during the day.

6. Analysis

Aural architecture can explain how sound and space can affect emotions, behavior, and experiences. If current research focuses only on architectural acoustics, aural architecture involves aspects of perception and culture that can provide a more comprehensive understanding of aural experience [2]. Physical aspects of aural architecture relate to measurable technical elements. Observations and measurements carried out provide an overview of the physical content of aural disorders in the church environment. Aspects of perception are related to the responses and perceptions of each individual. While the cultural aspects explain the aural experience of the environmental culture side that affect other aspects. Analysis of aural experiences and their disturbances based on physical, perceptual and cultural aspects are described simultaneously as a unity of understanding.

6.1. Components of Bahareque

Church surrounded by other buildings with a relatively close distance. The boundaries of the church site are:

North : Residence
East : Road, office, restaurant
South : School
West : Residence

At Sunday Services, the neighbors does not cause disturbing sounds. For example, the schools and offices that do not operate on Sundays, as well as restaurants that have not been operating in the morning. But it would be very disturbing if the worship took place on another day. Both school and restaurants have a peek time resulting in noise. One way to keep the sound from being heard is to keep the sound source away. If the point distance to the source of the sound multiplies 10 times, the intensity level is reduced by 20 dB. The distance between adjacent buildings precisely leads to the opposite.

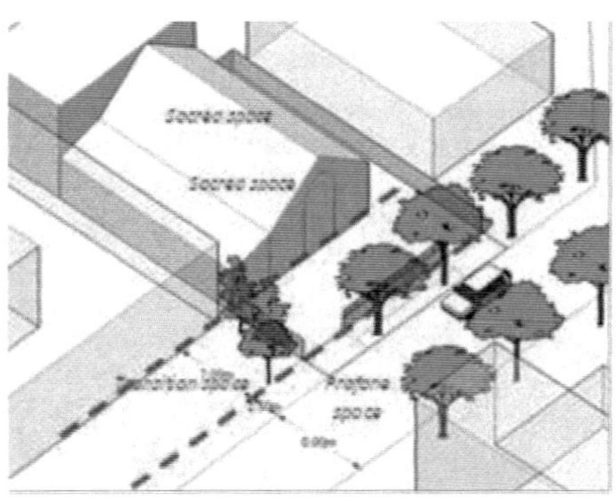

Figure 8. Decrease in the mean value of the measuring point

Ideally, the church should have enough courtyard and not only on the front but also on the left and right even behind it. In addition to the sound mileage factor, the presence of this yard will create a comfortable environment and enhance the impression of spirituality as a place of worship.

6.2. Sound Barrier

The condition of the land in contact with the road requires a barrier element so that the sounds that reach the building do not cause discomfort or disruption to the activity inside. According to Koenigsberger in Satwiko [3] states that the allowed noise level in the church is 35-40 dB. Trees and wall fence can be an option as a barrier of noise from the street. Vegetation can muffle sound up to 95% by absorbing sound waves by leaves, branches and twigs.

The church environment is pretty much overgrown with trees and other. At the front of the church there are trees with a height of approximately 10 m with canopy and wide leaves. In principle, trees with characters like this fit into a common barrier of sound that intersect with the road. However, based on the observation it is seen that these trees do not cover the bottom that can still be penetrated by the noise from the road.

Figure 9. A church still visible to be recognized from the street

There are elements of the church fence to help in dispelling the sound that comes with covering the bottom of the tree. The front fence of the church has a height of 2 m with a combination of stone materials and iron arrangement. A solid field is effective in reflecting sound. While the transparent field is not very effective. View and aesthetic considerations are the reasons for the design of this fence.

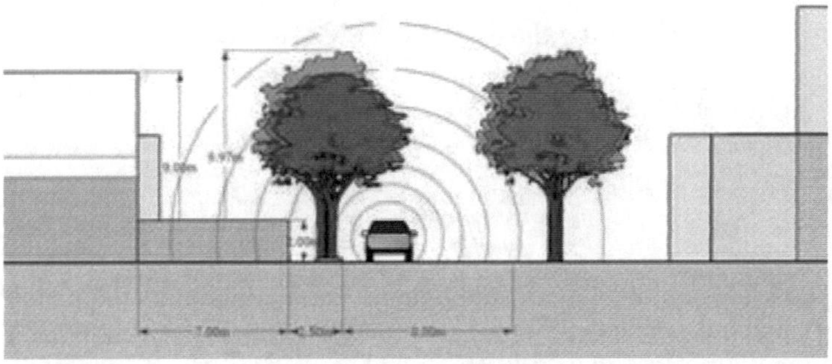

Figure 10. Street section

In Indonesia the fence has become a complement of buildings to boundary the area. Few people have reason to present the fence as a barrier of sound, but rather to give a boundary of land, privacy, or security considerations. But in this discussion the fence serves as a barrier of sound that unwittingly reduces the intensity of noise from the street.

Figure 11. Street section

The combination of trees and fences is a design strategy in controlling noise from the road. Design options can vary. However it is advisable in the selection of vegetation should be of various height. Short plants such as teh-tehan or soka flowers can be used as a sound barrier while beautifying the church environment. In addition to the transparent part of the fence can be modified by providing a place to be overgrown with plants.

7. Conclusions

Coming to a house of worship is an experience that is different from other places because it relates to the sanctity of the environment. A worship environment that requires sound aesthetics cannot be separated from the approach to the sound space in it. Experience involving aural and visual can determine the atmosphere built during worship. This experience can affect the emotions through perception of the sound.

To establish the sanctity of a place of worship should consider a good environment as it affects the atmosphere formed from its surroundings. There are voices that are allowed to be present in the church environment, some are not. Birds' voices are natural voices permitted to be present in the church environment.

Most of the sound is not recorded by the Sound Level Meter due to the small intensity of sound that only records the sound intensity above 30 dB. The observed sounds are not necessarily sound interference. If it sounds vague then it is not a sound interruption.

Sound barriers with vegetation need to be carefully considered so that other aspects such as the view towards the church are not blocked.

References

[1] Blesser B and Salter L -R 2007 Experiencing Aural Architecture (Massachusetts: MIT Press)
[2] Algargoosh A 2016 Review of Aspects That Shape The Aural Experience in Worship Space. *Proceedings of the 22nd International Congress on Acoustics* (Buenos Aires: ICA)
[3] Satwiko, P. (2008). Fisika Bangunan. Yogyakarta: Penerbit Andi

Acknowledgements

The research about the assessment of aural disturbances experienced by a church located in housing district still need to be developed. The obstacles that exist in this research in terms of data collection and the results may be refined in further research. The hope of this study can contribute to the development of science.

IOP Conf. Series: Earth and Environmental Science **152** (2018) 012025 doi:10.1088/1755-1315/152/1/012025

Typology Study of Urban Canyon in Residential Area and The Quality of Its Thermal Environment

A M Firdausah [1] and S Wonorahardjo[2]

[1]Master Student of Architecture, SAPPK ITB, Bandung, Indonesia
[2]Building Technology Research Group, SAPPK ITB, Bandung, Indonesia

E-mail : azzahra_miftahul@ymail.com

Abstract. The current physical development in urban area turns the cityscape into vertical growth with building that separated by narrow streets. Consequently, canyon formed in urban residential impact on thermal quality. Surfaces in canyon receive shortwave radiation and increase the Urban Heat Island (UHI) intensity. This paper discusses the quality of thermal quality in various urban canyons form which are influenced by several factors such as density, materials, topography,building arrangement and vegetations. The aims to categorize a number of residential canyon causes and its qualities of thermal environment. The initial result shows that building materials, canyon geometry and orientation gives significant role to their thermal environment.

1. Introduction

The availability of limited land makes the urban vertically and dense designed. The phenomenon of vertical residential design forms a building cliff called urban canyon phenomenon. Buildings that are made vertically also lead to heating due to solar radiation absorbed by urban surfaces such as roofs, walls of buildings and roads. However, the radiation cannot be released directly into the atmosphere due to urban structure density. Aksoy & Ekici [1] stated that vertical surfaces are increasingly important in terms of solar absorption due to increasing of vertical urbanization in recent years.

According to Nunez and Oke [2] 30% of solar thermal energy is absorbed and stored by canyon material and released at night with the result increasing the ambient temperature. The increasing of the air temperature in the urban at night and the temperature difference between urban and peri urban is an indication of heat island phenomenon. The difference in atmospheric stability between the urban and its surroundings is due to the greater heat capacity of the urban canyon structure (Kusuka and Kimura [3], Santamouris et.al [4] also found a high intensity of the heat island with an increase in air temperature of 13°C around the canyon at night due to anthropogenic heat and the use of heat-absorbing material.

Air circulation and air temperature distribution within urban canyon have significance for air quality inside buildings as well as outside environment. Buildings that are in canyons of low thermal quality use more energy for air conditioning and even more electricity for lighting [5]. High urban temperatures increase also decrease comfort in open spaces and reduce the potential for cooling by natural convection at night [6]. Akbari et al. [7] found that for every 1 °C rise in air temperature, 2 to 4% of electric energy is required for cooling as an effect of heat island. Based on these findings means that urban canyon has a bearing in the formation of heat islands and as an index to evaluate the quality of thermal environments.

Urban canyon plays an important role in the difference of air temperature between urban and peri urban. This is related to the amount of radiation absorbed by the surface of the canyon. The aspect ratio between the height and width of the canyon is the dominant factor in solar radiation on road and wall surfaces. In the field of shadowed buildings, the heat entering into space only occurs through the conduction process itself due to the difference in outside temperature and inside temperature. However, in the field of buildings exposed to sunlight (not exposed to the shadow), then the heat that enters into space can be through the process of radiation resulting from back radiation of the heat of the walls exposed to sunlight [8] Building orientation and ratio aspect between height of building and road width on canyon impact on urban air distribution have been proven by C. Georgakis and M. Santamouris [9], Takabayasi [10] Niachou et.al,[11]. Canyon geometry and surface orientation that absorb radiation also greatly affect the duration of time and the amount of energy stored and released by the surface of the walls and base of the canyon [2]. Niachou et.al. [11] suggested that the distribution of air temperature inside the canyon is in addition influenced by geometry and canyon orientation also influenced by the optical properties and thermal properties of building materials as well as micro climatic conditions.

Figure 1. Urban canyon phenomenon [12]

2. Urban Heat Island In Residential

Heat island defined as a higher urban air temperatures condition in urban centers compared to rural or suburban areas. Heat islands are formed in urban and suburban areas because many construction materials absorb and retain more solar thermal than natural materials. There are two main reasons for this warming. First, most urban building materials are watertight, so moisture is not available to remove the solar thermal. Second, dark material on concrete buildings with canyon configurations and pavement collecting and trapping more solar energy [13].

Changes in land use which settlements into trading has changed the physical condition of the spatial region. Building originally in the form of single level buildings transformed into a densely populated area with a vertical- narrow building. Wonorahardjo [14] analyzed the physical character of ten districts in Bandung Air temperature in old and well planned residential district is relatively higher than others (29.06°C). The trade center and settlement district had higher density of buildings than the other districts. Density aspect of the district could be explained as function of buiding material (wall, roof, street), in which case this aspect act an entrapment effect of hot air surrounding buildings. A dense district like residential district tend to trap the air inside the canyon and increase the air temperature compared with the rare district.

In Putrajaya, Malaysia, the UHI occurred in high-rise buildings are lower than low-rise buildings. The high-rise buildings are thermally comfortable, while low-rise buildings suffer from a heat stress. However, the low-rise building experienced the highest mean radiant temperature which exceeded 70 °C and it received the longest period of Tmrt around 10 h. The high-rise residential received a shorter period of Tmrt with the maximum value reaching 63 °C in high-rise building.

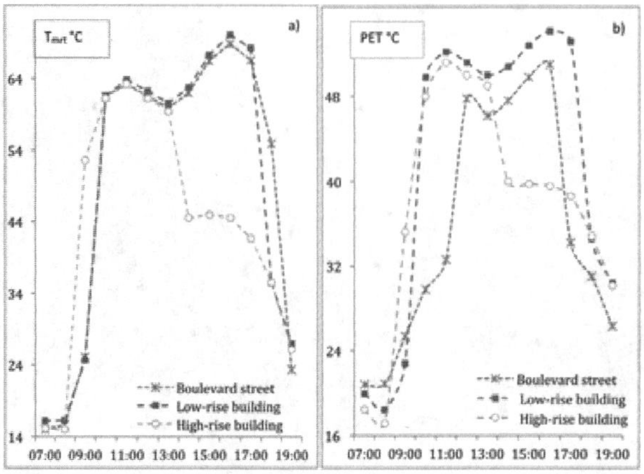

Figure 2 (a) Mean radiant temperature (Tmrt °C), (b) physiologically equivalent temperature (PET °C) [15].

Yang et al. [16] studied the UHI in three areas of high-rise residential buildings, the study found that the density, layout and greenery have impacts on UHI patterns. Site shading factor is closely related to day-time UHI. Variations in density and plot layout have different impacts on-site shading condition. The lowest in day-time UHI achieved with high building density, while long-near layout with low density is the highest. In practice TSF can be used for thermal environment assessment at design stage.

3. Typology Of Urban Canyon

The urban canyon's characteristics are places (usually roads) flanked by two buildings on both sides [2]. Air circulation and air temperature distribution within urban canyon have significance for air quality inside buildings as well as outside environment.

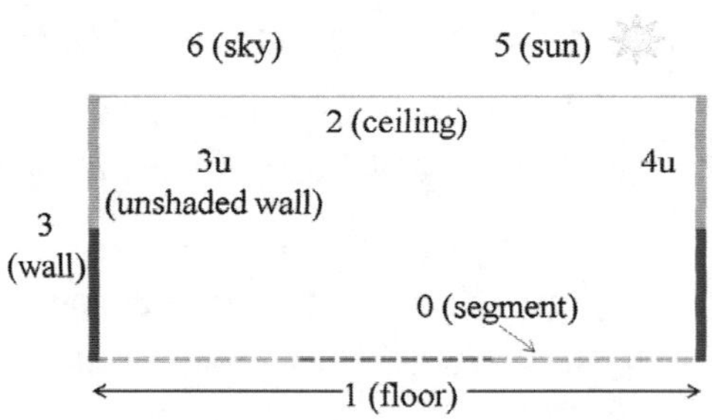

Figure 3. Elements of urban Canyon (surface 0-4) and light source (surface 5 and 6). [17]

Buildings located in canyons without high climate quality use more electricity for air conditioning and for lighting [5]. Akbari et al. [7] found that for every 1 ° C rise in air temperature, 2 to 4% of electric energy is required for cooling as a result of heat islands. Based on these findings means urban canyon has a bearing in the formation of heat islands and as an index to evaluate the thermal

IOP Conf. Series: Earth and Environmental Science **152** (2018) 012025 doi:10.1088/1755-1315/152/1/012025

environments quality. An understanding of the specific thermal characteristics is essential for understanding airflow within the canyon and for the study of natural ventilation and hybrid in urban environments [11]. This is related to the amount of energy absorbed by the urban surface. Santamouris et al [4] revealed that there is a temperature difference on the opposite wall of the building. This occurs because of the impact of the convection phenomenon of adjacent wall surfaces.

A number of studies in urban canyon have been conducted to improve thermal quality. Niachou et.al, [11] stated that the distribution of air temperature within the canyon of is a geometry function and canyon orientation, thermal properties of buildings and road materials and weather conditions.

There are three typical canyon patterns that presented by Shashua-Bar [18]: continous, separated form and colonnaded form (figure.4).

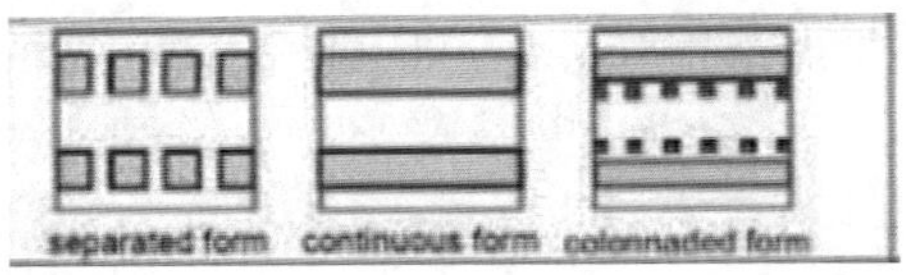

Figure 4. Three archetype streets forms [18]

3.1. Geometry of Canyon

Pearlmutter et al. [19] noted the importance of street orientation and aspect ratio in thermal quality control in the canyon. The distribution of air temperature within the canyon is influenced by geometry, canyon orientation, optical and thermal properties of buildings and materials.

Urban geometry and thermal properties of urban surface are found as two main parameters affecting the urban climate [20]. The height of the building (H) and the width between both of them (W) affects the number of both. Radiation in and out and also affects wind speed. Nocturnal heat island has been proved increasing with the ratio H = W because the outgoing wave radiation is reduced due to decreased sky view factor (SVF) [21].

Another proof is stated by Emmanuel [20] that deep canyons with higher aspect ratios (building height: road width, H / W), can provide more shade at street level during the day.

It can be concluded that the comparison between the height of the building (H) and the width of the street (W) will form a space in canyon and give shadow protection in the street canyon. Shadow protection decrease temperature, thereby providing thermal comfort to its users.

3.2. Canyon Orientation

Chatzidimitriou [23] used a simulation model to analyze the aspect ratio and canyon orientation. The results show that the most comfortable conditions in summer occur in north-south oriented to medium or high aspect ratio nuances while the East-West oriented requires repairs such as additional shade on the open north side. Street oriented in a north–south direction, shade can be provided by the buildings with high H/W in both the morning and the afternoon. However, around noon, shade cannot be provided by the buildings alone.

3.3. Material

3.3.1. Potential of urban surface in the improvement of thermal quality. The balance of urban thermal is defined by the materials optical characteristics that make up the urban envelope. The building envelope is responsible for the most significant load of external environmental quality. Materials in urban areas absorb solar and infrared radiation, and heat accumulation is transferred to buildings or released into the atmosphere.

The materials used in building envelopes and urban structures play a very important role in urban thermal balance. They absorb solar and infrared radiation and throw some of the heat accumulated through convective and radiation processes into the atmosphere that increases the ambient temperature. Thus, the technical characteristics of the materials used determine the level of energy consumption and the comfort conditions of individual buildings or even open spaces.

The thermal capacity of the canyon surface material also contributes to nocturnal heat islands because most of the incoming heat during the day is stored in such materials and is not released until the evening [21] Many research have been done to have better understanding of the optical characteristics and thermal material as well as its impact on the urban climate and systematic environment temperatures over different types of materials have been reported [24].

Therefore, improving optical properties (albedo and emissivity) of urban surfaces can reduce the consumption of cooling energy, due to the decrease of air and surface temperatures [25]. Highly reflective or cold materials are a cost-effective technique, environmental friendly and passive that contributes to achieve energy efficiency in buildings by lowering energy requirements for cooling and increasing the microclimate in the air by lowering surface and air temperatures. Cold material is characterized by:

a. Solar reflance (SR)

Solar reflance is capability of a surface material to reflect solar radiation and indicate total surface fluctuations, remember the reflection of hemispherical radiation, which is integrated through the solar spectrum, including specular reflections and difeksi. It is measured on a scale of 0 to 1 (or 0-100%) [24] The balance of the thermal material is determined primarily by their reflections to solar radiation and their emissivity against long-wave radiation during the day. Therefore the dark surface is warmer than the light-colored ones. The use of 'cold' material contributes to the reduction of air temperature due to heat transfer phenomenon. However the use of warm than cold materials is used for urban environmental structures. This use is due to economic and aesthetic reasons, or by poor environmental planning. As a result, temperatures in urban environments are increasing and demand for cooling loads in buildings is growing [26].

b. Infrared Emissivity (IRE)

Infrared emissivity is a measure of surface ability to release, absorb heat. This determines how well the surface radiates energy from itself compared to the black body operating at the same temperature. Infrared input is measured on a scale from 0 to 1 [22]. The value for emissivity (e) can be between 0 (perfect reflector mirror) and 1 (perfectly emitter black body). Consequently, emissivity plays an important role in constructing thermo graphic surveys and relies on temperature, wavelength and surface conditions. Surfaces with low emissivity values (eg aluminum, steel, etc.) serve as a mirror (high reflection).

3.3.2. Material Properties on the building surface. Solar radiation exposed by urban surfaces such as building facades, sidewalks, roofs and roads is not flattened back into the atmosphere due to urban structures density. It is also known that vertical surfaces are increasingly important in terms of solar absorption due to increased vertical urbanization in recent years [1]. The parameters of each building material are solar reflectance, Emissivity, thermal conductivity, heat capacity [10]

According to Rosenlund in Noerwasito and Santosa [27], the material resistance against heat that affects buildings, called thermal properties are: Density: has a unit of kg / m3, is the ratio between weight and volume, density holds a large role for thermal properties, the material has a mild density that has greater insulation power than large density materials. Conductivity: has a unit of W / mK, is the ability of the material to conduct heat conduction. Materials which have low conductivity have good insulator power; otherwise the material having high conductivity is a good conducting material. Specific heat: has a unit Wh / kgK, is indicating the material that has the ability to store some energy. High specificity heat means the material has the ability to store lots of heat (heat storage).

The combination of these three thermal properties materials results in what is called Time lag, the maximum time the wall uses to remove heat from the outside surface of the wall to the inside of the wall. The material also has thermal capacity; the amount of heat stored by the material, then releasing it.

The dynamic thermal performance of the building envelope depends on three basic properties; Heat capacity, density, and thermal conductivity. High heat capacity and density maximizes the amount of heat absorbed in every m of the material and the thermal conductivity required for the material to make useful heat capacity. A sufficient thermal conductivity allows the material to exchange heat with an air environment at an appropriate level. The thermal diffusivity of the material is a measure of how quickly the temperature of the material adjusts to the ambient temperature. The lower the diffusivity is, the greater time-lag material will be [28].

Thermal conductivity and specific heat capacity are important properties of building materials. These values are required, for example in thermal performance calculations. However, the resulting values for material properties are usually not accurate enough. The fact that certain products such as plaster, may have very diverse material properties. Since many of the material properties are listed commonly to the whole ingredient group, the value is determined from the actual product required to achieve a high degree of accuracy in the calculation for a particular case.

4. Discussion

In line with Niachou et.al [11] research, which stated that the distribution of air temperature within the urban canyon is a function of geometry and orientation, as well as the optical properties of buildings and road materials and weather conditions. This is in line with other studies conducted such as research in the desert town of El Oued Algeria, Bourbia and Awbi [30] reported that the maximum temperature tends to decrease as the H / W ratio increase. While in the hot, humid Dhaka environment of Bangladesh, Ahmed [31] found that the average daily maximum temperature decreased, by 4.5 K when the H / W ratio increased from 0.3 to 2.8. The results show a clear relationship between urban geometry and microclimate on the road level.

The surface temperature of the material also affects the air temperature. Correspondingly, Radhi et.al [32] the material temperature on the surface and air temperature depends on the color, thermophysical properties and radiation properties of the material. Doulos et al. [26] observed that materials which surface is darker are warmer than bright ones. The use of 'cold' material decrease air temperature due to heat transfer phenomenon. However the use of warm materials rather than cold materials is more widely used for urban environmental structures. This use is due to economic, aesthetic, or poor environmental planning reasons. As a result, the temperature in urban environments increases and is an indication of the low thermal quality.

The balance of material thermal determines the material's reflection ability and its emissivity against long-wave radiation during the day. The albedo effect of the building surface has the highest effect that is on the maximum at the day temperature while the emissivity has the highest effect at the minimum night time temperature. High use of albedo on the surface material of the canyon can lower air temperature, reduce energy use, achieve thermal comfort and decrease the intensity of heat islands

In surface temperature at the soil surface. The parameters of each building material are Solar Reflectance, Emissivity, Thermal Conductivity, Heat Capacity [10] The temperature of the canyon surface material is governed by the thermal balance [11].

5. Conclusion

Various studies have dealt with thermal environment in urban canyon. In urban areas where land is limited, the cityscape grows vertically. Consequently, it is formed the urban canyon that impact on thermal quality. The results of research on thermal quality in canyon are influenced by canyon geometry and materials. All of these conclusions are closely related to the geometry of the canyon and the angle of the sun. The aspect ratio between the height and width of the canyon is the dominant factor in solar radiation on road and wall surfaces. The key property of the urban canyon road is expressed by its W / H aspect ratio. In addition, it is also influenced by the orientation of the building, which is related to the direction of solar radiation.

The material thermal balance is determined the material's reflection ability and its emissivity against long-wave radiation during the day. The albedo effect of the building surface has the highest effect that is on the maximum temperature during the day while the emissivity has the highest effect at the minimum night time temperature. Surfaces in canyon that receive radiation can increase the Urban Heat Island (UHI) intensity. The parameters that describe the canyon are geometric dimensions (wall height, road width, and roof width), solar reflectance surface (albedo), infrared emmisivity, and thermal surface properties (heat capacity).

Acknowledgment

This publication and research supported by Research Schema of Desentralisasi ITB PUPT Ristekdikti Indonesia 2017, contract No. 009/SP2H/LT/ DRPM/IV/2017.

References

[1] Ekici Betul Bektas, Aksoy U Teoman 2011 *Prediction of building energy needs in early stage of design by using ANFIS Expert Systems with Applications* Sustainable Cities and Society 26

[2] Nunez M and T R Oke 1977 *The Energy Balance of an Urban Canyon* Journal Of Applied Meteorology Volume 16

[3] Kusuka Hiroyuki and Fujio Kimura, 2004 *Thermal Effects of Urban Canyon Structure on the Nocturnal Heat Island: Numerical Experiment Using a Mesoscale Model Coupled with an Urban Canopy Model* Journal Of Applied Meteorology Volume 43

[4] Santamouris M, N Papanikolaou, I Koronakis, I Livada, D Asimakopoulos 1999 *Thermal and air flow characteristics in a deep pedestrian canyon under hot weather conditions.* Atmospheric Environment Volume 33, Issue 27, November 1999, Pages 4503–4521

[5] A Bitan 1992 *The high climatic quality city of the future Atmospheric Environment.* Part B. Urban Atmosphere Volume 26Issue 3 Pages 313-329

[6] Stathopoulou G M, M Santamouris, H Bagiorgas 2008 *Impact of temperature on tropospheric ozone concentration levels in urban environments* Journal of earthsystem Science 117(3)

[7] Akbari H, M Pomerantz, H Taha 2001 *Cool surfaces and shade trees to reduce energy Use and improve air quality in urban areas.* (Solar Energy 70 (3) 295–310.

[8] Sukawi 2010 *Kaitan Desain Selubung Bangunan Terhadap Pemakaian Energi dalam Bangunan* Seminar Nasional Universita Wahid Hasim Semarang

[9] Georgakis C and M Santamouris 2009 The Vertical Stratification of Air Temperature in the Center of Athens. Journal of Applied Meteorology and Climatology Volume 49

[10] Takebayashi Hideki. Moriyama Masakazu 2012 *Relationships Between The Properties Of An Urban Street Canyon And Its Radiant Environment: Introduction Of Appropriate Urban Heat Island Mitigation Technologies*. Solar Energy Volume 86

[11] Niachou K, I Livada, M Santamouris 2008 *Experimental study of temperature and airflow distribution inside an urban street canyon during hot summer weather conditions*—Part I: Air and surface temperatures. Building and Environment 43 1383–1392

[12] https:// ryanari-wordpress.com

[13] L Gartland. 2008 *Heat Islands. Understanding And Mitigating Heat In Urban Areas* (Earthscan : London UK)

[14] Wonorahardjo Surjamanto 2012 *New Concepts in Districts Planning, Based on Heat Island Investigation* Procedia - Social and Behavioral Sciences, Volume 36

[15] Qaid, et.al 2017. A Qaid, H B Lamit, D R Ossen, R N R Shahminan. 2016 *Urban heat island and thermal comfort conditions at micro-climate scale in a tropical planned city* Energy and Buildings 133: 577-595

[16] Yang F, Lau S S Y, Qian F 2010 *Summertime heat island intensities in three high-rise housing quarters in inner-city Shanghai China: building layout, density and greenery* Build Environment 45(1):115–134.

[17] Rosado P J, G Ban-Weiss, A Mohegh, R M Levinson 2017 *Influence of street setbacks on solar reflection and air cooling by reflective streets in urban canyons.* Solar Energy 144 144–157

[18] Shashua-Bar L, Hoffman M E, and Y Tzamir 2006 *Integrated Thermal Effects of Generic Built Forms and Vegetation on the Ucl Microclimate* Building and Environment 41:343–54.

[19] D Pearlmutter,P Berliner, E Shaviv 2007 *Integrated modeling of pedestrian energy exchange and thermal comfort in urban street canyons*. Building and Environment. Volume 42, Issue 6, Pages 2396–2409

[20] Oke T R 1988 *Street design and urban canopy layer climate* Energy and Buildings Volume 11

[21] Oke T R, Johnson G T, Steyn D G, Watson ID 1991.*Simulation of surface urban heat islands under 'ideal' conditions at night—Part 2: Diagnosis and causation*. Boundary Layer Meteorol. 56

[22] Emmanuel R, Johansson E 2006 *Influence of urban morphology and sea breeze on hot humid microclimate: the case of Colombo, Sri Lanka*. Climate Res. 30:189–200.

[23] A Chatzidimitriou, Kleo Axarli 2017 *Street Canyon Geometry Effects on Microclimate and Comfort; a case study in Thessaloniki.* Procedia Environmental Sciences 38 (International Conference on Sustainable Synergies from Buildings to the Urban Scale)

[24] Santamouris, M Synnefa, A Karlessi T 2011 *Using advanced cool materials in the urban built environment to mitigate heat islands and improve thermal comfort conditions* Sol. Energy 85, 3085–3102.

[25] Alchapar, N L EN Correa 2015 *Comparison of the performance of different facade materials for reducing building cooling needs.* Eco-efficient Materials for Mitigating Building Cooling Needs (Woodhead Publishing)

[26] Doulos, L M, Santamouris, I Livada 2004 *Passive cooling of outdoor urban spaces. The role of materials*. Solar Energy 77 231–249

[27] Noerwasito T & Santosa M 2006 *Pengaruh " Thermal Properties " Material Bata Merah Dan Batako Sebagai Dinding , Terhadap Efisien Enerji Dalam Ruang Di Surabaya* Dimensi Teknik Arsitektur Vol. 34

[28] Mohammad Shaghayegh, Andrew Shea 2013 *Performance Evaluation of Modern Building Thermal Envelope Designs in the Semi-Arid Continental Climate of Tehran* Buildings

[29] Holman J P 1997 *Heat Transfer* 8th edition (Mcgraw-Hill Book Company New York) pp. 56–59.

[30] Bourbia F, Awbi HB 2004 *Building cluster and shading in urban canyon for hot dry climate. Part 1: Air and surface temperature measurements*. Renewable Energy ;29:249–62.

[31] Ahmed KS 1994 *A comparative analysis of the outdoor thermal environment of the urban vernacular and the contemporary development: case studies in Dhaka*. Architecture of the extremes (Proceedings of the 11th PLEA conference, Dead Sea, Israel). Ben-Gurion University of the Negev P. 341–8.

[32] Radhi Hassan, Essam Assem, Stephen Sharples 2014 *On the colours and properties of building surface materials to mitigate urban heat islands in highly productive solar regions* Building and Environment 72

IOP Conf. Series: Earth and Environmental Science **152** (2018) 012026 doi:10.1088/1755-1315/152/1/012026

The Application of the Environmentally Friendly Co-Housing Concept in the Residences at Desa Tegaldowo, Kecamatan Bantul, Yogyakarta

T W Murtini[1] and N Muladica[2]

[1] Lecturer Architecture Department, Faculty of Engineering Diponegoro University, Semarang, Indonesia

[2] Master Student of Architecture Engineering Departement Faculty of Engineering Diponegoro University, Semarang, Indonesia

E-mail: titien_wm@yahoo.com

Abstract. Humans' comfort when living in their residences is heavily influenced by their economic level, and it distinguishes the design of residences of the poor and the rich. Having limited position, architects and society should participate in order to solve the aforementioned problem. In participatory architecture, the client's special position that is non-profit and non-private, and in this case the clients are the communities and people in lower class should be considered. The participatory architecture is also supported by a concept called 'CoHousing' or 'Community Housing'. It is a community-based concept in Yogyakarta where its people are identical with communal work. The study case took place at Desa Tegaldowo, Kecamatan Bantul, Yogyakarta where the people still live in simple houses. The participatory architecture concept and 'CoHousing' concept will complement each other in involving the community that can give impact for their household. So we can apply the concept of eco-friendly Cohousing on settlements in Desa Tegaldowo based on the participation of community. This research used the method of design guidance. The result of the research was the guidance of settlement design with concept of design participation, co-housing' and permaculture concept using environmental friendly materials and structures at Desa Tegaldowo.

1. Introduction

The property developments, especially in sector of housing are increasing. This is related to the needs of the community for housing in the increasingly limited availability of land. So, the developers are competing to provide the best facilities to attract for consumers to be interested in housing offered. But, sometimes the design and facilities of housing are just "opening" at the beginning. These new housing in general have the character of industrial goods that are made in bulk so that not in accordance with the needs of its users and the climate in Indonesia. And also affected at the comfort and feelings do not feel at home.

Home is not just a collection of physical but also a social gathering. High demand and demand factors for housing, forcing most urban communities in Indonesia to become individualists. The increasingly fragile urban security level, plus the lack of interaction with neighbours with the fence of houses such as fortresses make a dead social interaction. The characteristics of housing as mentioned above is not in line with the general character of Indonesian citizens are no exception Yogyakarta city who like to socialize and cooperate. Actually, the problem of land limitations and social interaction in building a house is not just happening in our country alone. Abroad, this problem has long been happening so that there have been various concepts to overcome the problem. One of them is a concept called 'Cohousing' or short for 'Community Housing'. A community-based housing concept as one of the answers to various issues related to housing today. This concept has been started by some communities such as "Rumah Bersama Community".

Implementation of this concept is suitable when viewing the condition of the city of Yogyakarta as one of the friendly city between its citizens. This is seen in one of the villages in Yogyakarta, namely in Desa Tegaldowo, Bantul. As a village that has tourism potential, this village belongs to an independent village because it has various businesses in agriculture, farming, and also in handicraft, which belong to micro, small and medium enterprises.

1.1. Formulation of the Problem

Housing is a residential area inhabited by certain types of houses with large residential capacity. In urban areas have emerged various models of housing with fancy concepts. However, the character of the housing is actually not in line with the general character of Indonesian citizens are no exception the city of Yogyakarta in the village of Tegaldowo, Bantul District who love to socialize and cooperate. So the research question arises as follows:

1. What is the potential of Tegaldowo Village community in Kecamatan Bantul that can increase their occupancy value ?
2. How to connect the community potency of Tegaldowo Village, Bantul District with the application of Permaculture and Co-housing concept that is environmentally friendly on the dwelling ?

1.2. Research Purposes

The purpose of The Application of the Environmentally Friendly Co-Housing Concept in the Residences at Desa Tegaldowo, Kecamatan Bantul, Yogyakarta to be able to identify the potential of the community and apply the concept of eco-friendly Cohousing on settlements in Desa Tegaldowo, Bantul, Yogyakarta city based on the participation and ability of the village community.

1.3. Research Outcomes

The result of this research is a house design guidelines of settlement for the society with the concept of co-housing and participatory architectural approach using environmentally friendly materials and structures in Tegaldowo Village, Bantul District, Yogyakarta.

1.4. Research Methods

This research used the method of design guidance with the concept that refers to the way people build residential with his basic theory using the design participation, the concept of co-housing and permaculture.

2. Literature Review

2.1. Co Housing

Cohousing is a concept of residence. This concept is emphasizes "common space" called common house on housing. The core of the concept of Cohousing i.e. in one housing there are spaces that can be shared by all residents of the house in the housing [1]. The difference with ordinary housing, in housing cohousing residents of houses already formed the community first before the housing itself was built. Then they commit to live in a housing together and come to design the housing design. According to McCamant and Durrett [2] there are several characteristics of Co-Housing concept, among others:

1. Participatory process, prospective occupants are involved from the beginning in designing settlements and community forms that suit your needs
2. Shared Facilities, designed for everyday use to save land & expenses.
3. Design Housing, must support community life. Usually cluster-shaped, motor-free and allows each member to always interact with each other.
4. Managed by Occupants, there is a division of tasks among fellow residents to manage the settlement area.
5. Decision-making is done by consensus, not hierarchy.
6. The economic system is not communal, in which each occupant has his own source of income that does not come from the community.

2.2. Permaculture

Permaculture derived from the word "permanent" and "agriculture" can be interpreted as the management of agriculture, animal husbandry, and fisheries by improving the quality of ecosystems so fulfilled human needs that remain sustainable into the future. In fact, according to Yayasan IDEP (one of the institutions that develop Permaculture in Indonesia) in the Permaculture Handbook, the permaculture can also be meant "permanent culture" which is to preserve, support and cooperate with local culture and environment, and grow together at the same time [3]. Work with nature and people and learn from them instead of fighting or competing with them. These ethics and principles speak of responsibility for our own lives, our environment, and the future. It also helps us plan for a safe future for our families, nature, and culture [4].

According to Permatil [5] there are characteristics of the concept of permaculture

1. Permaculture is one of the most holistic, integrated systems analysis and design methodology found in the world.
2. Permaculture can be applied to create a productive ecosystem from a human perspective using or to help restore the health of degraded and wild ecosystems. Permaculture can be applied to any ecosystem, no matter how degraded.
3. Permaculture is the value of validating traditional knowledge and experience. Permaculture combines sustainable farming practices and land management techniques and strategies from around the world. Permaculture is a bridge between traditional culture and emerging earth cultures.
4. Permaculture aims to maximize symbiotic and synergistic relationships between the components of the site.
5. Permaculture is urban planning and land design in the countryside.

2.3. Participatory Planning

The development through community participation is one of the efforts to empower the community potency in planning development related to local resource potential based on consultation study, that is increasing the aspiration in the form of desire and real need that exist in the society, increasing the motivation and role of community group in development process, And increased sense of belonging to community groups on program activities that have been developed. Development planning process based on community participation must pay attention to the interest of the people which aims to improve the welfare of the community. The working principle of development through community participation is as follows:

IOP Conf. Series: Earth and Environmental Science **152** (2018) 012026 doi:10.1088/1755-1315/152/1/012026

1. Work programs are presented openly to the community by conducting participatory communication in order to get community support,
2. Work program implemented through cooperation and working together groups between communities, village officials and all citizens in order to minimize barriers in the program
3. The work program does not lead to certain groups in society or groups in order not to cause division,
4. During the program, coordination is always done vertically or horizontally,
5. No need to be superior or "feel most know" in every opportunity of implementation of work program,
6. No need to promise to anyone but the seriousness of work within the context of a work program that has been determined.

Community Development with all its activities in development should avoid the "doing for the community" work method, but adopt the "doing with the community" work method. The working method of doing for, will make the community become passive, less creative and helpless, even educate people to rely on government assistance or voluntary aid organizations. In contrast, the working method of doing with, stimulating the community to be active and dynamic and able to identify where the needs of its nature - real needs, felt needs and expected need. The working method of doing with, closely matches the great ideas of KI Hajar Dewantara about education leadership in Indonesia - ing ngarso sung tulodo, ing madyo mangun karso, and tut wuri handayani - which focuses on the need for participative independence in the development process [6].

3. Discussion

The location of this research is located in desa Tegaldowo Kecamatan Bantul kota Yogyakarta. The village is located on the border of the sub-district Pajangan and the subdistrict of Sewon.

Figure 1. Site location of Tegaldowo Village

Tourism Tegaldowo Village has a variety of interesting tourism potential to be visited by tourists and folk art that is still preserved. As geographic this land area is located in Bantul regency located on the south side of Yogyakarta Special Province. This land has an area of 7,327 m2. With a relatively large number of residents and most of the livelihood as private employees and laborers.

IOP Conf. Series: Earth and Environmental Science **152** (2018) 012026 doi:10.1088/1755-1315/152/1/012026

Figure 2. The condition of housing in Tegaldowo village

With the growing population of villagers, coupled with immigrants like the Pringgomukti community of Pinggokusuman in early 2017 to settle for a long time that determines their upcoming life plans. It shows that there will be a development of residential space for them. The development of this dwelling refers to their life plan based on their needs as residents. The development of this dwelling will not take a long time, because in the empirical conditions of Tegaldowo village has the characteristic of people who like mutual cooperation and tolerance. In addition, the development of houses in the village Tegaldowo itself evolved from the concept of Java house with the division of three main rooms namely the front room, living room, back room. This proves that their culture is still preserved.

Therefore, the application of residential design concept in Tegaldowo village still refers to the concept of Javanese house with the shape and material adapted to the ability of the community. And the development of residential space tailored to the needs of residents who are the same majority of the community in accordance with the concept of co-housing. The role of Tegaldowo villagers themselves is vital because the attitude of mutual participation among the community is so high that the concept of co-housing is very appropriately applied in the village. In addition, land use that can be used as a cultivation area as a basis for applying the concept of permaculture to use it as an area of cultivation and alternative resources

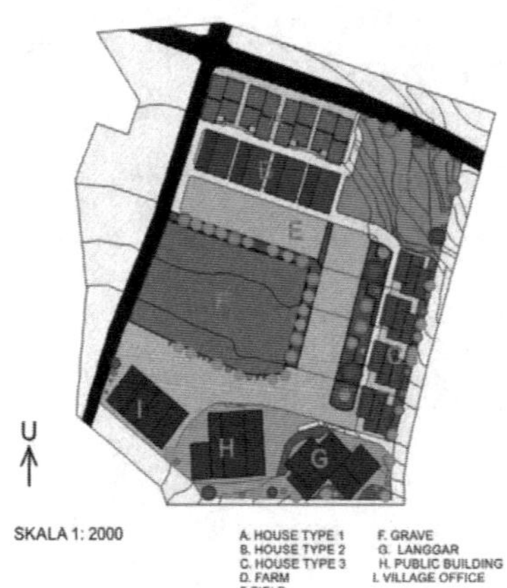

Figure 3. Part of siteplan in Tegaldowo Village

The design process starts from the area of the village. The design includes the location of the cemetery, rice fields, houses, gardens, village hall and langgar. Each placement is based on the hierarchy of the place. In Tegaldowo village the burial area has the highest hierarchy so that the cemetery area is placed some distance from the residential area. The boundary between the cemetery area is limited by the rice fields surrounding the cemetery. New settlement and public place of the village. As explained before, with the development of villagers affected by the immigrants there will be changes in residential space. Based on the results of meetings with the residents, there are several aspects that can be taken as the life plan of residents if already settled in Tegaldowo, namely health, religious, harmony, comfort, environment, economic growth, productive, new life, happy children. The concept of house refers to the way villagers from Tegaldowo village build their house with the concept of traditional Javanese house by dividing the house into 3 parts (Javanese traditional house concept of the front room, living room, back room / servise) which is also applied to the design guide. The design guidelines include:

1. Create a private space as a stimulant for space restrictions.
2. The development of space according to the needs of residents of the front room and living room in accordance with the concept of traditional javanese house.
3. The addition of service space on the back of the house.
4. Use of insubstantial spaces between spaces to maximize the existing shelter.
5. Use some natural materials such as wood as walls and bamboo as a roof frame and columns.
6. Periodic maintenance of all materiasl used in that house.
7. Use of environmentally friendly materials that use woven bamboo coated cement on the outer wall of the dwelling house.

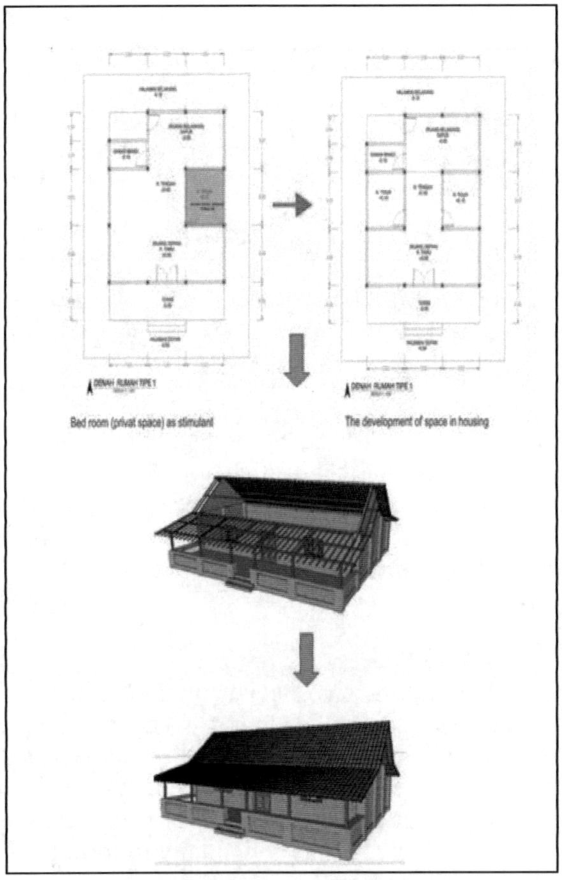

Figure 4. The steps of space housing development in Tegaldowo village

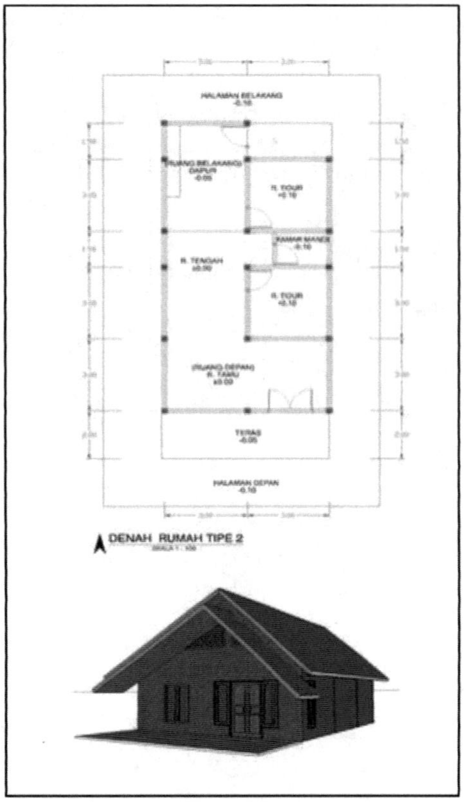

Figure 5. Alternative design 2 (floor plan and 3D model)

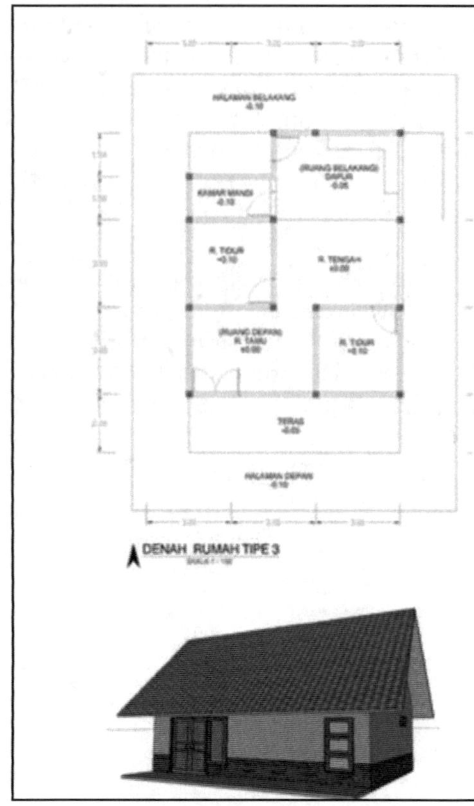

Figure 6. Alternative design 3 (floor plan and 3D model)

In the process of residential design in Tegaldowo village, the material and the shape of the house adjust the ability of the community. So that the participation of village communities is needed in the planning process. There are 2 alternative home design models that have the same concept on the basis of Javanese house concept (consist of front room, living room and back room). But using the saddle roof and bamboo material according to the ability of the villagers.

4. Conclusion

In Desa Tegaldowo, Bantul, Yogyakarta has the potential of settlements that can be developed according to the needs of residents and the community. The development of settlements through residential design where the space within it develops according to the needs of its inhabitants. The role of the tegaldowo village community is crucial in the residential planning process that suits their ability. So that the design and design model can be easily understood and applied because the proposed design is adapted to the activities of the community to form their house with the application of the concept of co-housing and permaculture.

5. References

[1] McCamant K and Durrett C 2011 *Creating Cohousing: Building Sustainable Communities*, New Society Publishers ISBN 978-0-86571-672-8
[2] McCamant, Kathryn M.; Durrett, Charles. 1994, Cohousing: *A Contemporary Approach to Housing Ourselves (2nd Edition)*, Ten Speed Press, ISBN 0-89815-539-8.

[3] McKenzie L 2006 *Buku Panduan Permakultur* Yayasan IDEP Bali ISBN 979-15305-0-5
[4] Mollison B 1979 *PERMACULTURE: Practical Design for Town and Country in Permanent Agriculture* (Australia: Tagari) ISBN 090-82280-0-7
[5] Permatil 2011 *Buku Panduan untuk Permakultur: Menuju Hidup Lestari* Yayasan IDEP ISBN 979-15305-0-5
[6] Tampobulon M 2006 *Pendidikan Pola Pemberdayaan Masyarakat Dan Pemberdayaan Partisipasi Masyarakat Dalam Pembangunan Sesuai Tuntutan Otonomi Daerah* (Fakultas Ilmu Pendidikan. Universitas Negeri Medan. Sumatera Utara)

A Review on Mitigation Technologies for Controlling Urban Heat Island Effect in Housing and Settlement Areas

Heri Andoni[1] and Surjamanto Wonorahardjo[2]

[1]Master Student of Architecture Research, Department of Architecture, SAPPK ITB, Indonesia
[2]Building Technology Research Group, Department of Architecture, SAPPK ITB, Indonesia
E-mail : heri.andoni.ha@gmail.com and wonorahardjo@yahoo.com

Abstract. Urban Heat Island (UHI) has many harm impacts to urban and human life, the examples are increased building and energy consumption, increased air pollutants emissions, compromised human comfort and health, and many more. Despite UHI phenomenon and those impacts has been realized long enough by the experts along with development of major cities in the world, but its mitigation technology has not been well developed. This paper discusses few mitigation technologies for UHI phenomenon that have been developed by experts. Some mitigation technologies such as double skin façade, shading strategies, and many more are discussed in this study, include the strengths and the weaknesses of each strategy. The focus of the study is mainly on the potential of building skin engineering in which thermal conductivity, infrared emissivity, and specific heat factors. The selected case discussed are buildings made of heavy weight materials. The result in this study reveals a potential map where thermal insulation is one potential strategy to reduce the intensity of UHI through the reduction of heat emissions of heavy building materials components. By this study, improvement of the urban life in its relation to UHI are expected to occur, especially in housing and settlements area.

1. Introduction

UHI is a phenomenon closely related to the improvement of cities and also its expansions [1]. The term 'heat island' represents a region with a higher temperature (urban) than non-urbanized area around it [2]. Urban centers have lower solar reflectivity, higher solar absorption, and greater thermal capacity compared to the rural areas due to their less vegetated surfaces and darker surfaces [3]. Wanphen & Nagano [4] in their report found the temperature difference could be as high as 5-15° C between urban and its surroundings (rural). Experts and scientists has realized about Urban Heat Island impacts and have been trying to find solution to mitigate it but the mitigation technology has not been developed much. Accordingly, this study is motivated to look the mitigation technologies for Urban Heat Island, especially on Building Skin Engineering. This paper is a preliminary study which is aimed to increase urban environment because of Urban Heat Island and to discover a new alternative solution for UHI mitigation. Urban Heat Island can be detected by surface or air temperature. Figure 1 showed that commercial and residential areas are the third most areas which have the hottest temperature. It means the Urban Heat Island intensity in this area is high, and that's why we need to mitigate UHI in housing and settlement.

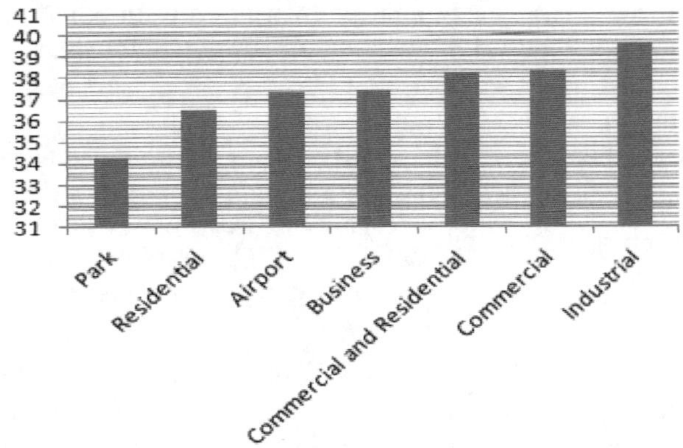

Figure 1. Daytime temperature in various land use areas [5].

The causes of UHI are identified [6] :

- Greenhouse gas emissions

Vehicles, industrial processes and the heating of buildings with fossil fuels are main sources of gas emissions, which trap solar energy in the atmosphere and warm the atmosphere.

- Gradual loss of urban forest cover

Vegetation has roles in preventing heat through evapotranspiration and shading of the ground and buildings. Loss of vegetation means loss of cooling in urban area.

- Impereability of materials

The change of ground covers from natural soils to asphalt made a problem because they can't provide water filtration or absorpsion functions, modify the path of stormwater [7, 8, 9].

- Thermal properties of materials

Impermeable surfaces and building materials influence the microclimate and thermal comfort conditions, since they absorb considerable heat during the day, which they release back into the atmosphere at night, thus contributing to the urban heat island effect [10].

- Urban morphology and city size

Urban morphology, which relates to the three-dimensional form, orientation and spacing of buildings in a city, also plays a role in the forming of Urban Heat Islands [11]. Large buildings and narrow streets can hamper good ventilation of urban centres because they create canyons where the heat generated by solar radiation and human activities accumulates and remains trapped [8]. In fact, the reduction of the sky view factor limits net radiative losses of buildings and streets [12]. In addition, urban morphology can also influence vehicle traffic and thus promote inputs of heat and air pollution from this mode of transportation [13].

- Anthropogenic heat

The production of anthropogenic heat such as heat emitted by vehicles, air conditioners and industrial activity is another factor that contributes to the development of heat islands, particularly in dense urban areas where activities are concentrated [11]. The best way to mitigate Urban Heat Island first is to know about the causes of the phenomenon explained above. After we know the causes, we could explore how to mitigate the phenomenon itself.

2. Selection of research studies

This paper reviews recent research about some various strategies to mitigate UHI effect in the urban environment. The main objective of such strategies is to map mitigation technologies for Urban Heat Island, and to find new innovative solution for UHI mitigation to improve urban life. The article had to be deep study of the heat island effect to be included in this review, its characteristics, influential factors, consequences, mitigation strategies, etc. The result of this study will reveal the possible effective strategies to mitigate UHI effect. However, it should be underlined, because of the corelation and interrelation, some of these strategies might be overlap to each other, and it is also because some of these researches tend to imply several of the strategies simultaneously.

3. UHI mitigation strategies

For decades, the experts has been developed some mitigation strategies to face the Urban Heat Island phenomenon. Some of the strategies really has a big impact to environment. The strategies implied are focusing on the air temperature reduction or surface temperature reduction. The common mitigation strategies are explained below.

3.1. Green spaces

Based on Shashua-Bar research, the combination effects of tree shading and evapotranspiration cause a big decrease in temperature, and even produce what are referred to as cool islands within the city [14]. Varied studies have analyzed the temperature in parks and beneath trees. the overall conclusion was that green areas were cooler than areas with none greenery [15, 16]. In another research, Eliasson found air temperature difference between the city and the park can be as high as 4°C [17]. Ca et al. [18] collected field knowledge to quantify the result of a park on the summer climate in a nearby area. They analyzed ways in which to cut back the energy consumed for air con. At varied locations, the air temperature, relative humidity, and different meteorological factors were measured. The study was dispensed within the park in Tama ny, a coastal city within the metropolitan area of Tokyo. The results obtained indicated that vegetation considerably changed the city climate. At noon, the cooling was the maximum amount as 1.5°C in an exceedingly busy commercial area one km down wind [18]. The analysis analyzed over that parks and green areas might help to mitigate urban heat island effects and reduce cooling energy consumption in summer.

What is more, it had been found that such green areas additionally reduced the temperature changes made by building materials, stabilizing the temperature fluctuation caused by building materials [19, 20]. However, it had been not clear specifically however park affected the formation of the heat island. The cooling result looked as if it would rely on the dimensions of the park and therefore the seasonal radiation conditions, however there was no linear relation between the dimensions of the park and therefore the intensity of the cool island. This intensity was primarily determined by the realm occupied by the trees and shrubbery within the park in addition as by the form of the park.

3.2. Vegetations

Rosenfield analyzed with the use of vegetations, lighter colors of paving, and cooler roofs can reduce the air temperature as much as 3° C [21]. The study above combined all three aspects to reduce the air temperature. Another studies analyzed the use of vegetations and planting only can reduce air temperature between 1.3 – 1.6° C [22,23].

3.3. Green roofs

The study analyzed by Li et al showed green roof fraction has to be close to 90% and the cool roof (albedo = 0.7) fraction has to be close to 95% in order to reduce the near-surface UHI by 0.5 °C.

3.4. Albedo

Albedo is Latin, meaning whiteness. The albedo of a surface is a shard of sunlight reflected by the surface [24]. Taha in his research found by the increase in albedo and vegetation, it has impact to the reduction in the surface temperature and near-surface air temperature. This study demonstrated that the use of materials with high albedo properties decreased the radiation absorbed by building and urban structures. By keeping these sur- faces cooler, the intensity of long wave radiation was reduced. From this study, it can be concluded that by modifying albedo of the surface (0.25 to 0,40), air temperatures can be decreased as much as 4°C [25].

3.5. Pavements

Thermal balance in pavement depends of the amount of solar radiation absorbed by the pavement, the emitted infrared radiation, the heat transferred by convection to the atmospheric air, the heat stored into the mass of the material and the heat conducted to the ground [26]. Synnefa et al. [27] report that the measured surface temperature differences due to the differences in solar reflectance, reached 12 C (between the off-white thin asphalt layer sample and the conventional black asphalt).

4. Building skin engineering to reduce air temperature

The mitigation strategies above specifically developed in macro and meso scale of urban space. The reduction of the air temperature varies between 1-3° C. In order to mitigate the heat island more detail, the mitigation strategies could be developed in micro scale of urban space, or the building scale, which is more related to architecture scope. Building skin or envelope absorbed the heat from solar radiation, and emitted the heat to the environment. The potential of building skin engineering in which thermal conductivity, specific heat, and infrared emissivity factors could lead to the maximum reduction of air temperature and UHI intensity. Building skin engineering for UHI mitigation strategies which have been explored by experts are explained below specifically their effects to reduce air temperature.

4.1. Double skin facade

Al Kaabi [28]on his thesis examined the interaction between the urban canopy and a building with a Double Skin Façade (DSF). It has been found that the heat gains from the building envelope can be reduced by applying the DSF, the reduction of heat gains causes an enhanced energy performance and lower cooling loads. Additionally, the application of DSF can help achieve lower pavement temperature, which increases the outdoor conditions for pedestrians. In terms of UHI intensity, the application of DSF have resulted in a reduction of about 1.4°C which is a significant enhancement to the urban environment

4.2. Tree shading strategies

Tan et al [29] indicates that the tree canopies are able to maintain the surface cooler than the air above. Ta is cooled by 1 ◦C at the pedestrian level (1.5 m) in the morning and by 1.5 ◦C in the afternoon.

4.3. Vertical greeneries : green facade and living walls

Vertical greeneries consist of two types : living walls and green facades. Green façade is a kind of green walls in which a simple structure (such as metal frames,square panels and cable systems for retaining structure) for scaffolding attached to the wall of building ,and acts such as anchor for creeping and climbing plants: vine, ivy and etc. Green facades can rely on fences and columns or be built as an independent structure. Green facade scaffolding structure can be different materials such as wood, steel (galvanized, stainless, coated), plastic or aluminum, etc., each of this materials will follow the usage , aesthetic aspects and different functions [30].

A research about cooling effect of green façade has been done by Jeffrey Price in 2010, tested the effect on West and South walls of four small buildings. The results showed the significant reduction of surface temperature, air temperature inside, building's ambient air, and also heat flux. On hot, sunny days, the green façade on the south and west sides cooled their respective ambient air temperatures by as much as 1.12 °C at 12:00 pm DST and 2.97 °C at 7:50 pm DST , respectively.

4.4. Reflective coatings

Synnefa et al in 2007 made a research which shown a cool coating applied on a rood (SR 0.65) could reduce the temperature (1.2-3.7°C). Look at figure 2 below for the comparison of building engineering method for reducing air temperature.

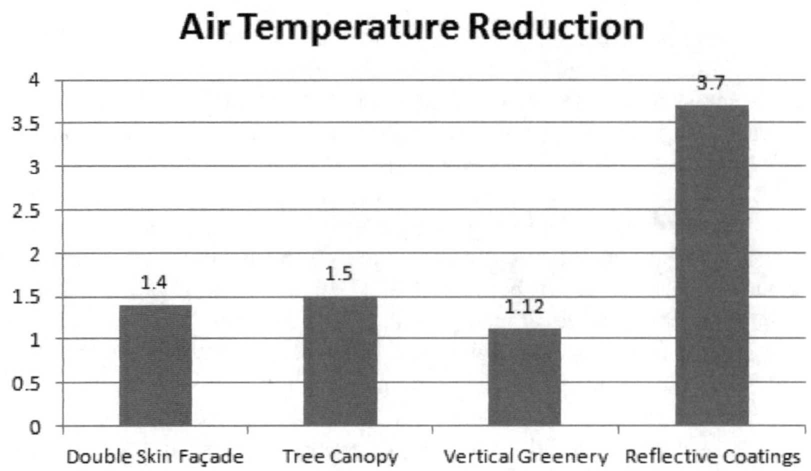

Figure 2. Air temperature reduction with several building skin engineering method.

5. Building skin engineering to reduce surface temperature

To measure Urban Heat Island, experts usually use air temperature or surface temperature. Correlated with building envelope, it is hard to use air temperature as a factor to be examined, so it would be easier to use surface temperature as a factor to be examined.

5.1. Vertical greenery system

Wong et.al [31] have examined by the use of vertical greenery system, the reduction of wall surface temperature could be maximized to 11.58 °C.

5.2. Tree canopy system

Berry et.al [32] have examined by the use of tree canopy system, the reduction of wall surface temperature could be maximized to 9 °C.

5.3. Reflective coatings

Shen et al [33] have examined by the use of reflective coatings, the reduction of wall surface temperature could be maximized to 19.9° C.

5.4. Double skin facade

Kinnane [34] have examined by the use of DSF, the reduction of wall surface temperature could be maximized to 7° C.

5.5. Thermal insulation

Ibrahim [35] experiment on uninsulated and insulated envelopes so far resulted great outcomes. With the use of thermal insulation, west wall surface temperature could be reduced up to 26°C (from 49°C to 23°C). This result means thermal insulation is one of great solution to reduce surface temperature. Look at figure 3 for comparison building skin method for reducing surface temperature.

From figure 3 below, we can conclude that thermal insulation is a great solution to reduce surface temperature. Surface temperature reduction is up to 26°C.

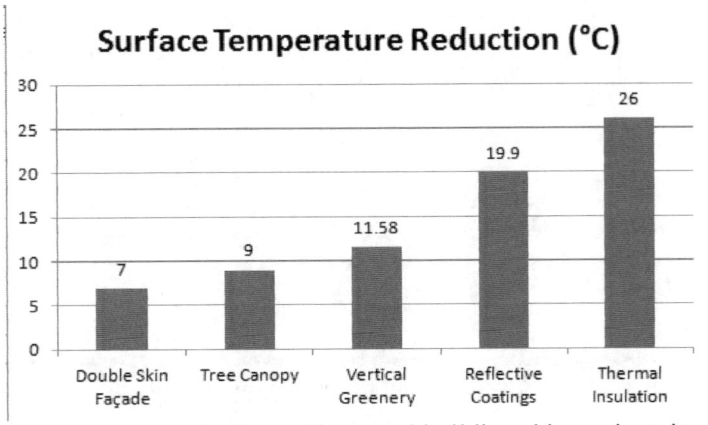

Figure 3. Surface temperature reduction with several building skin engineering method. Source : Analyzed by literature reviews.

6. Disadvantages of building skin engineering that has existed

Despite of the great reduction of ambient air temperature resulted by those building skin engineering (1.4 – 3.7° C) and great reduction of surface temperature (7-19.9°C) , they have few disadavantages. Double Skin Façade has higher construction costs, which Oesterle et al [36] describe "noone would dispute double skin façades are more expensive than single skin forms". DSF has no clear information, positive or negatif about how its reaction in case of fire. DSF also have additional maintenance and operational cost, and there is not a very efficient way to estimate the costs [36]. DSF also have overheating problems. If the DSF system is not properly designed, air temperature in the cavity might increase and make interior space overheating. Other disadvantages of Double Skin Façade are increased structural weight and the reduction of light entering the building, because of its large size [37].

For tree shading strategies, it takes a lot of time to wait for the tree can shade the building. If the tree used for shading are imported, there are some factors which need to be considered, such as the strength of the soils, the façade of the building which is hindered by the trees. It should be considered about the type of the vegetation used as well.

Vertical greeneries also have some disadvantages. Vertical greeneries change the building façade, which affects the design of the building. From the economic perspective, the cost of vertical greeneries is expensive. The cost is between $900 - $1,500 per square m. Another disadvantage, it will attract biodiversity (fauna) which could damages the infrastructures. It is also need extra maintenance. Another disadvantages are : Incorrect dimensioning of the support of the vegetation, both in relation to the size of the plant and the loads will cost incident on the structure; lack of space for the plant to develop, mainly regarding the thickness of the branches, faults and poor quality of the coating, which can cause damage to the building even if the choice of the plant and the support has been appropriate [38, 39]

7. Thermal insulation as good UHI mitigation technology opportunities

With all disadvantages of building skin engineering above, we could see that we still have some room for the improvements of building skin engineering for UHI mitigation. Thermal insulation could be explored as good UHI mitigation technology opportunities. How thermal insulation could be one of potential strategy for UHI mitigation is explained below.

Physical development in big cities generally use heavy material in building envelope such as brick and concrete. This heavy material component has a high heat capacity that stores a lot of solar heat (see figure 1). This has a detrimental effect on the thermal environment, where the heat stored in the morning and afternoon is released back in the afternoon and evening in the form of heat re-emissions resulting in a hot city environment becoming hotter (see figure 2). This extreme rise in urban temperatures resulted in the emergence of the Urban Heat Island phenomenon where city temperatures differ significantly compared to rural temperatures. This heat island phenomenon significantly affects the activity, productivity, and quality of life of urban residents so that mitigation is needed to prevent further damage [40].

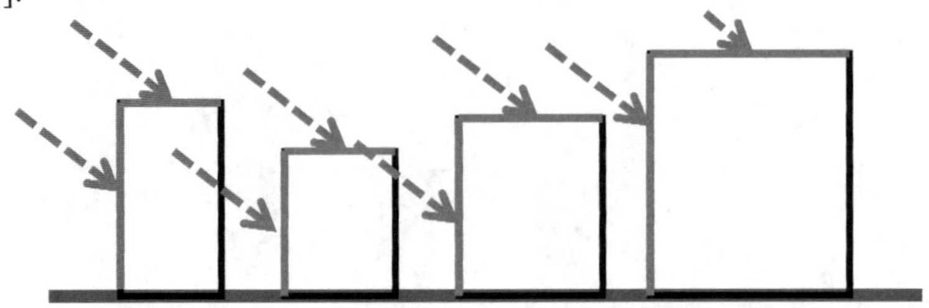

Figure 4. Heavy materials store heat from the sun on the building envelope in the morning and afternoon.

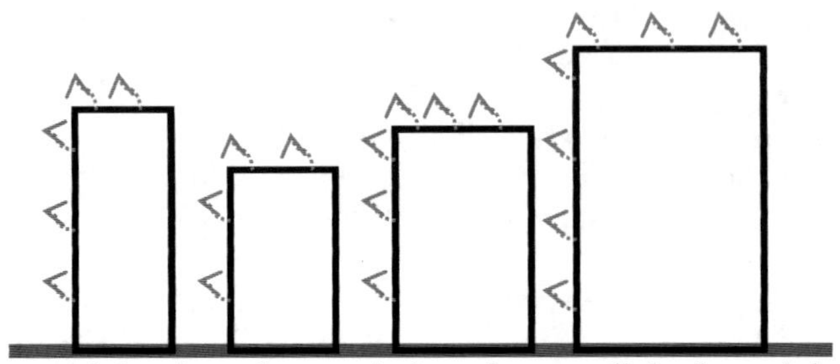

Figure 5. Heavy materials release heat emissions from sunlight through building envelopes

Existing mitigation strategies generally do not take into account the release of heat emissions (IRE / infrared emissivity) from buildings, whereas the release of heat emissions is one of the most important aspects in the formation of heat islands where the release of heat emissions by heavy building material increases the urban temperatures of hot cities In the afternoons and evenings. For that it is necessary heat island mitigation that can intervene before heat is absorbed into the building that will break the path of heat before it is released back outside the building. In addition, there are considerations of heat island mitigation in terms of economy, time, and the appearance of the facade. As explained above, the use of double skin facade is expensive and heavy, the use of vegetation takes a long time until the tree can give the effect of shading, while the use of green façade will change the appearance of the facade that will affect the design aspects of the building. Therefore, this research tries to present heat island mitigation strategy that is fast, cheap, does not change the facade, but able to intervene before heat is absorbed by building envelope, which will influence to decrease of heat emission which is released by the building. The potential strategy proposed in this study is by providing thermal insulation attached to heavy building material sheath (see figure 3).

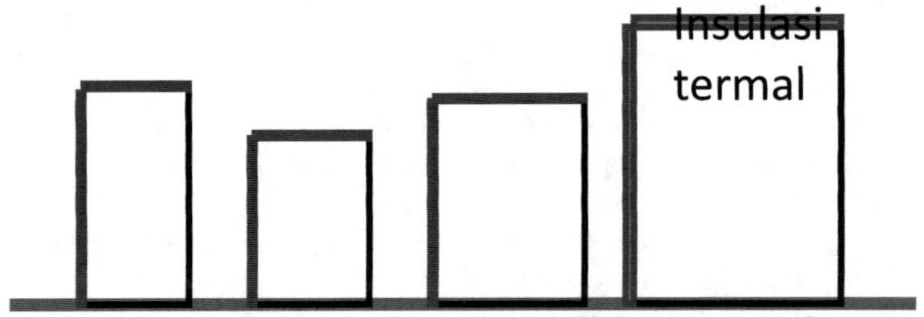

Figure 6. Thermal insulation as an intervention on the building envelope for the reduction of heat emissions of heavy building materials

From the explanation, we can conclude there are 2 important heat factors that needs to be considered for thermal insulation strategy. The factors are heat resistance and heat re-emission.

7.1. Heat resistence

Heat ressistance is about how the thermal insulation could resist from the heat from solar radiation before the heat enters the heavy weight materials (m^2K/W). The more thickness means the less heat flow through the building and so does a lower conductivity. High thermal resistance means a good insulator, and one with a low thermal resistance is a bad insulator. The equation is Thermal Resistance (m^2K/W) = Thickness (m) / Conductivity (W/mK)

Thermal Conductivity measures the heat flow by conduction through a material. Conduction is the main sort of heat transfer through insulation. The lower the thermal conductivity, the performance of the material as insulation will increase. The density refers to the mass (or 'weight') per unit volume of a material and is measured in kg/m3.

The Specific Heat Capacity of a material is the amount of heat needed to raise the temperature of 1kg of the material by 1K (or by 1°C) . A good insulator has a higher Specific Heat Capacity because it takes time to absorb more heat before it actually heats up (temperature rising) to transfer the heat.

7.2. Heat re-emission

Factors related to heat re-emission is albedo and emissivity.Albedo is the ratio between incident solar energy and reflected solar energy. The higher the albedo, the better the performance. Albedo rate depends on the combination of surfaces, material, pavements and color. Materials with lighter color will have high albedo and it will absorb less energy than materials with darker color. Materials with dark color will absorb more heat during day time and it will have high reflectance at night. Another important factor of heat re-emission is thermal emittance. It means surface ability to emit IRE (infrared radiation). Surfaces which have high emittance value will release more heat and remain cool. Many construction materials except metal have high thermal emittance value [41].

8. Conclusions

From all the reviews above we can conclude that thermal insulation could be the potential strategy to mitigate Urban Heat Island effects, by large amount of surface temperature reduction (26°C) with the use of thermal insulation [35]. With large reduction of surface temperature, the heat will not absorbed much into heavy weight materials and it will not reradiate heat much into environment as well. So, it is why thermal insulation could be potential to mitigate Urban Heat Island, by increasing the thermal resistance of the building envelope. Thermal insulation is one of the easy and effective ways of energy conservation and heat reduction available today. It offers a number of applications in residential sectors, and also commercial and industrial sectors.

To find the best thermal insulation to mitigate Urban Heat Island effects, we need to understand about 2 most important factors of thermal insulation : heat resistance and heat re-emission, and their subfactors such as thermal conductivity, thermal difussivity, density, thickness, specific heat capacity, albedo, and emissivity. We need to explore the most combination of the factors with most advantages to reduce maximum surface temperature and air temperature.

One of the most developed thermal insulation is bio-massed insulation or natural fiber, because it is cheap and it is everywhere [42], it is easy to dispose as waste, so it will reduce negative impact to environment [43], and it has a very low thermal conductivity. So, we can begin the exploration of thermal insulation to mitigate UHI effects from pure bio-massed insulation, or mixed bio-massed insulation : lighweight concrete insulation – agriculture waste based thermal insulation.

9. Acknowledgements

This publication and research is supported by the Research Scheme of Desentralisasi ITB PUPT Ristekdikti Indonesia 2017, contract No.009/SP2H/LT/DRPM/IV/2017.

References

[1] Mirzaei P 2015 *Recent challenges in modeling of urban heat island* Sustainable Cities and Society vol 19 pp : 200–206

[2] Arifwidodo S and Orana Chandrasiri 2015 *Urban Heat Island and Its Effects to Health and Well-being in Bangkok, Thailand* Procedia Environmental Sciences

[3] Kaloustin and Diab 2015 *Effects of urbanization on the urban heat island in Beirut* Urban Climate vol 14 pp 154–165

[4] Wanphen, S and Nagano 2009 *Experimental Study of the Performance of Porous Materials to Moderate the Roof Surface Temperature by its Evaporative Cooling Effect* Building and Environment vol 44 pp 338– 351

[5] Jusuf S 2007 *The influence of land use on the urban heat island in Singapore* Habitat International vol 31p 232-242

[6] Giguere M 2012 *Literature Review of Urban Heat Island Mitigation Strategies* (Quebec : Institut national de santé publique du Québec)

[7] Rushton B 2001 *Low-impact parking lot design reduces runoff and pollutant loads* Journal Water Resource Plng. and Mgmt vol. 127 pp 172-179

[8] Coutts A, Beringer J, and Tapper N 2008 *Changing urban climate and CO2 emissions: implications for the development of policies for sustainable cities* Urban Policy and Research, In Press

[9] Mailhot A and Duchesne S 2005 *Impacts et enjeux liés aux changements climatiques en matière de gestion des eaux en milieu urbain* VertigO Hors-série no 1 pp 1-9

[10] Asaeda A, Ca V.T and Akio Wake A 1994 *Heat storage of pavement and its effect on the lower atmosphere* Atmospheric environment vol 30 pp 413-427

[11] EPA's Report on the Environment (Roe) 2008 Final Report US Environmental Protection Agency Washington, D.C.

[12] Pigeon, G., Moscicki, M. A., Voogt, J. A., and Masson V. 2008. *Simulation of fall and winter surface energy balance over a dense urban area using the TEB scheme*. Meteorology and Atmospheric Physics 102 (3-4) pp: 159-171

[13] Oke T. R. 1988 *The urban energy balance* Prog. Phys. Geogr vol 12 pp 471–508 Crossref

[14] Shashua-Bar L and Hoffman ME 2000 *Vegetation as a climatic component in the design of an urban street An empirical model for predicting the cooling effect of urban green areas with tress* Energy and Buildings vol 31 pp : 221–235

[15] Parker JH 1983 *The effectiveness of vegetation on residential cooling* Passive Solar Journal vol 2 pp : 123–32

[16] McPerson EG et al 1989 *Effect of three landscape treatments on residential energy and water use in Tucson* Energy and Buildings vol 13 pp : 127–138

[17] Eliasson I 1996 *Urban nocturnal temperatures, street geometry and land use* Atmospheric Environment pp : 379–392

[18] Ca VT et al 1998 *Reductions in air conditioning energy caused by nearby park* Energy and Buildings vol 29 pp :83–92

[19] Yu C and Hien WN 2006 *Thermal benefits of city parks* Energy and Buildings vol 38 pp :105–120.

[20] Cao X et al 2010 *Quantifying the cool island intensity of urban parks using ASTER and IKONOS data* Landscape and Urban Planning vol 96 pp : 224–31

[21] Rosenfeld A H et al 1998 *Cool communities: Strategies for heat island mitigation and smog reduction* Energy and Buildings vol 28 pp : 51–62

[22] Ashie Y and Asaeda T 1999 *Building canopy model for the analysis of urban climate* Journal of Wind Engineering and Industrial Aerodynamics vol 81 pp : 237–248.

[23] Tong H et al 2005 *Numerical simulation of the urban boundary layer over the complex terrain of Hong Kong* Atmospheric Environment vol 39 pp : 3549–3563

[24] Coakley J A 2003 *Reflectance And Albedo Surface* Oregon State University Corvallis USA

[25] Taha H 1997 *Urban climates and heat islands : albedo, evapotranspiration and anthropogenic heat* Energy and Buildings vol 25 pp : 99–103

[26] Santamouris M 2013 *Using cool pavements as a mitigation strategy to fight urban heat island –*

a review of the actual developments Renewable and Sustainable Energy Reviews vol 26 pp 224–240

[27] Synnefa et al 2007 *Estimating the effect of using cool coatings on energy loads and thermal comfort in residential buildings in various climatic conditions* Energy and Buildings vol 39 pp : 1167–1174

[28] Al Kaabi M 2016 *Double skin façade as an Urban Heat Island mitigation strategy - case study of a health care facility in Abu Dhabi* Thesis : Masdar Institute of Science and Technology

[29] Tan Z et al 2015 *Urban tree design approaches for mitigating daytime urban heat island effects in a high-density urban environment* Energy and Buildings Journal vol 114 pp : 265-274

[30] Perini K 2012 *The Integration of Vegetation in Architecture ,Vertical & Horizontal Greened Surfaces* International Journal of Biology vol 4 p 81

[31] Wong NH et al 2009 *Thermal evaluation of vertical greenery systems for building walls* Building and Environment Journal vol 45 pp : 663-672

[32] Berry R et al 2013 *Tree canopy shade impacts on solar irradiance received by building walls and their surface temperature* Building and Environment Journal vol 69 pp : 91-100

[33] Shen et al 2011 *The Effect of Reflective Coatings on Building Surface Temperatures, Indoor Environment and Energy consumption An Experimental Stud.* Energy and Buildings vol 43 pp : 573–580

[34] Kinnane O 2014 *Assessment of the double-skin façade passive thermal buffer effect* PLEA 2014

[35] Ibrahim S 1987 *The thermal behavior of thermally insulated and uninsulated buildings* Energy Journal vol 12 pp : 615-622

[36] Oesterle W, Lieb E, Lutz, and Heusler 2001 *Double skin facades – Integrated planning* Prestel Verlag Munich

[37] Poirazis H 2006. Double Skin Façade a Literature Review. Lund University, Lund Institute of Technology Department of Architecture and Built Environment

[38] Dunnett N and Kingsbury 2004 *Planting Green Roofs and Living Walls* Portland: Timber Press

[39] SHARP R et al 2008 *Introduction to Green Walls: technology, benefits & design* Toronto: Green Roofs for Healthy Cities

[40] Wonorahardjo S 2012 *New Concept in New Concepts in Districts Planning, Based on Heat Island Investigation* (ASEAN Conference on Environment-Behaviour Studies. Procedia - Social and Behavioral Sciences) vol 36 pp : 235 – 242

[41] Bhargava L and Bhargava S 2017 *Urban Heat Island Effect: It's Relevance in Urban Planning* Journal of Biodiversity and Endangered Species vol 5 p : 187

[42] Guilbert S, Guillaume C, Gontard N 2011 *New packaging materials based on renewable resources: properties, applications and prospects* Food Engineering Series pp 619-630

[43] Manohar K 2012 *Renewable building thermal insulation – oil palm fibre* International Journal of Engineering and Technology vol 2 pp 475-479

Comparison of Liveable City of Three Cities in Indonesia through Index of Happiness Data from Social Media and Urban Structure

Munawir[1,2], B J Dewancker[1] and M D Koerniawan[3]

[1] The University of Kitakyushu, Japan
[2] The University of Cokroaminoto Palopo, Indonesia
[3] Bandung Institute of Technology, Indonesia

Abstract. Social media has contributed to providing information from the users. Users can expressing and sharing their thoughts and opinions on all kinds of topics and events freely. Twitter offers organizations fast and effective way to monitor the users' feelings towards their mood in the cities. In twitter there are the big information data that be able to download freely, this study collected data from twitter and make the classification based on users' tweets to categorize the index of happiness of the city. The urban structure data are collected from the urban guideline development from each city. The index of happiness becomes the prosperity level paradigm of a city, it becomes a tool to assess and compare how the city is liveable. This study will analyse from the users' tweets on a regular basis in three cities of Indonesia, to the parameters of happiness that will be compared with the urban structure development. Based on data of the index of happiness from social media can offer better information for city planners and developers that can be used to improve planning and quality of life in the city and future of urban development.

1. Introduction

The Urban and regional analysis involves the use of a wide range of approaches to understand and manage complex sectors, such as transportation, environment, health, housing, the built environment, and the economy. The goals of urban research are many, and include the theoretical understanding of infrastructural, physical and socioeconomic systems; developing approaches to improve urban operations and management; long-range plan making, and impact assessments of urban policy [1].

Indonesians were early adopters of Twitter and are among the most prolific Twitter users. In 2010, the comScore report ranked Indonesia as the country with the highest Twitter penetration, with 20.8% of the Internet-using population visiting Twitter in the month of June. In early 2012 a market research firm put the Indonesian Twitter-using population at 29.4 million, the fifth largest in the world in raw numbers [2].

In this paper, we try to introduce the feelings of public on sentiment analysis into the field of computational sustainability. We collected data about their feelings from the twitter and we use tweets classification based on the feelings of the user that is positive, negative and neutral than we can know it is happy or unhappy. Here we use Twitter to study the fine-grained geography and dynamics of sentiment in the greatest of three cities in Indonesia Bandung, Makassar, and Bandar Lampung Bandar Lampung areas, identifying areas and times of positive and negative sentiment.

This paper we currently use data from twitter, due to its wealth and volume. To understand the demographic information of the city area, we can improve the process of making it cheaper and easier to infer existing demographic information and investigating whether Twitter volume or the volume of a particular discussion topic, correlates with the level of happiness or quality of life. If so, then we can use social media as a proxy for concrete action, and thus collect that information at a low price and in real time.

In this study, we try to find the condition of the three cities based on data from tweets. We want to analyze that their share social in media and for the future data can be a reference to determine the index happiness city and provide that the city that livable or not

The aim of the research is how to use data from social media about their feelings and expected to be a tool to determine the characteristics of the city. Investigate how geographic place correlates with influences societal levels of happiness. The index of happiness becomes the prosperity level paradigm of a city, it becomes a tool to assess and compare how the city is livable, and so it can be used as a reference for future urban development and sustainable urban planning.

2. Methods

2.1. Data Collection

We used data from Twitter. We selected three cities in Indonesia are Bandung, Makassar, and Bandar Lampung. The three cities are 10 largest cities in Indonesia. Table 1 shows the condition of the three regions of each city.

Table 1. The population of three cities, Bandung, Makassar, and Bandar Lampung City

No.	City	Total Area	Population Census 2010	Coordinate
1	Bandung	167.67 km2 (64.74 sq. mi)	2.394.873	6°54′43″S 107°36′35″E
2	Makassar	199.3 km2 (77.0 sq. mi)	1.338.663	5°8′S 119°25′E
3	Bandar Lampung	167.67 km2 (64.74 sq. mi)	881.801	5°27′0″S 105°16′0″E

2.2. Sentiment Analysis

We use twitter to study of the cities based on geography and dynamics of sentiment in the greater area of the city. The relevant tweets can be filtrated basic on feelings or topic and followed by an assessment of whether the tweets are objective or subjective before identification tweets is positive, negative or neutral. The number of words in each text tweets with positive meaning is then compared to the number of words with a negative meaning to give an overall sentiment score [3, 4]. The framework is shown in figure 1, a typical sentiment analysis model that grants access to historical content and to various tools for searching, analysing and reporting based on data.

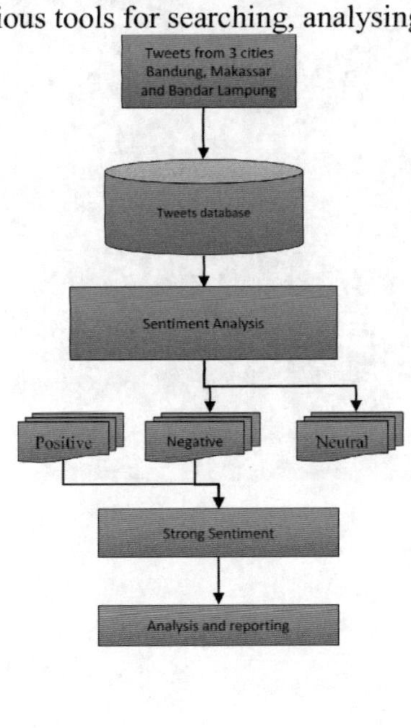

Figure 1. The Frameworks of sentiment analysis model

2.3. *Text Mining*

In this research, we use Naïve Bayes classification for text classification. Naive Bayes classifier is a simple model for classification. It is simple and works well on text classification. It is a probabilistic classifier based on applying Bayes' theorem with strong independence assumptions.

This is the simplest form of Bayesian Network, in which all attributes are independent given the value of the class variable. This is called conditional independence. It assumes each feature is conditional independent to other features given the class. A Naive Bayes classifier is a technique that applies to a certain class of problems, namely those that phrased as associating an object with a discrete category. We make two classes positive and negative tweets [5]

A conditional probability is a probability that event X will occur, given the evidence. So, our initial formula looks like this:

$$P(sentiment|sentence) = \frac{P(sentiment)P(sentence|sentiment)}{P(sentence)}$$

Process stemming is each tweet was stemmed into the group Indonesian words, a match of each word was searched in the lexicon database, scoring Positive and negative matches were summed to define a score of each tweet and polarity: (P-N)/ (P+N), where P=total sum of positive sentiment words; N=total sum of negative sentiment words.

3. Analysis and Discussion

We collected twitter data from three cities in Indonesia are Bandung, Makassar, and Bandar Lampung during September and October 2015. Table 2 shows the total of tweets from the three cities.

Table 2. The Total of tweets data from three cities during September and October 2015

No.	Location	Total Tweets	Total Users
1	Bandung	544,791	100,417
2	Makassar	222,353	27,802
3	Bandar Lampung	50,456	9,076

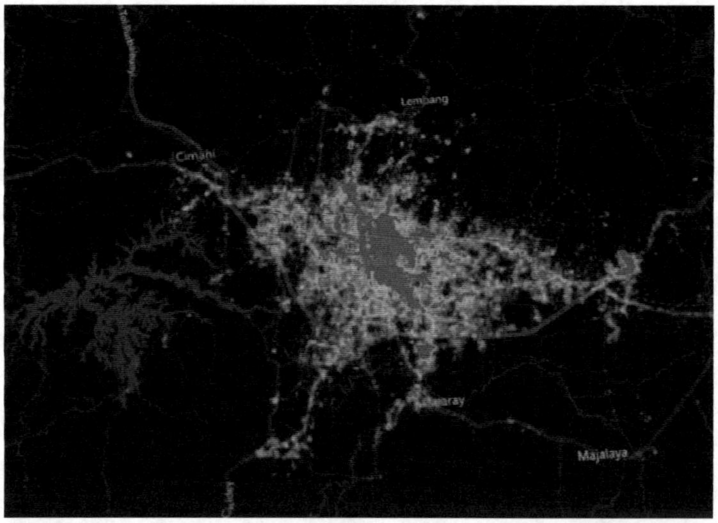

Figure 2. Twitter activity of Bandung city. Each point by geolocated of users. Maps were created using

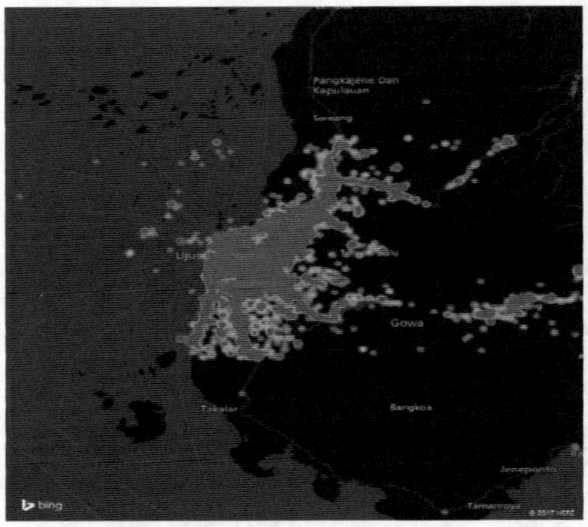

Figure 3. Twitter activity of Makassar city, each point by geolocated of users. Maps were created using Power Map.

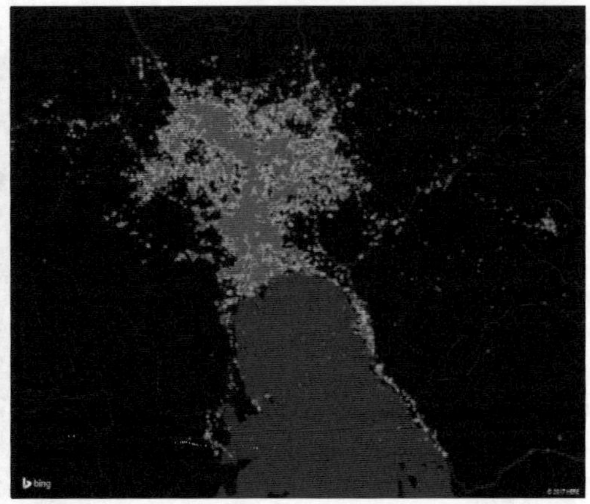

Figure 4. Twitter activity of Bandar Lampung city, each point by geolocated of users. Maps were created using Power Map.

We collected tweets by name of user, text tweets, date, time, longitude, and latitude. From this data we classification which tweets positive or negative based on geolocation. From the figure 2, 3, and 4, we can see the tweet activity reflects population density each cities. Most of the all city very active in social media twitter.

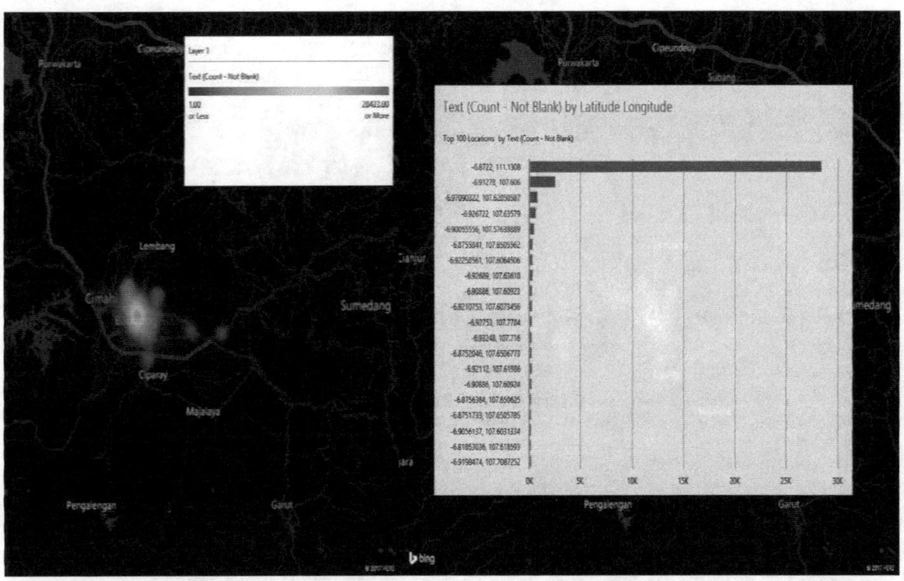

Figure 5. The frequency of tweets by geolocated, Bandung City

We calculated user's position by the number of tweets. In figure 5, 6, and 7 shows there are positions where users often upload their tweets at coordinate points. It is that points we can analysis about their feelings are more concerned their city. We can select any location by looking at the number of frequencies tweets and more detail what is their feelings.

finally, we found the total of sentiment classification from three cities, Bandung, Makassar, and Bandar Lampung. Figure 8, 9, and 10 shows the total of sentiment each city. We can see that from tweets

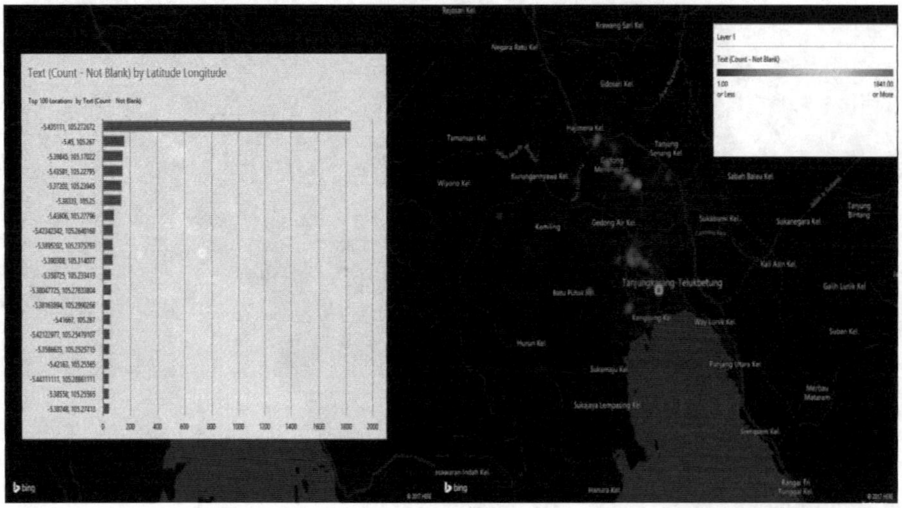

Figure 6. The frequency of tweets by geolocated, Bandar Lampung City

The totally from three cities the percentage of positive tweets more than negative tweets. For example in Makassar city the total of tweets collected 222,353 tweets and the positive tweets is 46391(21%), negative sentiment is 3776(2%) and 77 % is neutral. In this analysis why the neutral sentiment is very high because the system can not calculate half of the tweets the meaning, sometimes the characteristic different write of tweets, users used their local language or symbol.

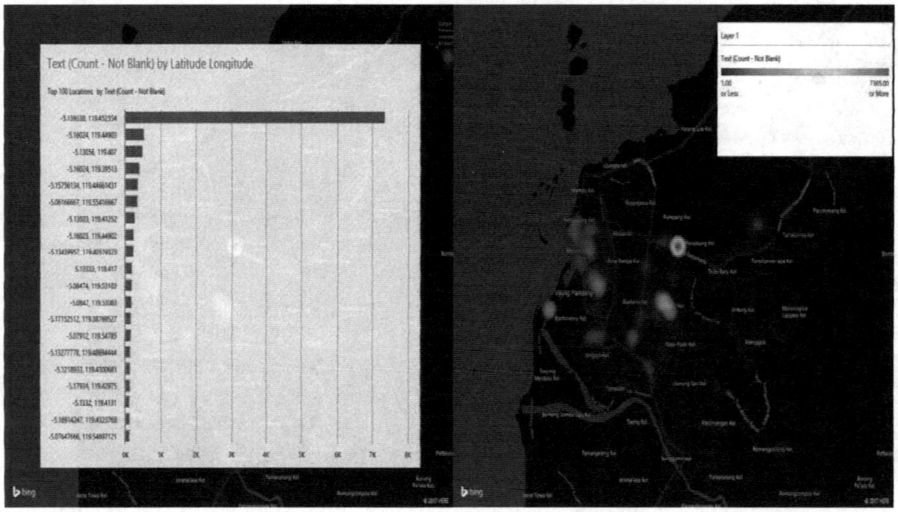

Figure 7. The frequency of tweets by geolocated, Makassar City

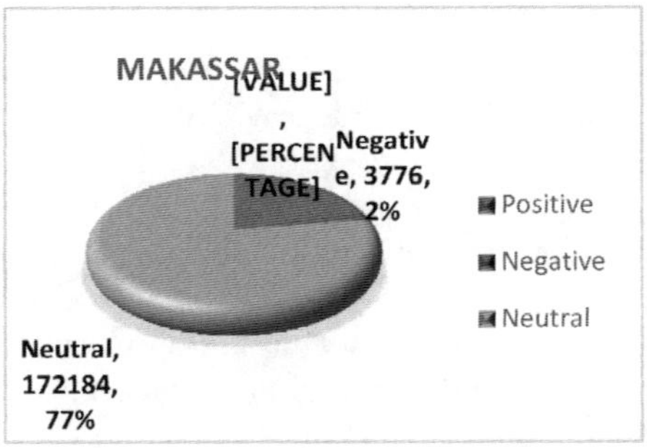

Figure 8. Overall sentiment classification tweets of Makassar.

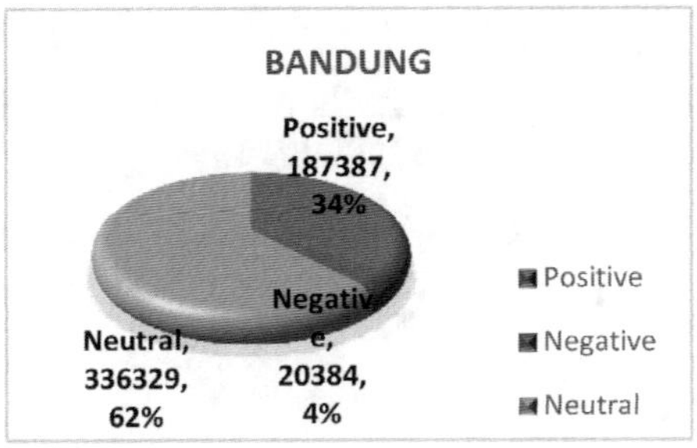

Figure 9. Overall sentiment classification tweets of Bandung

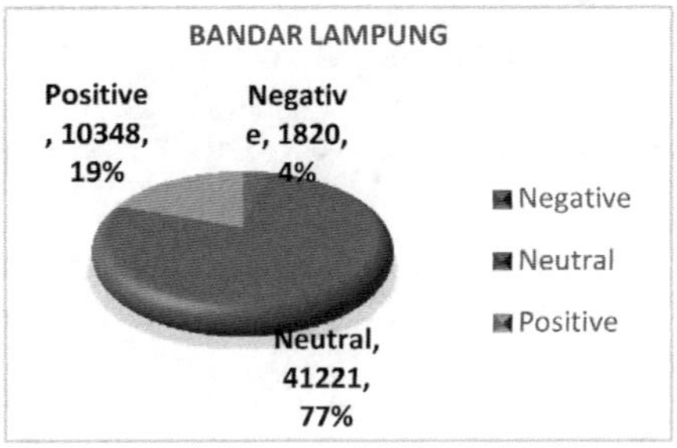

Figure 10. Overall sentiment classification tweets of Bandar Lampung.

4. Conclusions and Future Works

This study is an initial attempt to investigate citizen feelings in the city, determine the index of happiness city based on social media data. Potential of tweets identification their feelings by geolocation on sentiment analysis, as well as offering possible new direction for calculating the well-being of the city and parameter index of happiness city. Using an opinion mining to determine sentiment and impact on the decision-making process can be used and to better information for a city planner.

The report could be effective in generating participation but faced substantial technical, analytical, and communication barriers to influencing decision making. Social media it is very popular in this cities, Bandung, Makassar and Bandar Lampung, that we can see the population of the user using the Twitter is very high. The overall sentiment classification from three cities Bandung, Makassar, and Bandar Lampung shows that positive sentiment more than negative sentiment, it is mean the majority of people happy their life.

For the future work, based on the data we can analyze more detail what the issue make their happy and unhappy related to the city. It is hoped that the data obtained provide better information, understand the people and area of the city are based on the feelings of their mood so that they can accelerate in the city planning process itself. The data can be utilized for the development of sustainable urban and operates efficiently.

Acknowledgments

This study was supported by The University of Cokroaminoto Palopo Indonesia, The University of Kitakyushu, and Directorate General of Higher Education of Indonesia (DIKTI).

References

[1] Thakuriah, P., N. Tilahun and M. Zellner (2015). Big Data and Urban Informatics: Innovations and Challenges to Urban Planning and Knowledge Discovery. In Proc. of NSF Workshop on Big Data and Urban Informatics, pp. 4-3

[2] Kathleen M. Carley, Momin Malik, Mike Kowalchuk Jürgen Pfeffer, Peter Landwehr "Twitter Usage in Indonesia"Center for the Computational Analysis of Social and Organizational Systems.2015

[3] Geetika, V., & Sangharsh, T. (2014). Facebook as a Corpus for Emoticons-Based Sentiment Analysis. International Journal of Emerging Technology and Advanced Engineering, 904-908.

[4] Jasleen, K., & Jatinderkumar, R. S. (2014). On Classifying Sentiments and Mining Opinions. International Journal of Emerging Technologies in Computational and Applied Sciences (IJETCAS), 210-214.

[5] Pablo Gamallo and Marcos Garcia "Citius: A Naive-Bayes Strategy for Sentiment Analysis on English Tweets" Proceedings of the 8th International Workshop on Semantic Evaluation (SemEval 2014), pages 171–175, Dublin, Ireland, August 23-24 2014

Transforming Shell and Society Elements in Human Settlements for Sustainable Tourism Development: Setu Babakan, South Jakarta, Indonesia

R M P Koncara[1], R Tiarasari[1], and W D Pratiwi[1,2]

[1]School of Architecture, Planning and Policy Development, Institut Teknologi Bandung, 40132, Indonesia

wdpratiwi@ar.itb.ac.id[2]

Abstract. The government has established the settlements around Setu Babakan as Betawi Cultural Village (PBB). The goal of PBB Setu Babakan is to preserve Betawi culture which is fading, and realized by making the location as a tourism destination in South Jakarta. This paper discusses the transformation on the two ekistics elements in human settlements around Setu Babakan; shell and society in the development of tourism. Shell as a physical element is reflected in the citizen's houses which are adapted into traditional Betawi houses. While the society refers to Setu Babakan's citizen who participated as economic actors by selling traditional Betawi food and beverages, as well as a performers at the Betawi traditional art performances. In its application, the adaptation of citizen's houses into traditional Betawi house cannot be done completely. The adaptation that can be done to the residents' houses is by beautification in the appearance of the building. The application of typical Betawi ornaments makes the settlement feels like Betawi traditional village. This is one of the attractions for tourists. They can enjoy the atmosphere of Betawi village, enjoy Betawi food and beverages, watch Betawi art performance, and enjoy the lake tour at PBB Setu Babakan.

Keywords: transformation; ekistic's elements; tourism development; Betawi culture

1. Introduction

Jakarta as the capital of the Republic of Indonesia is the center of all activities; those are government, economy, trade, and others. This is related to *Pembangunan Lima Tahun* (PELITA) program which makes Jakarta and its surrounding areas as a pilot in various aspects of development such as industrial area, port area, trade center area, central office area, and education center. The development and expansion of the city have brought changes to Jakarta not only in terms of physical but also the form of a city that was originally a region with ethnic Betawi culture into a metropolitan city culture.

The Governor of DKI Jakarta once issued a decree to make Condet area as a cultural heritage area and agro tourism area [1]. However, the efforts of DKI Jakarta provincial government in reviving Betawi culture in Condet area were unsuccessful due to diverse development interest in the increasingly complex area of Jakarta City. The efforts of Jakarta provincial government to revive Betawi culture as the identity of Jakarta Province did not stop there; supported by *Badan Musyawarah*

Betawi (BAMUS Betawi) was built *Perkampungan Budaya Betawi* (PBB) by developing the region as well as designated the area as recreation and tourism place.

Culture and tourism are closely intertwined in the development of PBB Setu Babakan. This is because the tourism magnet in Setu Babakan is Betawi culture by utilizing elements in the surrounding settlements. The development of this area is functionally implemented by *Unit Pelaksana Teknis* (UPT) of Setu Babakan.

In this paper, we discuss what changes are made to the two settlement elements around Setu Babakan which are shell elements and society. The shell element refers to citizen's houses which are adapted into Betawi traditional houses. The transformation was done on the house of the citizens' aims to make the atmosphere of settlement to have impression like Betawi village again. The society refers to the resident who lives around Setu Babakan. Citizens are invited to become economic actors in the tourism industry in Setu Babakan by selling Betawi food, beverages, or souvenirs. In addition, residents are also invited to be performances in traditional Betawi arts performances which are held every weekend. In addition, to attract tourists, it is also expected to raise the economy of society.

2. Human Settlement's elements as tourism

Developing the area around Setu Babakan is also means to develop the pre-existing areas including settlements. Settlements belong to an ecological system made up of five elements and interactions between the elements in it. The five elements can be referred to as the ecological element, the nature as a system, the human as individuals, the society as a system, the shells that are all types of buildings, and networks of all sorts [2].

In the planning of tourism in Setu Babakan which makes the residential area itself as an object, the five elements play a very important role. Therefore, some changes or maximize the potential of these elements are things that can be done. In this paper, the transformations that will be discussed are related to two elements: the shell and society.

Shell or can be called also built environment is not solely a physical form of housing but also a reflection of the culture of local communities. How Betawi people's daily behavior and philosophy of Betawi people life is reflected in their traditional house. For tourists, it can be very interesting because they can see and feel directly how the daily Betawi society. Then, settlement tourism can be conceived of as a non-verbal cultural dialogue through which local cultural identity is communicated to outsiders [3].

Figure 1. Community Development Increases Capacity for Decision Making on Society [4]

In addition to the shell, another interesting element to be developed is the society. The development of society brings together an idea to society in its development including the association of groups that combine action together [4]. This is in line with Cavaye's [5] exposure that the community itself is involved in a process aimed at improving the social, economic, and environmental stewardship by adopting a good decision using labor, infrastructure, and knowledge as a resource (Figure 1).

Studies suggest that tourism can contribute very well to the quality of community life, it is indicated by the activity of tourists either directly or indirectly contribute to the community through the tourism industry. The tourism industry provides opportunities directly to the creation of employment so that local people have income derived from tourism activities [6]. In addition, tourism also provides opportunities to create small and medium enterprises such as souvenir production, homestay provision, to guide tour services. Tasci, et al [7] explained that tourism activities open the way for the integration between local entrepreneurs and the creation of workers who previously did not have skill in the field of tourism, in addition to tourism potential to create employment, poverty alleviation, human development, and sustainable environment.

3. Shell Element's Transformation

Residential buildings built by the community both migrants and original Setu Babakan most of the modern buildings that spatial arrangement does not follow the rules of spatial arrangement in accordance with the philosophy of Betawi architecture. This is due to the limitations of different space in each house and some standards of Betawi traditional architecture that are not relevant to modern society such as the location of the well in front of the house and family graves on the side of the house. In addition, the use of materials is also a consideration because of the use of wood materials are high maintenance costs and are vulnerable to weather.

Therefore, the transformation carried out on the residents' homes is the beautification of the visible buildings. Beautification is done by applying the ornaments typical of Betawi, so that although the building is not in accordance with the Betawi custom, the environment still feels like the original Betawi village. Here are some ornaments typical of Betawi that applied to the residents' homes:

1. *Langkan/Ginggang*

Langkan is a substitute for the yard fence contained on the terrace or front porch of traditional Betawi houses usually made of wood or bamboo, thus blocking outward views into the home and maintaining the privacy of the homeowner. *Langkan* is made as high as waist that has a symbol like a human statue that also has a moral message, that is good ethics in the visit must pass through the front yard, because when a visit through the back or side of the house is a bad ethic for the Betawi people.

Figure 2. *Langkan/Ginggang*
Source: Fieldwork, 2017

2. Lisplank *Gigi Balang*

Lisplank or lis *Gigi Balang* is made of wooden material board carved with triangular ornaments lined up resembling grasshoppers that symbolize that life must always be honest, diligent, tenacious, and patient, because grasshoppers can only break the wood if done continuously and usually within the time frame can be categorized long but overall can mean a strong defense.

Figure 3. Lis *Gigi Balang*
Source: Fieldwork, 2017

3. *Banji* Swastika (Sunflower Ornament)

Besides *Gigi Balang*, ornaments in other Betawi houses are *Banji*. *Banji* has a rectangular pattern, this pattern is developed from the basic ornaments *Swastika* which is the influence of Hindu culture which means dynamic. *Banji* ornaments are often combined with elements of plants. This ornament in the form of a translucent engraving that is usually located at the top of the door of this living room symbolizes that the homeowner's life should be an inspiration for the surrounding community, and use sunflower because the sun is symbolized as a source of life and light, sunlight here means that homeowners must always have bright thoughts and minds.

Figure 4. Use of *Banji* Swastika or Sunflower Ornament on the top of front door
Source: Fieldwork, 2017

4. Jasmine Ornament

The jasmine ornament is on the side of the front roof cover, this ornament is a symbol of joy, fragrance, and hospitality towards anyone, no wonder if the Betawi people are always open for anyone who wants to visit his village. Like jasmine flowers, *cempaka* flower ornament is a symbol that the life of the home owner must always be fragrant and harmonious.

Figure 5. Jasmine Ornament
Source: Fieldwork, 2017

3.1. Beautification of Betawi Character in residential buildings

In a residential building in Setu Babakan, the beautification of local Betawi characters is not a requirement and there are no specific rules yet, but for residents who want to maintain this local character will be assisted and financed, so there are still many houses that do not apply. In addition, the surrounding community does not have its own initiative and assume that the application of local character is the responsibility of the government solely, especially *Dinas Permukiman dan Perumahan*. Because so far ornaments in the form of fascia board and balustrades are financed and provided by *Dinas Permukiman dan Perumahan* for the homes of citizens, especially for houses that are on the side of the road, close to the place of performances and the lake.

Some problems are also faced in the effort to maintain the local character of this Betawi. For example the use of *gigi balang* that is financed by *Dinas Permukiman dan Perumahan* including durable, it is just that many citizens who then abuse it and make it quickly damaged and destroyed. The holes in this *gigi balang* are used by residents such as to hang clothes and hang merchandise for those who sell. It is these heavy loads that cause damage to the *gigi balang*.

Figure 6. Beautification in residential buildings
Source: Fieldwork, 2017

3.2. Beautification of Betawi character in public buildings

Beautification of Betawi characters in public buildings in Setu Babakan which are applied amongst other buildings such as gate, mosque, and main place of the pilot tour of Betawi traditional house.

Figure 7. Beautification in public buildings
Source: Fieldwork, 2017

At the Bang Pitung gate located on the edge of Mohammad Kahfi II road which is the main entrance to the PBB Setu Babakan area, there are several ornaments of the Betawi culture, such as *Gigi Balang* ornaments in fascia board which also serves as the cover of the roof structure. In addition to the use of *Gigi Balang* ornaments, there are also *Ginggang* or *Langkan* ornaments that adorn the post that also serves as a support. *Ginggang* ornaments used to replicate the form of fence from the ornament of traditional house Betawi which is usually used as a guardrail found on the terrace with wood materials. But at Gapura Bang Pitung, this *Ginggang* ornament is made from concrete painted with brown color resembling wood.

In mosque and musholla buildings, ornaments or elements typical of Betawi traditional house that is applied that is *Ginggang* and lis *Gigi Balang*. Waist ornament is applied to the guardrail between the mosque hall and the outer space, while the *Gigi Balang* is applied to the fascia board. For *Banji* swastika or sunflower ornaments found only in one mosque. In embryo zone area which is made and designated as a pilot area of traditional Betawi house, it has several buildings that function as PBB office, PBB guesthouse, gallery and meeting room, and open stage. These buildings are made to represent the Betawi traditional house so that the use of ornaments is complete and in accordance with the original Betawi traditional house.

4. Society Element's Transformation

The development planning of Setu Babakan related to society element is to invite local people to become economic agents in tourism development. Basically, the community itself has great potential in Betawi culture such as traditional Betawi traditional food and beverage expertise, as well as Betawi art.

Besides to promoting tourism, invite the society on the development also helped promote the economy of the surrounding society. The community took advantage of the opportunity by opening food, drinks, and Betawi traditional snacks' stall scattered around Setu Babakan and the surrounding areas.

Figure 8. Society activities in promoting tourism
Source: Fielwork, 2017

In addition to the culinary field, economic activities of the society in utilizing the potential of tourism at PBB Setu Babakan is by opening a souvenir shop, *delman* rental, *odong-odong* rental, and dragon boat rental, and parking management. However, in tourism development planning, the community is not directly involved. The planning process uses a top-down scheme which was originally as a whole conducted by *Dinas Kebudayaan dan Pariwisata (Disbudpar)* Jakarta Selatan after which the community was involved in its development. The community referred to here is a society represented by *Badan Musyawarah (Bamus)* Betawi as the representation of the entire Betawi organization and *Lembaga Kebudayaan Betawi* (LKB) as an institution that encompasses all Betawi cultural activities.

The arts and cultural community have a significant role and influence for tourism in PBB Setu Babakan. Community is incorporated in several studios, such as studio *lenong* (theater), dance studio, music studio (*keroncong, tanjidor, gambus*), *pencak silat* studio, and community of souvenir artisans and street vendors community. These studios are regularly and alternately invited by the UPK PBB Setu Babakan as a weekly performer. Schedule of activities and performances for three months already installed on the bulletin board, making it easier for people who want to come to see a particular show. The tourists come very diverse, ranging from students, youth, and families or even researchers and foreign tourists.

5. Transformation Impact on Tourism in Setu Babakan

The development undertaken on the ecological elements of settlements in Setu Babakan has a positive impact on the tourism industry. The transformation of the shell element provides an attraction for tourists to enjoy the Betawi shades, though only on the outside. Nevertheless, tourists can still see how the design of traditional Betawi houses in the pilot zone.

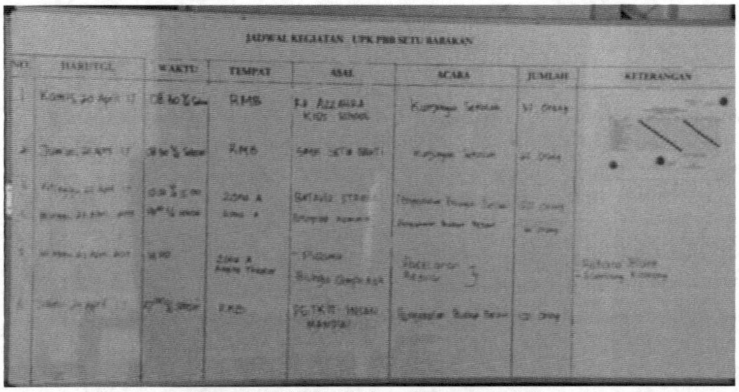

Figure 9. Schedule of tourist visits
Source: Fieldwork, 2017

Figure 10. Tourists in Setu Babakan
Source: Fieldwork, 2017

In addition can be enjoyed by tourists to take pictures, traditional house Betawi also be an object to learn. Many groups from schools and universities are visiting in order to learn Betawi culture. Schedule weekly visits are always fully charged each month. Usually, they make visits on weekends so that they can watch the performance at once.

The existence of Betawi food and beverage stalls around Setu Babakan also become a very interesting thing for tourists. In the midst of the hustle and bustle of Jakarta, Setu Babakan becomes alternative tourism other than usual. Tourists can enjoy the lake tour by renting various kinds of boats, fishing, or just sitting enjoying the lake view. In addition, there are also a variety of traditional Betawi food and beverage options at very affordable prices. On weekends, Jalan Srengseng Sawah around Setu Babakan is often crowded with tourists. Parking vehicles full of both motorcycles and cars, food stalls and streets were fairly solid.

The public facilities available around Setu Babakan are complete. In addition to tourist facilities, dining areas, toilets, and worship area was available with fairly comfortable conditions. Therefore, tourists are not hesitant to spend a day in there.

6. Conclusions

The transformation of elements in Setu Babakan which aims to promote tourism while preserving Betawi culture has shown a positive sign. Some of the deficiencies that still occur are because the government still has no direct interaction with the citizens, so most citizens do not know how actually the big plan of Setu Babakan development. More citizens only use the facilities that have been provided just like the kiosk to sell, as well as time and place to hold the show.

For beautification in their own homes, although many people have taken the initiative to use Betawi ornaments in their homes, they still feel confused about what to do. People are still afraid to initiate because they still do not know the plan of the continuation of development from the government, so that will be done will be pointless if not in line with the long term development plan. The suggested solution in this case is mainly communication with the community. The community as one of the great potential in the preservation of culture and also the development of tourism, it is very important to know how the development planning in Setu Babakan will proceed. In addition, good communication with citizens will also, increase support so that the development will proceed well.

References

[1] Saputra, Yahya A., et al. (2014). Sejarah Perkampungan Budaya Betawi Setu Babakan. South Jakarta: Suku Dinas Kebudayaan Kota Administrasi Jakarta Selatan.

[2] Doxiadis, C. A. (1975). Ecology and Ekistics. Colorado: Westview Press.

[3] Pratiwi, Wiwik D. (2011). Betawi Traditional Dwelling as a Place for Tourism and Creative Industry. Proceedings of the Indonesia International Conference on Innovation, Enterpreneurship, and Small Business "Creative and Cultural Enterpreneuship in the New Era". Bandung: CIEL-SBM ITB.

[4] Pratiwi, Wiwik D. (2011). Multicultural heritages in a city as productive tourism places. *ASEAN Journal on Hospitality and Tourism*, 10(1), 51-62.

[5] Cavaye, J. (2006). Understanding Community Development. Queensland: Cavaye Community Development.

[6] Pratiwi, Wiwik D. (2008) Heritage tourism and local creative economy, *Proceedings of the International Conference Arte-Polis*, ITB, Bandung

[7] Tasci, A. D. A., Semrad, K. J., & Yilmaz, S. S. (2013): Community based tourism: finding the equilibrium in the COMCEC context-setting the pathway for the future. Ankara: COMCEC Coordination Office.

IOP Conf. Series: Earth and Environmental Science **152** (2018) 012029 doi:10.1088/1755-1315/152/1/012029

Acknowledgements
This paper is a publication for research entitled "*Transformasi Ruang Luar Publik Permukiman dengan Kegiatan Wisata di Kawasan Bersejarah Perkotaan*" conducted within Housing Settlement Research Group, ITB http://dosen.ar.itb.ac.id/pp/, SAPPK, Institut Teknologi Bandung and funded by *Program Penelitian, Pengabdian kepada Masyarakat, dan Inovasi* (P3MI) LPPM ITB. The authors would like to thank all of the research team (M. Harits Insan Kamil, M. Iqbal Katik R. E., Rakotonindrina Onjaharimanana, Shandra Rama P. W., Sri Lilianti K., Hazrati Sharfina, Dennis Octaviano, Ayunda Putri P., Siti Hawa Fitrah E., Bunga Astari, Maulani Faradina, and Nakanwagi Orashida), for their effort in data collections and writing the fieldwork report.

IOP Conf. Series: Earth and Environmental Science **152** (2018) 012030 doi:10.1088/1755-1315/152/1/012030

Fire Safety Parameters of High-Rise Residential Building: A Literature Review of Performance-Based Analysis Method

T A Kurniawan[1,2], **L Tambunan**[1], and **L N Imaniar**[1]

School of Architecture, Planning and Policy Development, Institut Teknologi Bandung, Ganeca Street No 10, 40132, Bandung, Indonesia

[2]asktitus@gmail.com

Abstract. The basic concept in occupant safety assessments in high-rise building residential against fire hazard is the determination of the time the occupants can be safe/escape before the fire danger conditions occur. This concept is also called performance-based appraisal. It is known by another name Available Safety Egress Time > Required Safety Egress Time (ASET> RSET). The concept of performance-based fire safety assessment has been widely used in the concept of fire safety. Discussions related to this literature study are focused on strength and weaknesses in ASET and RSET methodologies, which are related to the parameters. This paper provides a brief overview of the methodologies and parameters of ASET and RSET. The results of this literature study provide a clear picture that the simulation of fire hazards in high-rise residential buildings can be done accurately by considering the parameters of the ASET and RSET. However the lack of resources and experts in using this related software make this performance-based method is difficult to achieve.

1. Introduction

During this time, one of the ways to meet fire safety building criteria is to follow the requirements stated in the fire safety regulations. Each building element is designed according to the specifications defined by the rules, then checked for compliance to assess whether the required level of safety has been met. This procedure is known as a prescriptive based evaluation method. However, the approach used in fire safety regulations over the past decade has begun to shift from prescriptive to performance-based.

This change is triggered by dissatisfaction with prescriptive-based rules that cannot be used as a reference for assessing buildings with special functions and complex building designs. In addition, the safety levels set by prescriptive rules are overvalued so that to meet them are relatively expensive. A performance-based approach has been popular since decades ago. This approach provides an opportunity for building designers to produce innovative and cost-effective, equivalent or better designs of regulatory solutions [1].

The main difference between the two approaches lies in its substance. Performance-based approaches emphasis on goals and do not specify solutions, while prescriptive-based approaches focus on standard and regulations. Performance-based rules place greater emphasis on fire safety goals. That's why performance-based rules are considered more flexible, more functional, and easier to apply. Through performance-based regulation can also be obtained optimum design with better safety and lower cost [2, 3]. Performance-based evaluation can be done by modeling and simulation. The simulation can be done by using computer analysis.

A modeling and simulation analysis is used to predict the behavior of occupants in the event of a fire. This analysis has evolved since the 1960s, with a model which is then applied to computer software where the risk of fire hazard can be predicted. One of the earliest known is Available Safe Egress Time (ASET) Model [4] that is one of the 2 part methods of ASET and RSET. The ASET model provides a good estimation of assessment time for this safe exit to introduce the fire safety concept by defining Available Safe Egress Time (ASET) as the time interval between detection time and hazard object. Meanwhile, to ensure that the building is considered safe for the occupants then there is a certain period of time used to know the movement of occupants to a safe place, that is known as Required Safety Egress Time (RSET). In RSET fire prevention concept should be less than ASET specified by space formation. RSET is only considered the time it takes to get to the exit. But in his theory, Cooper did not provide any guidance on the determination of RSET. Then some aspects of the RSET are enhanced by Sime [5] by adding the countermeasures phase, in early detection and exit. The coping phase may be more commonly referred to as a pre-movement time, to allow for various delays before starting to move towards the exit, following the moment of awareness of the fire cue.

2. ASET and RSET

The performance evaluation of buildings against fire hazard for buildings or other places of human occupancy, in the modern era has sought how to construct quantitative performance metrics [6]. Quantification is done by comparing two or more alternative strategies to be made. The ASET / RSET method is used to demonstrate occupant safety against fire hazard by taking time as an indicator.

The process of fire engineering analysis usually involves assessment of fire development, smoke distribution and evacuation of occupants. The concept of this method is that if a simple ASET > RSET relationship is satisfied, it is assumed that the performance of the building against fire hazard is fully adequate, in terms of the evacuation safety of the occupants. RSET (Required Safe Egress Time) is the time to evacuate occupants. Potential sufficient time for occupants to get out of the place where fire appears is the most important factor in the security aspect of the occupants. ASET (Available Safe Egress Time), is the time for unsustainable conditions. The ASSET is determined from the moment the fire effect reaches the limit of tenability specified in the acceptance criteria. RSET is determined from the analysis of time taken by the occupants to escape from the fire effect to a safe place. Therefore, the assessment of acceptance criteria is ultimately made relatively that residents can safely escape the presence of fire hazards.

The ASET / RSET parameter relies on a number of key variables as shown in Figure 1 [7]. Safety is defined by the Assessment of a set of agreed criteria that are largely expressed in terms of potentially exposed fires. Simply put, if the occupants can be shown to escape from fire hazard with acceptable margins, the solution is deemed to have met the performance requirements. Hall [8] states while escaping with enough time, residents will not face a very thick smoke and will see the right way out. The children will need help from adults in any case to get out. But if enough time is available, there will be time for adults to gather and guard the children and lead them to a safe place.

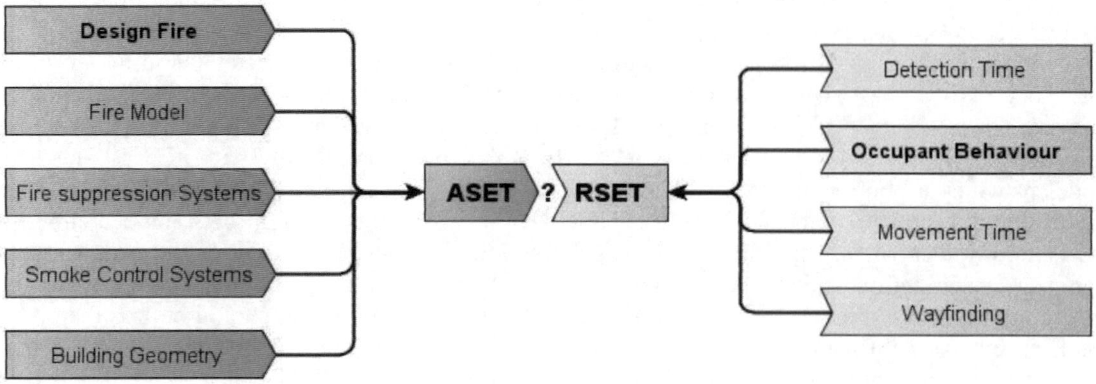

Figure 1. Dependent variables affecting ASET and RSET (most significant variable shown in bold) [7]

3. Computer Simulation Model

New research in evacuation behavior has developed a more comprehensive method of analysis and is able to model complex behavior during evacuation. Fire safety codes called performance-based codes are used to validate the fire safety of buildings. Parameters that influence the efficiency of performance-based evacuation can be categorized into four main groups: the physical characteristics of the building, the behavioral characteristics of the occupants, the physiological characteristics of the occupants and the characteristics of environmental fire [9].

- Model of physical characteristics of buildings. Building physical characteristics such as corridors, stair widths, landing areas are modeled. As well as obstacles and physical disturbance of other architectures such as furniture and building layouts are also taken into account in the evacuation route.
- Modeling characteristics of occupant behavior. Differences in occupant behavior affect evacuation time in two different stages of pre-movement time and movement time. The diversion behavior of the occupants or persistence in finding the route when facing the smoke-filled area is another characteristic of the behavior that can be simulated.
- Modeling physiological characteristics of the occupants. The physical fitness of the occupant, sex, age, height and weight affects the speed of movement as well as his / her ability to tolerate gas and heat. The agility of the inhabitants to jump over small obstacles like furniture can also affect the evacuation time.
- Model of fire environment characteristics. The speed of the evacuation process is hampered by exposure from smoke. A variety of toxic gases can be generated during a fire depending on the burning object. However, there is sufficient information about the effects of these gases on the mobility parameters. The most commonly known effects are toxic gases (CO, CO2, & HCN) and lack of oxygen can be seen in the evacuation analysis.

4. Discussion

From the parameter classification of ASET and RSET based on the software, related to the building function (Table 1) some parameters that affect the simulation performed can be seen. These parameters can affect ASET and RSET in the simulation process.

ASET is affected by the Type of Smoke and Fire parameters, Environment Condition, Ventilation, Wind Condition, Passive Protection, Active Protection, and Room Type.

- Type of Smoke and Fire relates to the height of smoke in the room, how poisonous the smoke, and how big of fire in the worst conditions.

- Enviroment Condition relates to temperature inside or outside room.
- Ventilation relates on length and width of doors and windows and their positions on the wall.
- Wind Condition associates with the wind speed in the room can affect the rapid or slow fire spread.
- Passive Protection includes some passive handling so the fire does not spread such as making a wall or ceiling with resistant for fire.
- Active Protection includes active handling which effectively and automatically can help reduce the fire spread, for example is sprinkler.
- Room Type is more emphasizes on the type of room and how large the room is. Room type can determine how long the fire can be spread to the next room.
- RSET is influenced by several parameters including Physical, Behavior Attributes, Person Mode, Type of Travel, and Number of Occupants.
- Physical relates to the physical factors of the occupants, Includes gender, height, weight and age.
- Behavior Attributes relates whether the occupants are well informed to handle fire.
- Person Mode relates to how the occupants react to the danger of fire, whether they are walking normally, walking fast, or half running.
- Type of Travel Time relates whether the occupants are moving horizontally or vertically.
- Number of Occupants covers how many occupants are in one building

Table 1 Parameter Grouping ASET(Available Safety Egress Time) and RSET (Required Safety Egress Time) based on software related building function.

No.	Building Function	Researcher	Software		Information
			ASET	RSET	
1	High-rise Student Apartment	Zhong-an, Jiang., et.al. [10]	-	**EXODUS** (Parameter: Physical, Behavior Attributes, Type of time travel, and Number of Occupants	1. Use Experimental Method 2. Comparing experimental and simulation evacuation methods without considering ASET
2	High Rise Building Flats	Sujatmiko W., et. al. [11]	**CFAST** (Parameter: Ventilation, Wind condition, passive and active protection)	**FDS-EAC** (physical, behavior, attributes, person mode, and number of occupants)	
3	High Rise Residential Building	Zhixiang X and Yong T [12]	**CFAST** (Parameter: Type of smoke and fire, passive and active protection, room type)	**EXODUS** (physical, behavior, attributes, and number of occupants)	
4	6-Storey residential building	Li, Xiao [13], Zhixiang X and Yong T [12] and Hadjisophocleous, G.V. et. al [3]	**CURisk** (Parameter: Type of smoke and fire, passive and active protection, room type)	-	1. Evacuation on buildings is not simulated but using existing data
5	Multistory building	Sujatmiko W., et. al. [11]	**CFD-FDS** (Parameter: Type of smoke and fire, environment condition. Ventilation, wind condition, room type)	-	

5. Conclusion

In some software, not all have same parameters so that we must determine the right software to help analyze the data desired. Assessment of building performance against fire hazard can be done well using computer simulation by considering some parameters related to ASET and RSET. The disadvantage of performance appraisal using computer simulations is that there are software that can combine ASET and RSET simulations. It also requires precision in determining the parameters before starting the simulation. Not many experts can use software and perform computational analysis related to performance appraisal of buildings against fire hazard

References

[1] Ramachandran,G 1995 Probability-based Building Design for Fire Safety Part 2.
[2] Richardson, J.K. 1993. *Changing The Regulatory System To Accept Fire Safety Engineering Methods*, Journal of Fire Protection Engineering, 135-140
[3] Hadjisophocleous, G.V. et. al. 1995. *Literature Review of Performance-based Fire Codes and Design Environment*, Journal of Fire Protection Engineering 7(2), pp 12-40
[4] Cooper, L. Y., and D.W. Stroup 1982 *Calculating Safe Egress Time (ASET)- A Computer Program and User's Guide* NBSIR 82-2578, National Bureu od Standards, (Washington. DC)
[5] Sime, Jonathan D. 1986. *Perceived Time Available: The Margin of Safety in Fires* Pp. 561-570 in Fire Safety Science -- Proceedings of the First International Symposium, edited by C. Grant and P. Pagni. (London, England: Interscience Communications Ltd)
[6] Babrauskas1,Vytenis, Fleming, Joseph M. Russell. B. Don 2010 *RSET/ASET, a flawed concept for fire safety assessment*. Fire and Materials. 34:341–355
[7] Poon, S.L 2014 A Dynamic Approach to ASET/RSET Assessment in Performance Based Design Procedia Engineering 71 173 – 181
[8] Hall JR 2004 How many people can be saved from home fires if given more time to escape? Fire Technol 40(2):117–126
[9] Oven, V.A., Cakici, N. 2009 Modelling the evacuation of a high-rise office building in Istanbul. Fire Safety Journal 44 : 1–15
[10] Zhong-an J, Mei-ling C, Xiao-hua W 2011 Experiment and Simulation Study on High-rise Student Apartment Fire Personal Evacuation in the Campus Procedia Engineering 11 p: 156–161
[11] Sujatmiko W, Dipojono H K, Soelami, F X N, Soegijanto, 2014 *Performance-based fire safety evacuation in high-rise building flats in Indonesia – a case study in Bandung*. Procedia Environmental Sciences 20 : 116 – 125
[12] Zhixiang X, Yong T 2012 *Simulation of fire and evacuation in high-rise building*. Procedia Engineering 45 705 – 7
[13] Li X 2013 Fire risk analysis of a 6-storey residential building using CUrisk. Procedia Engineering 62 609 – 617

IOP Conf. Series: Earth and Environmental Science **152** (2018) 012031 doi:10.1088/1755-1315/152/1/012031

Towards sustainable transportation: identification of the spatial configuration of rental housing area using space syntax method

Irfan Irwanuddin[1], Heri Andoni[2], and Allis Nurdini[3]

[1,2]Architecture Master Student, School of Architecture, Planning, and Policy Development, Institut Teknologi Bandung, Indonesia
[3]Architecture Program, School of Architecture, Planning, and Policy Development, Institut Teknologi Bandung, Indonesia

E-mail: irfanirwanuddin@gmail.com

Abstract. The emergence of several higher education infrastructures in Bandung city caused the significant increase of inhabitants in the certain region. This resulting in the fast growth of rental housing, while the region's structure itself is mostly organic and having some cul-de-sac areas. This kind of structure is considered as unideal for a residential area in the context of public transportation accessibility. Based on that, this study aimed to identify the structure of a rental housing complex in Bandung regarding its accessibility for public transport, using space syntax method. There were three main findings revealed from this study. The most segregated area was also the least integrated. There was a significant range of spatial hierarchy between the shallowest and the deepest area. There was an only small area in the case study that considered as accessible regarding the distance to public transport. This condition may lead to certain urban problems such as traffic congestion or excessive fossil fuel usage. These findings could act as a suggestion to the government for policy regarding the regulation for rental housing areas in order to minimize the growth rate of private vehicle possession so as to support the development of sustainable transportation of a city.

1. Introduction

Before impacted by urbanization, housing in most area in northern Bandung were commonly occupied for family living. But nowadays, as several universities like UNIKOM, ITB, UNPAR, UNPAD started to occupy this area, the houses were affected accordingly. Several houses that used to be a regular housing for family has now became a rental housing. Thus, it increased the inhabitant's population while there is no significant adaptation from the existing structure of the region.

This densification also affecting its transportation needs. By the time a population of a city become denser, there is another issues aroused which is transportation. As a result, accessibilities to public transportation became an issue for the housing area, especially those with a high density population such as rental housing. Accessibility from a housing area to public transportation is now becoming more important considering its impact to the inhabitant's willingness to use public transportation [1] as an attempt to reduce the city's traffic congestion. While at the same time, the housing structure itself should meet the criteria of walk for transport.

Space syntax, as an analysis tools, is having an ability to identify social phenomena in a built environment such as accessibility by examining its spatial configuration. This research used Tubagus Ismail region in Bandung as a study case since this region is meeting the criteria of the above described phenomena such as rental housing area, density, and accessibility. Tubagus Ismail is a sample of housing region that previously was built as regular housing which at certain period undergone a transformation to be a rental housing.

The aim of this research is to identify the spatial configuration of Tubagus Ismail as a rental housing region and its relation with its accessibility to nearest public transportation transit.

2. Literature Review

2.1. *Space Syntax*

Space syntax is a set of tools to represent, quantify, and interpret a spatial configuration in a building, city, interior, or landscape [2, 3]. The main focus of space syntax is the relationship between human and the space being occupied. This theory is based on a characteristics that uniquely presented in a society inside a spatial system, and the knowledge that adhere to this society is described through the space itself, as well as through the spatial configuration [2].

In its theory, the relation between the spaces is what considered as the spatial configuration while the idea about how specific form of a space could affects the behavior of its inhabitant is what considered as the social knowledge. The goal of space syntax research is to elaborate the social value from the configuration of the occupied space. This could be used to develop practical knowledge about the effect of particular space towards society with varied social or cultural variables. The basis of space syntax is to understand the configured space, specifically about the forming process and the social value [3]. In short, space syntax is an attempt to build a spatial configuration in architecture by producing theoretical understanding toward how human create as well as using spatial configuration. In other words, as a tool to identify how some spatial configuration could express the social and cultural value. And how this spatial configuration resulting a social interaction and built environment.

Several researches and publications proves that space syntax in the early period was focusing on a real world environment and had trying to identify the most basic attributes of built environment. By consistently developing the techniques for representing and analyzing spatial pattern, current space syntax research has reaching on an attempt to stimulate the spatial configuration of a design, as well as predicting the social pattern of the space being designed.

2.2. *Rental Housing Region*

Rental housing in this context is described as an informal rented house in Indonesia that commonly called as "Kost". The difference between "Kost" and other rental housing facilities is that this type of rental housing come in various conditions and arrangements, so the tenants are provided with a lot of choices in the market that may suit their preferences, economic capabilities and cultural issues. In addition, a "Kost" tends to be paid on a monthly basis and has a flexible contract period [4].

Recently, there is no any housing public policy in Indonesia that specifically administering the rental housing region's structure. The growth of such region are usually occurred in an organic way [5]. The morphology of this region are commonly started as a regular housing that transformed as a rental housing. Many of the housing areas in Bandung are having a structure that contains lot of intersections, narrow alleys, and cul-de-sac. These circumstances are what make the rental housing in Bandung, especially those in northern area, could be a unique region in terms of social characteristics related to its spatial configuration.

2.3. *Walk for Transports*

To reveal the properness of a certain region in terms of walk for transport, there should be a clear maximum standard distance requirement as the parameter. Walk distance is important because walk is the primary access mode for trips from home to public transport and because walking distance has a significant impact on public transport use [6]. For this, there are several publications that had identify the maximum standard distance between a residential region and a public transportation transit point within the reach of walking activity.

There is a publication from Malaysia that successfully revealing the maximum distance of people in Malaysia in walk for transport which is 500 meter [1]. Another publication from Maryland, US has had identify the maximum distance of walk for transport as 400 meter which is roughly equal to 5 minutes walking [7]. In Helsinki, the maximum distance of walk for transport is 300 meter [8] while in Perth, Australia, the maximum number is same as in Malaysia, which is 500 meter [9].

The standard number revealed from those publications are certainly pretty much affected by the physiological condition of the inhabitants from each region. In addition, every region has its unique characteristics in terms of climate and landscape. While in this research, the location of case study and its inhabitant's characteristics are almost similar to those in Malaysia since both Malaysia and Indonesia are sharing the same climate and same physiology condition based on their race similarity as Melanesian. From that, 500 meter distance is determined as the standard of walk for transport in this research.

3. Method

3.1. Data Collection Method

The region data used in this research is based on the 2015 AutoCAD photogrammetry map of Bandung city, while the mapping of rental housing unit location was based on field observation (Figure 1).

Figure 1. Location of rental housing units in Tubagus Ismail region.

3.2. Analysis Method

Based on the convex map from the AutoCAD data, an axial map firstly had to be created as the basis for the next step of space syntax analysis. From this axial map, the syntax structure of the region was revealed topologically. Then, in order to analyze the accessibility of certain region, there were several steps should be undertaken in space syntax method. First, integration analysis should be done to classify some area with the highest and the lowest density of accessibility. Second, depth analysis should be done to classify the range and the variation of depth occurred in the study case region. Besides, this depth analysis was conducted to identify the level of accessibility to the nearest public transport transit area. Third, there was segment analysis in order to identify the quality of region's structure in terms of walk for transport accessibility using the previously mentioned standard as the benchmark. By using this segment analysis, any areas that meet the walk for transport standard distance and the areas that does not would be identified.

These series of steps in space syntax analysis was conducted by the assistance of space syntax software namely DepthMap developed by University of College London.

4. Analysis

According to the map data obtained, firstly a convex map should be created as the boundary of the axial line going to be built (Figure 2). In this step, the convex map created was restricted only inside the Tubagus Ismail region and a big street namely Jalan Raya Juanda which was the nearest street to the Tubagus Ismail area that had a dense traffic of public transportation.

From this convex map, an axial line map could be then generated (Figure 3).

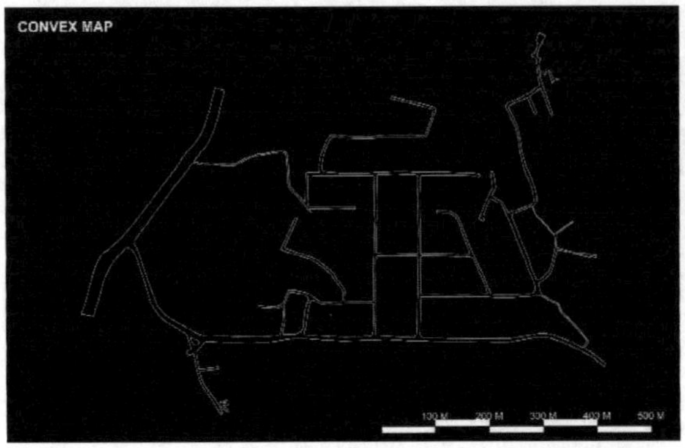

Figure 2. Convex map of Tubagus Ismail region.

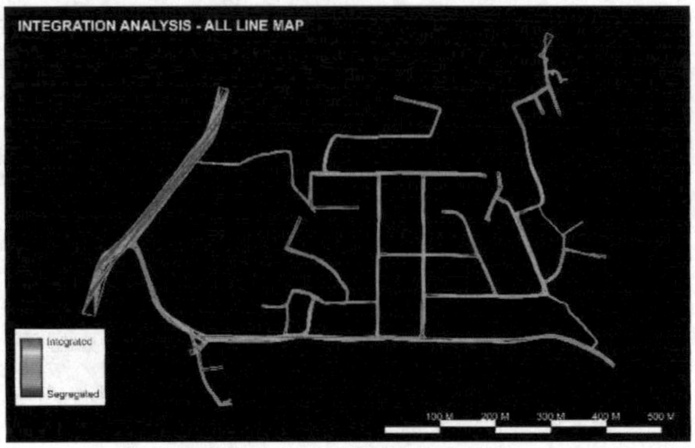

Figure 3. Axial map result.

4.1. *Integration Analysis*

This axial map was then reduced to be a more simplified version (Figure 4). This was due to facilitate a clear and easily quantified version of axial map. This simplified axial map was then used as the basis of integration analysis to identify the density of accessibility in this region in certain area.

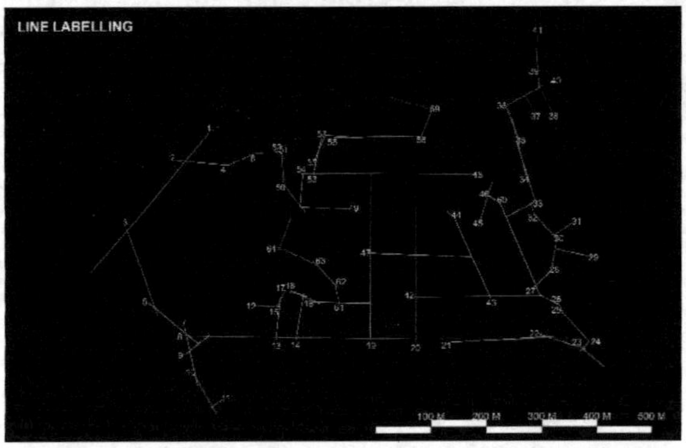

Figure 4. Simplified axial line map result.

According to integration analysis, there were several areas detected as the most integrated and the most segregated. The red colour on the map was illustrating the integrated areas and blue colour was illustrating the segregated areas. Every single line on this map were then given reference number on each of them to ease the identification process. It was clear from the colours that the most segregated area were line number 59 and 41, while the most integrated area are line number 8, 20, and 42.

4.2. *Depth Analysis*

According to the identified integration value, a depth analysis was then operated. This analysis require one line as the "root" to be the point where the other lines distance to it would be calculated. In this case, Jalan Raya Juanda (line 1) became the root. This depth analysis was more of a topological calculation rather than metrical. In determining the nearest access, depth analysis is pretty much affected by the integration value which has defined previously. The higher the value of a line, the more this line is going to be chosen as the most accessible line during this depth analysis process.

The characteristic of depth analysis is the higher the depth number of a line, the more inaccessible this line is. So as the opposite.

Through Figure 5, how the depth condition resulted from the Tubagus Ismail housing's structure was demonstrated. Blue line in Figure 5 is indicating the shallowest area, while the red is indicating the deepest area. From this analysis, it was found that line 40 and 41 were the deepest areas. But, if the line with the rental housing unit was included in this process, the deepest areas were on the line 59 and 35. This was due to the absence of rental housing units in line 40 and 41.

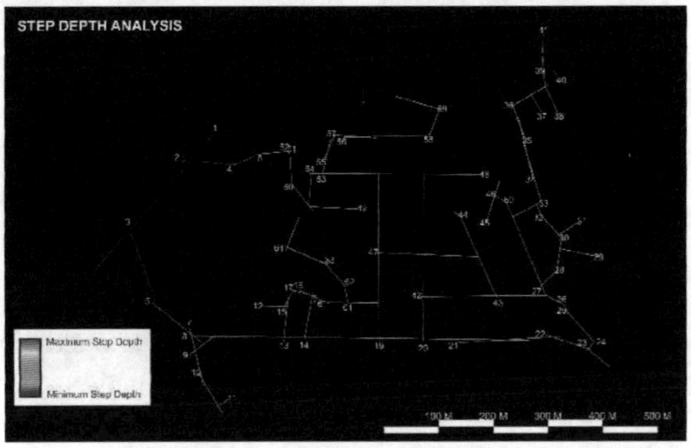

Figure 5. Depth analysis result.

According to the depth analysis, a justified permeability graph (JPG) was constructed. JPG is very useful to help identifying the range of depth occurred in this region. By identifying the range of depth, the hierarchy was then could be visualized (Figure 6).

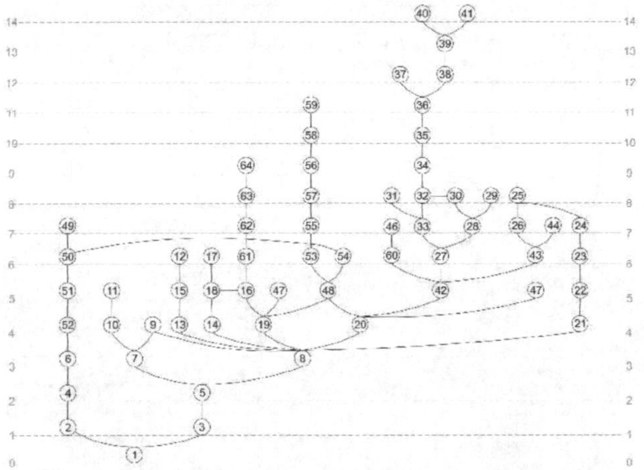

Figure 6. Justified permeability graph.

By examining the JPG, the range of depth hierarchy occurred in the whole region was revealed. The lines involved in this JPG construction were restricted only to those that occupied by the rental housing units. This region was proven to have 14 levels of depth hierarchy which was indicating a big gap of accessibility level throughout the Tubagus Ismail rental housing region.

4.3. Segment analysis

Segment analysis was then conducted so that the metric distance of the axial line could be read clearly. This was due to identifying any areas of rental housing in the case study that meet the standard

IOP Conf. Series: Earth and Environmental Science **152** (2018) 012031 doi:10.1088/1755-1315/152/1/012031

distance of walk for transport and those which are not. In this case, line 1 (Jalan Raya Juanda) was still works as the root of the calculation. But this time, the calculation was done metrically, not topologically. For a detailed result, the distance calculation was operated using every single rental housing unit in the location as the starting point to reach the root line (Figure 7).

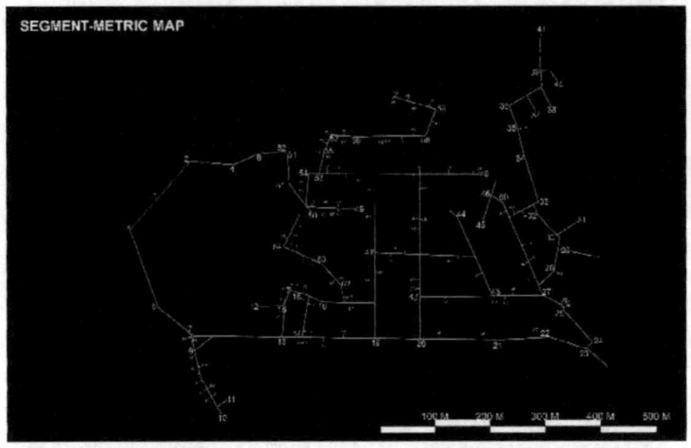

Figure 7. Segment-line map.

Through metric distance calculation on the segment-line map, by using the standard distance of walk for transport discussed earlier, there were some areas within the region that were considered as accessible and some that were not. In Table 1, the green coloured rows were indicating the rental house unit that meet the standard distance.

Table 1. Metric distance per housing unit.

Tubagus Ismail Raya

No.	House Unit	Line Number	Depth Level	Metric Distance (m)
1	AA	8	3	421
2	AB	8	3	430
3	AC	8	3	466
4	AD	8	3	502
5	AE	8	3	675
6	AF	8	3	683
7	AG	21	4	778

Tubagus Ismail 1

No.	House Unit	Line Number	Depth Level	Metric Distance (m)
1	BA	7	3	233
2	BB	7	3	269
3	BC	7	3	289
4	BD	7	3	289
5	BE	7	3	301
6	BF	10	4	313
7	BG	10	4	331
8	BH	10	4	333
9	BI	10	4	348
10	BJ	10	4	349

Tubagus Ismail 3

No.	House Unit	Line Number	Depth Level	Metric Distance (m)
1	CA	14	4	445
2	CB	14	4	495
3	CC	14	4	500
4	CD	13	4	433
5	CE	15	5	464

Tubagus Ismail 4

No.	House Unit	Line Number	Depth Level	Metric Distance (m)
1	DA	17	6	484
2	DB	17	6	494
3	DC	18	6	507
4	DD	16	5	592
5	DE	16	5	601
6	DF	16	5	600

Tubagus Ismail 5

No.	House Unit	Line Number	Depth Level	Metric Distance (m)
1	EA	62	7	634
2	EB	62	7	631
3	EC	63	8	691
4	ED	63	8	708
5	EE	63	8	741
6	EF	63	8	730
7	EG	64	9	768
8	EH	64	9	780
9	EI	64	9	799
10	EJ	64	9	812

Tubagus Ismail 6

No.	House Unit	Line Number	Depth Level	Metric Distance (m)
1	FA	19	4	666
2	FB	19	4	679
3	FC	19	4	678
4	FD	19	4	691
5	FE	19	4	756
6	FF	19	4	767

7	FG	19	4	782
8	FH	19	4	832

Tubagus Ismail 7

No.	House Unit	Line Number	Depth Level	Metric Distance (m)
1	GA	20	4	651
2	GB	20	4	667
3	GC	20	4	761
4	GD	20	4	779
5	GE	20	4	777
6	GF	20	4	834

Tubagus Ismail 8

No.	House Unit	Line Number	Depth Level	Metric Distance (m)
1	HA	42	5	806
2	HB	42	5	895
3	HC	42	5	915
4	HD	42	5	929
5	HE	60	6	1022
6	HF	25	8	1028
7	HG	27	6	1043
8	HH	60	6	1052
9	HI	60	6	1054
10	HJ	60	6	1084
11	HK	33	7	1167
12	HL	60	6	1153
13	HM	60	6	1153
14	HN	34	9	1325

Tubagus Ismail 9

No.	House Unit	Line Number	Depth Level	Metric Distance (m)
1	IA	47	5	752
2	IB	47	5	762
3	IC	47	5	778
4	ID	47	5	860
5	IE	47	5	876
6	IF	47	5	871

Tubagus Ismail 12

No.	House Unit	Line Number	Depth Level	Metric Distance (m)
1	JA	43	6	905

Tubagus Ismail 14

No.	House Unit	Line Number	Depth Level	Metric Distance (m)
1	KA	53	6	1006
2	KB	53	6	1008
3	KC	55	7	1028
4	KD	55	7	1045
5	KE	57	8	1068
6	KF	57	8	1069
7	KG	56	9	1134
8	KH	56	9	1147
9	KI	56	9	1174
10	KJ	56	9	1195
11	KK	58	10	1247
12	KL	59	11	1283
13	KM	59	11	1311
14	KN	59	11	1324
15	KO	59	11	1348
16	KP	59	11	1349

Tubagus Ismail 15

No.	House Unit	Line Number	Depth Level	Metric Distance (m)
1	LA	48	5	956
2	LB	48	5	935
3	LC	48	5	924
4	LD	48	5	901
5	LE	48	5	932
6	LF	48	5	941
7	LG	48	5	952
8	LH	48	5	1005
9	LI	48	5	1034
10	LJ	48	5	1049
11	LK	48	5	1059

Tubagus Ismail 17

No.	House Unit	Line Number	Depth Level	Metric Distance (m)
1	MA	54	6	1006
2	MB	54	6	1022
3	MC	54	6	1061
4	MD	49	7	1081
5	ME	49	7	1088
6	MF	49	7	1097
7	MG	49	7	1109
8	MH	49	7	1108

Based on the metric distance calculation resulted in Table 1, it was evident that only 20 units of rental housing that meet the standard distance of walk for transport, while the rest 88 units were below standard. In fact, there were 81,5% from the total housing units that did not meet the standard distance.

5. Result

According to the analysis, there were three important findings revealed as follows:

- Some areas with the lowest integration value had the deepest structure. The point is, this area had the most unideal structure for a rental housing region. An area with very low integration value and very deep structure is indicating that it is isolated and having low accessibility.

- The hierarchy range between the shallowest and the deepest areas were considerably high (14 levels). This interval was showing a significant depth levels throughout the Tubagus Ismail region's structure.
- According to the standard of walk for transport, most of the area in this region was considered as below standard. Therefore, the properness of Tubagus Ismail as a rental housing area surely needs to be reviewed.

6. Conclusion

With the mismatch between the rental housing area and the integrated public transportation system, there will be a raising on the private vehicle's possession inside the neighbourhood. This, may lead to another problem that could interfere the sustainable development of a city such as traffic congestion and exceeded carbon emission produced by the vehicles. Moreover, compared to the regular housing, rental housing region is having more contribution to this problem considering its higher population density per area. Hence, the properness requirement of rental housing region regarding its accessibility should be reviewed by the authority, so that the problems above could be minimized.

Despite the findings revealed, there were also several drawbacks in this research. In order to set a more accurate standard toward walk for distance accessibility in a residential scale, there should be an involvement from the inhabitant's perception toward the maximum distance for public transportation. Besides, another factor affected the sustainability assessment such as traffic, road physical condition, and walkability should be considered for a more holistic result. Therefore, further study to identify a more contextual perception about public transportation accessibility from the study case region is needed to be done.

7. References

[1] M. A. M. Din, "The Aspect Of Walking Accessibility In The Development of GIS-Based Transit System Modelling In Kuala Lumpur," *International Journal of Digital Earth,* 2009.

[2] P. Dursun and G. Saglamer, "Spatial Analysis of Different Home Environments in the City of Trabzon," in *4th International Space Syntax Symposium*, London, 2003.

[3] S. Bafna, "The Geometrical Intuition of Genotypes," in *3rd International Space Syntax Symposium*, Atlanta, USA, 2001.

[4] D. W. Wulandari and S. Mori, "Nature and Operation of Kost Private Rental Housing in Urban Settlement Development of Jakarta, indonesia," *Journal of Civil Engineering and Architecture,* pp. 1362-1369, 2015.

[5] P. Widjaja, Kampung - Kota Bandung, Bandung: Graha Ilmu, 2013.

[6] R. Daniels and C. Mulley, "Explaining walking distance to public transport: the dominance of public transport supply," *The Journal of Transport And Land Use,* vol. 6, pp. 5-20, 2013.

[7] L. Aurbach, "TND Design Rating Standards," United States Environmental Protection Agency, Maryland, US, 2005.

[8] HKL, "Public Transport Planning Guidelines in Helsinki, HKL publication series A," HKL Planning Unit, helsinki, 2008.

[9] Public Transport Authority, "Design and Planning Guidelines for Public Transport Infrastructure: Bus Route Planning and Transit Streets," Public Transport Authority, Western Australia, 2003.

IOP Conf. Series: Earth and Environmental Science **152** (2018) 012032 doi:10.1088/1755-1315/152/1/012032

Local Wisdom of The Native Settlement as A Main Gate in The Northern Axis of Javanese City Center In Semarang

R. Siti Rukayah[1], Dr. Dhanang Respati Puguh[2], Endang Sri Susilo[3], Mohammad Sahid Indraswara[4]

[1]Architecture Departement Faculty of Engineering Diponegoro University,
[2]History Departement Faculty of Culture Diponegoro University
[3]Faculty of Marine Sciene Diponegoro University &
[4]Architecture Departement Faculty of Engineering Diponegoro University

[1]tututrsiti@yahoo.com

Abstract. Similar to the city structure of Islamic cities in Java, Semarang also has multi-ethnic kampongs located around *alun-alun* / traditional square. There is a uniqueness kampong ethnic village that was located in the coastal city of Semarang. Now there are no longer in coastal area. Based on historical method using some old maps and the old images and field study resulted that the old city centre of Semarang had a north axis connecting the seaport (Kampung Darat - 14th century) to the *alun-alun* as administrative city centre (16th century). The axis was once a commercial corridor with settlements arrays behind the commercial corridor. There is some local wisdom in the settlement arrangement. The economic and religious harmonies live together along the corridor. The ethnic Chinese and Arabs as a merchant set up shop houses in the corridor that had a strategic value. Meanwhile, Malay, Arab, Cirebon and Banjar settlements located behind commercial corridors. This corridor present that the economic activity dominated done by Arabic and Chinese trader. Local wisdom in urban design heritage along North axis functioned as main gate to traditional city centre. The local wisdom is potential to be integrated with recent situation and replicated in another region

1. The first section in your paper

This paper is a continuation of research that has been done since the doctoral thesis 2010 that focused on traditional cities in Java, research funded by Sumitomo foundation [1], the similarity of urban patterns in Java and Japan research funded by Diponegoro University in 2016 [2]. Those studies revealed a strong Islamic and Chinese influence in Javanese urban design in the past. The influence of trade and Islamic broadcast that has been done by Cheng Ho from 1405 to 1433 (during his seven expeditions overseas) [3]. Cheng Ho participated in spreading Islam In Java. So, It is not surprising to find many Islamic sultanates in Java that have a building character with architectural style from China. [4, 5].

Semarang city is the capital of Central Java province. With an area of 373.67 km, originated from a city that grew from a small village in the area of Bergota [6]. Due to the sedimentation that occurred on the north side of the island of Java since the 10th century [7] then the center of the city turned to the area near to the seaport in the old boom area (Late 14th century). The area of seaport grew into a multi ethnic village because it became a meeting place for traders from various nations [8]. Recently, the old village has a potential historical value and has significance in the formation of the city of Semarang, namely Kampung Melayu. Kampung Melayu is one of the oldest kampung in Semarang city that has characteristics as multi ethnic village such as Arabic, Chinese, Pakistani, Indian, Cirebonan and Banjar.

Rukayah and Sahid [8] found that the village of multi ethnic was located exactly right on the north side of *Alun-alun*. Its existence is now not directly faced with the *alun-alun* because the *alun-alun* has been closed by the building. It is estimated that since the *alun-alun* and the old Semarang city center in Kanjengan were cut off by the posweg street in 1809/1811 the north side of the area seemed to have no connection with the administrative center in Kanjengan [2]. Thus, it is hard to understanding the historical form of the old Semarang city by future generations

Hartono [10] states that an understanding of the history of the city is very important for planners and urban designers. This is because most urban planning techniques depend on extrapolation from the history of the city itself.

The oldest urban planning in Java is the Majapahit palace (1293 to 1500). The planning of cities in Java after the Majapahit era refers to the unwritten rules on the concept of the city layout in Majapahit. The planning of the city center concist of *Alun-alun*, palace, mosque and North-South axis with houses on the North side [11].

The structure and pattern of cities in Java can be recognized by the dominance of North-South axis [12].The dominance of the North South axis is also found in the pattern and structure of Islamic cities in Java [11, 13, 14]. Similarly, the old city of Semarang. In contrast to the urban structural pattern in inland Java, Semarang is a coastal city with a pattern and a city structure that has a northern axis [9] (see Figure 1). At the end of North Axis there are multi ethnic settlements. To unveil the past condition, this research will be done using historical methods. How was the local wisdom of a multi-ethnic village located in the North Axis of the indigenous government at that time?

2. Research Methods

The research location is the settlemet and corridor in northern axis of the old city center Kanjengan Semarang. The conditions in the past will explored by historical method. The old maps and photographs of the city of Semarang in 1695,1719, 1875 and old photos obtained from kitlv have revealed the formation of a traditional city center with the composition of paseban (square), Javanese temple (mosque), administrativ government (kanjengan office), pasaar (market) and Javanese nagorij (Javanese settlement) on the North side. In old photo from kitlv, there is information of kampong Layur as center of old Semarang city. Through some old images and maps the researchers will do the reserach by using 'seeing by believing' approach from Pole [15]. According to Pole [15], pictures and maps speak more than a narrative. Overlapping of some ancient maps in current conditions (google earth) and manual sketches will be done to analyze urban design patterns. The type old buildings to be analyzed the architecture and function of the building at that time. The photo will be compared with the old photo to strengthen the analysis of the function of the area at that time.

3. The City in Java

Kuta (city-Jawa) means a residential area protected by a wall that built around a city center [16]. He stated that city in Java will be related to the context of the state as the center of government since Hindu-Buddhist era to Mataram-Islam civilization era. Unlike the city concepts in Europe, the physical form of city in Java is not absolute the result of the composition of buildings and spaces, but it was closer to the idea of community coaching consisting of peasants, traders and prijajis that centered on a public figure [14].

3.1. The City Pattern

In the old view of the Indonesian people, the king is considered a figure identified as a god. At the time of the influence of Islam, the Sultan is also considered a figure who controls the living community and can connect it with the unseen society [17]. He revealed that the establishment of the palace as the core of the royal city in Southeast Asia (including Indonesia) during the influence of Islam is associated with the symbol of Meru as in Hindu mythology. According to Tjandrasasmita [18], the city was made traditionally and planned by the supreme ruler. The ruler arranging the city with the composition of the square in the middle, the mosque in the west, and the palace on the south. Streets and other transportation routes made straight intersected form a square to the center of the Javanese Royal Palace. The center of the palace has a function as Imago Mundi (a world image), a center with its concentric circles [19].

The pattern of old city planning in Java is suspected to be found in the former ruins of the capital city of Majapahi. Rukayah [11] states the concept of urban planning from the era of Majapahit kingdom continued in the era of Islamic sultanate (Demak, Cirebon, Banten). In the era of Islamic sultanate still maintain the concept of north-south axis. In the end of the north axis was the sea. Thus, the city had a waterfront concept [20].

The old city Semarang in Kampung Melayu had a boom/ seaport located exactly right on the end of the north side of the central square of the indigenous government. On the old map found a line connecting the port with the square. The line on the map is now named the Jalan Layur. In the area of Jalan Layur written the name of the village of Java (Javanese nagoirij). The pattern of urban planning with the North-axis of the square, is also found on the old map of coastal cities in Banten [21], Cirebon [22] and Surakarta sultanate [23].

3.2. The Boom Area on North Side of Alun-alun

There is an interesting about the sultanate city pattern and structure along the northern coastal of Java and in the inner-city of Java that facing the river / riverfrontcity (Surakarta sultanate). Some cities in coastal and inland Java have a port or boom that are directly connected to the *alun-alun* as the center of the city. Literally the meaning of the word 'boom' equals 'haven' in Dutch or pelabuhan in Indonesian [10]. In Tuban, city, the boom was built long before the Dutch came. The physical form of boom, is the land that protrudes to the sea. Geographically located just a few hundred meters in the northern side of the *alun-alun* Tuban [10].

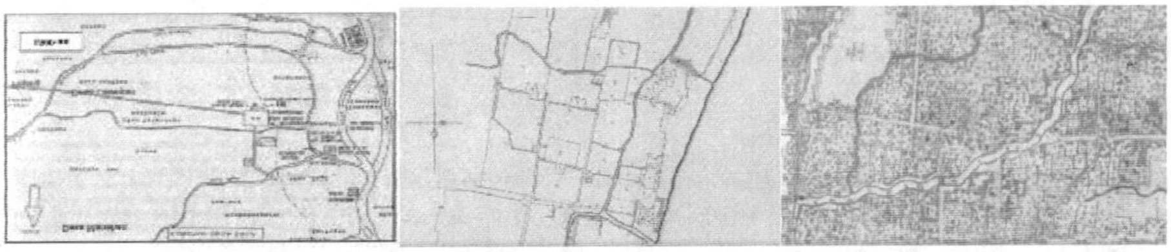

Figure. 1. Old Maps of Surakarta, Cirebon and Banten. North axis connecting *alun-alun* and port [24].

Solo City which is located on the riverside in the interior of Java also has a port called a bandar. The city of Solo was originally formed by the coolie community living on the edge of Bengawan Solo, near the port where they worked for their employers in Kadipaten Pajang (1530s). They formed a river bank settlement [23]. Similarly, the pattern and structure of the sultanates of Banten and Cirebon. The two cities became the center of the world market network and the local market. Both cities have a port, multi ethnic settlements and markets on the northern side of the *alun-alun*.

3.3. The Toponyms and Settlements in Ancient Cities in Java

The old city spatial pattern in Java consists of the structure of *alun-alun, kraton*, building of worship and market. The settlements are clustered and located around the core of the city. Settlements that develop outside the center of power are usually villages that are not guided by the street pattern but on the territorial integrity from the center to the sub center.

The road was not built earlier than the settlement, but was a liaison between the existing centers [16]. The grouping of the Javanese population depicts various social groups or types of occupations.

IOP Conf. Series: Earth and Environmental Science **152** (2018) 012032 doi:10.1088/1755-1315/152/1/012032

The existence of such toponyms can be used to track the existence of various physical artifacts that are now no longer visible. For example: group of craftsman, group of services. Groups based on social status, groups of educators, groups based on race, ethnicity or place of origin [13].

4. Layur Corridor

4.1. Layur Corridor in Old Map

The existence of the Layur corridor has appeared on old maps from atlas of mutual heritage in 1719 (see Figure2.). Layur Corridor is a road in Kampung Melayu Semarang formed by Chinatown and Arabic shophouses that served that corridor as a commercial area. Formerly this area was a port and entrance to the city of Semarang.

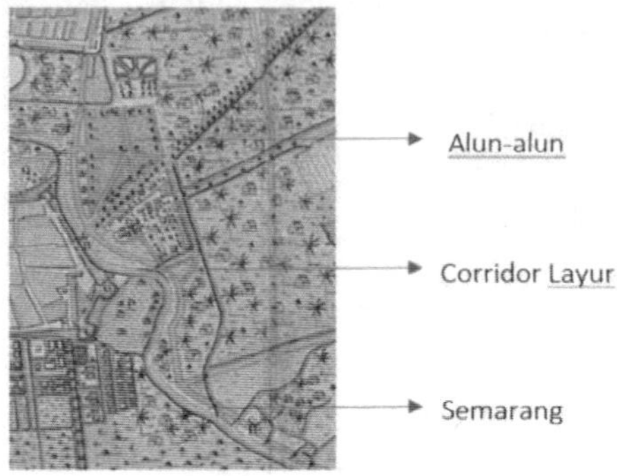

Figure. 2. Semarang old map 1787. Corridor Layur connecting alun-alun Semarang and rifer to the sea [24]

The uniqueness that occurs in the Layur Corridor is the meeting of two ethnicities along the path, Arab and Chinese. The existence of these two ethnic groups is marked physically by the existence of the building of worship, the Menara Layur Mosque was built in 1802 AD and the temple Dewa Bumi built in 1900 AD Arabian merchant houses are behind the Layur Corridor.

4.2. The Traders in the Layur Corridor

The existence of Corridor Layur at this time is very apprehensive, looks shabby, and prone to crime. Many buildings that are not functioning, damaged and collapsed while the Arabian merchant house also not worked. Problems that arise due to several factors and aspects in relation to the development of Semarang City. The existence the Layur Corridor has been abandoned because it has been replaced with trading buildings equipped with modem facilities.

The tower of the mosque had served as a light tower has been officially designated as a cultural heritage building since 1992.

Figure 3. a), Chinese and b), Arabic c), Malay shophouses along the corridor Layur

Figure 4. Some houses behind the corridor Layur a) corridor layur b) old house c) arabic houses

5. ANALYSIS

The process of Islamization in Indonesia was possible done by traders from China and Arabic The existence of Chinese and Arabic artifacts as shophouses and old mosques Masjid Menara Layur in the corridor in the northern side *alun-alun* Semarang proof that there are many local wisdoms found in settlement lay out. The Mosque building and Klenteng is a symbol of socio-cultural wealth in the glorious era of kampung layur. The Chinese and Rabs shophouses arranged along corridor, meanwhile the houses settlement arranged as the background.

China's influence in some cities in Java became clearly. This finding strengthens the statement of Rukayah, 2014, that found a Chinese influence on urban patterns in the traditional cities in Java. The city formation look similar with traditional city in Japan that consist of North-South axis, grid pattern and position of worship building on the west side, central government on the south side. There was a sustainable Chinese concept due to the city lay out in Kyoto was a copy paste of the center of Chang an as the kingdom city in China.

Figure 5. a) Corridor Layur in 1927 and b) recent time (2017).

Figure 6. a) Corridor Layur in 1930 and b) recent time (2017).

IOP Conf. Series: Earth and Environmental Science **152** (2018) 012032 doi:10.1088/1755-1315/152/1/012032

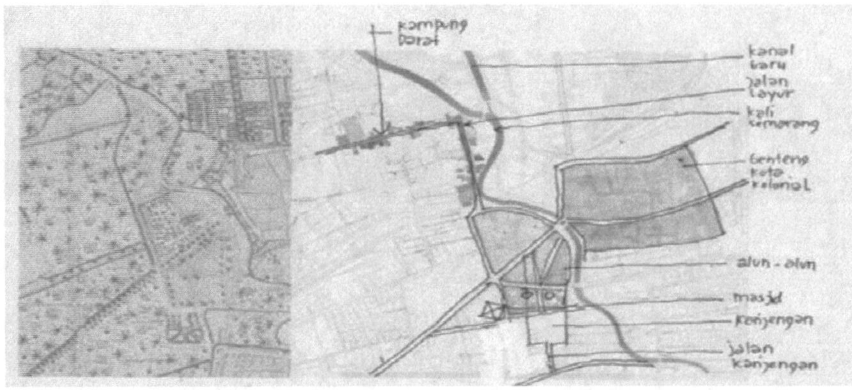

Figure 7. comparing old map and manual sketch about corridor Layur as main gate to alun-alun Semarang

In the old coastal city of Semarang, there was a strong Chinese-Arabic influences. The rows of store houses with Chinese and Arabic architectural styles, mosque buildings and temples show a harmonious life between Muslims and China in the 14th century. From the review of old cities of Java, it can be concluded that the spatial pattern of Semarang city follows an unwritten rule form the old traditional city center in Java. The city center consists of: *alun-alun*, government building, religious building and market. The North axis connects the *alun-alun* and sea port.

The physical structure of multi ethnic settlements in the north axis in the corridor Layur Semarang is a settlement that develops outside the center of power (central government of Kanjengan). However, the settlement pattern cannot be said to be based on road pattern. Existing road is used as the northern axis of the central square / government center.

The remains of old buildings along the northern axis indicate the uniqueness of an old shopping building. Now there are some buildings that still function as a shop. Below is a comparison of Layur road conditions in the past and present. (see fig.5)

Comparing the old photographs with the recent condition there is an indication of old buildings that show its function as a shophouse, or a home and shop. Figure 7. below shows that the corridor layur in the past served as a commercial corridor. Commercial corridor connecting the port in the village of Darat to the *alun-alun* Kanjengan. Here are juxtaposed maps in the present and in the past (in 1719) (see figure 7)

The development of the Layur corridor has a layer period

- Corridor as orientation of ethnic settlements at the outside the center of indigenous government.
- The development of the North axis of the Layur has increasingly become the direction of the orientation of buildings and villages since the construction of Groote Postweg path in 1809-1811 which is on the South side. Since then the Semarang River was no longer the orientation direction.
- The old city of Semarang lay out- grid pattern- is no longer visible since the construction of the postweg.
- The influence of China and Arab Islam dominates the shopping character of the Layur corridor.

6. CONCLUSION AND RECOMMENDATION

6.1. Conclusion

North axis of the old city center of Semarang towards the seaport at that time has an architectural and urban design that is dominated by Arabic and Chinese architectural style. There is local wisdom in the settlement arrangement that locates Chinese and Arab traders. This area illustrates the process of Islamization in Indonesia made possible through the trade channels by both ethnic groups.

The city's architecture follows the pattern of urban planning in Java where the north corridor functions as a commercial corridor. This is marked by the discovery of the form of building with the function of the building as a light house. Meanwhile the residential area is located behind the commercial corridor. The layout of the city-based multi-ethnic settlement establishes enclaves behind commercial corridors.

The existence of mosques, Islamic boarding schools, temples in commercial corridors and arrangement of villages based on the region of origin, describing that the place were function as the coastal city at that time. The Local wisdom found in the area are harmony in social, economy and religious

6.2. Recommendation
Local wisdom in urban design heritage is need to be appreciated more as a traditional technology in city lay out especially along North-South axis as main gate to traditional city center in Java. The local wisdom is potential to be integrated with recent situation and replicated in another region. Recommendations are directed to the Tourism office and Trade office to revive commercial corridor areas and develop them as coastal trade that potential for tourism.

Acknowledgements
We would like to thank to Department Architecture Engineering Faculty Diponegoro University which has provided research grants in 2017 and Spatial Planning Office of Semarang City that has lent the map of Semarang city.

References
[1] Rukayah, Siti, R, Tomohiko, Wijayanti 2014 A Comparative Study About Space and Place for All On Cosmic City In Kyoto And Yogyakarta As A World Heritage Tourisme. not published.
[2] Rukayah, Siti,R, Respati, P. D & Susilo. 2016. S Morphology of Traditional City Center in Semarang: Towards Adaptive re- use in urban heritage. E. S. *Environment-Behavior Proceedings Journal*, **1** 109–118.
[3] Yuanzhi, K. 2000 *Cheng Ho Muslim Tionghoa. Misteri Perjalanan Muhibah di Nusantara.* (Jakarta: Yayasan Pustaka Obor Indonesia).
[4] Rukayah, Siti, R 2005. *Satu Lagi Potensi Wisata Sejarah, Temuan Typologi Konsep Alun-alun di Jawa, Kreasi Muslim Cina Anak Buah Cheng Ho.* Pemenang ke 3 Lomba Penulisan Tingkat Nasional Cheng Ho dan Kaitannnya dengan Indonesia, dalam rangka peringatan 600 tahun pendaratan Cheng
[5] Muljana, S 2005 *Runtuhnya Kerajaan Hindu–Jawa dan Timbulnya Negara-negara Islam di Nusantara*. Yogyakarta: LkiS.
[6] Soekirno 1956 *Semarang. Djawatan Penerangan Kota Besar Semarang.*
[7] Bemmelen, R. Van *1948 Geology of Indonesia Vol-IA General*. Government Printing Office.
[8] Tio, J 2004 *Kota Semarang dalam Kenangan*. Semarang.
[9] Rukayah, Siti, R; Sahid, M. 2017 *Menemukan Sumbu Utara Selatan Alun-alun Lama Kanjengan Sebagai Dasar Perencanaan Pusat Kota Semarang Berdasarkan Heritage*
[10] Hartono S. 2005. Alun-alun Dan Revitalisasi Identitas Kota Tuban. Dimensi - journal of architecture and built environment. 33 131–141
[11] Rukayah, Siti,R. Soetomo, Joesron 2010 *Simbiosis Di Ruang Terbuka Kota Simpang Lima Di Kota Semarang Simbiosis Di Ruang Terbuka Kota Simpang Lima Di Kota Semarang*. Phd Dissertation, not published.
[12] Handinoto. 1992 *Alun-alun* sebagai identitas kota jawa, dulu dan sekarang . (September), ***Dimensi*** *- journal of architecture and built environment* 1–15.

[13] Adrisijanti, I 2000 *Arkeologi Perkotaan Mataram Islam*. Penerbit Jendela Yogyakarta.
[14] Wahyudi, M. A 2006 *Korelasi tata ruang rumah kuno di krajan kulon terhadap tata ruang kota kaliwungu.* Semarang: Tesis, not published.
[15] Pole, C. J. 2004. Seeing Is Believing? Approaches To Edited By. *Studies in Qualitative Methodology.* **7** 5-6
[16] Wiryomartono, A. B. P 1995 *Seni Bangunan dan Seni Bina Kota di Indonesia*. (Jakarta: PT. Gramedia Pustaka)
[17] Geldern, H. R 1982 *Konsepsi Tentang Negara dan Kedudukan Raja di Asia Tenggara*. (Jakarta: CV. Rajawali)
[18] Tjandrasasmita, U 2009 *Arkeologi Islam Nusantara*. (Jakarta: Gramedia)
[19] Lombard, D 2005 *Nusa Jawa Silang Budaya.* Gramedia Pustaka Utama.
[20] Rukayah, Siti,R. & Malik, A 2012 Between Colonial, Moslem, and Post-Independence Era, Which Layer of Urban Patterns should be Conserved? *Procedia - Social and Behavioral Sciences.* **68** 775–789.
[21] Wessing, R 1992 Robert Wessing A Tiger In The Heart : The Javanese Rampok Macani, *kitlv.nl* **148** 287–308.
[22] Hendro, E. P. 2004 Perkembangan morfologi kota cirebon dari masa kerajaan hingga akhir masa kolonial. *Paramita: Historical Studies Journal*. **24** 1
[23] Prayitno, B, & Qomarun, Q 2007 Morfologi Kota Solo (Tahun 1500-2000). ***Dimensi*** *- journal of architecture and built environment.* **1** 80-87
[24] Atlas Mutual Heritage retrieved from http://www.atlasofmutualheritage.nl/en/

IOP Conf. Series: Earth and Environmental Science **152** (2018) 012033 doi:10.1088/1755-1315/152/1/012033

The Model of Landscape Development in Big Cities Of Central Java

E. Darmawan1* and T. W. Murtini2

1,2Architecture Department, Engineering Faculty, Diponegoro University

edy_darmawan@yahoo.com

Abstract. The existence of urban parks as a part of urban green space is very important for the environment and the citizen of the city, and inseparable part from the urban landscape. In its development, the existence of an urban parks could create a safe, comfortable, productive, and visually aesthetical environment. The problem arising now is a view that the urban parks are often unsuitable with the surrounding. Therefore, the parks are not functional and does not have significant visual. So that, this research is aimed to reveal model of landscape development in big cities in Central Java. The method used is descriptive qualitative that can describe the detail of problem, in determining the plan to overcome the problem. The research location will focused on big cities in Central Java with potential landscape that can be improved. The results of the research will be composed in an international scientific journals and is expected to be a reference in the field of urban landscape arrangement.

1. Introduction

The dynamic development of the city has brought consequences on the needs of space/ land. The high number of inhabitants in urban areas demands the provision of facilities, infrastructure, and services for life and its activities. Cities with limited resources has attempted to provide a fair environment for the citizens. However, population projection on city planning is often missed, resulting in an imbalance of the number of residents with the provision of facilities and infrastructure of the city. This condition leads to the various problems, one of them is the decrease in the environment quality. The natural ecosystems in urban areas has change slowly into artificial ecosystem. A lot of vacant in urban areas become built lands. The open space areas functions have been switched into trading, settlement, industrial, transportation network, and other facilities and infrastructure. Function changing certainly brings some negative impacts such as the rise in temperature, pollution in urban areas, the decrease of ground level, the danger of flooding, landslides, coastalabration, etc.

Basically, the negative effects of growth of a city can be reduced if the government can implement Law No. 26 Years 2007 about The Space Management by developing green open space area at least 30% of city's total area (20% public green open space and 10% private green open space). Green open space area is advantageus to keep the balance, the continuity, health, comfortability, sustainability, and enhancement of city's environment quality. Correct arrangement green open space will exactly able to perform to increase improve the quality of the city atmosphere, provide fresh air, decrease pollution, and reduce noise. Green open space filled with plants will affect the quality of city air because the plants creates the micro climate which cool down temperature, maintain air humidity, and increase the level of oxygen [1].

IOP Conf. Series: Earth and Environmental Science **152** (2018) 012033 doi:10.1088/1755-1315/152/1/012033

Based on the data from the office of city planning Central Java, the provision of public green open space in Central Java is only meet 10% of the target where it should 20%. Therefore, the development of open city parks should constinuously held. From 60 districts, there are 15 districts that have meet the target. These districts are mainly located in dry areas such as city of Semarang, Demak, Kudus, Pekalongan, Tegal, Pati, etc.

Over these past few years, those cities has conducted rehabilitation and built some parks to meet the target. However, Central Java does not have guidelines in spatial planning, especially in the development of city park. Hence, the researchers of this article interested to assess model of landscape development in big cities in Central Java.

2. Related Study

According to the Act of Republic Indonesia No. 26 Year 2007 [2] about The Space Management, green open space is defined as an elongated and/or group area/path that has open function such as a place for plants to growing while public green open space is a green open space that is owned and managed by the government and it is generally used for public purposes. The existence of a green open space significantly improves the quality of life and the sustainability of environment by improving citizen's prosperity through a better urban environment and healthier. According to Hakim [1], green open spaces have many functions such as:

a) Bio-ecological function (physical) that ensures the construction of green open spaces as the part of the air circulation system (lungs of cities) and as the micro climate controller, so the air and water circulation system can naturally work as a shade, an oxygen producer, a rain absorber, a habitat for animals, a pollutant absorber, and as a wind barrier.
b) Social, economic (productive) and cultural functions to describe local cultural expression as communication media for citizens, recreational facility, and research and education facility.
c) Urban ecosystem,
d) Aesthetical function.

Darmawan [3] stated that an interesting green open space will attract many visitors from various ethnic, ages, education levels, socio-economic levels, and interest.

Landscape is a part of green urban open space. Rob [4] defined landscape as an integrated system of biotic and abiotic components including a human influence component. According to Naveh on Restyanto [5], landscape is connected to all physical, ecological, and geographical integration of human's and nature's processes and patterns. In urban areas, the landscape is formed due to the change of natural landscape that is degraded and formed into the shape of a city caused by human activities. Over the time, the landscape will continually change due to human activities. Therefore, it needs proper management in order to maintain its components. Danier [6] stated that the regulation of space and the management of landscape should contribute for creating a unique character of a landscape. Designing landscape should consider principles of landscape design and elements of landscape. According to Hakim [1], the principles of landscape design consist of balance, rhythm and repetition, stress and accentuation, simplicity, contrast, proportion, and unity. While landscape elements that should be concerned are lines, fields, spaces, textures, and colors.

City parks is a support for urban landscapes that is able to shape the character and keep the sustainability of cities. According to Carr [7], the city park is the part of public space and the place for social life. The city park portrays dynamic life of community, it also serve as a destination for many people to do various activities. Without city parks, the citizen of a city will suffer mental strain because city parks are physically able to protect the surrounding environments. Susiloarifin [8] stated that awell managed city parks will be beautiful, comfortable, and safe for their users. Inimproving a park to be a beautiful place, things like materials (soft material and hard material), scales, circulation, parking facilities, lighting, tile patterns, comfort, drainages should be the focus of city park design, there are also supporting elements of park such as display front/ landmark, benches, trash bin, lamps post, vegetations, pavements, parking lot, stalls, children playground and facilities, and sports facilities [1].

3. Research Method

This research are carried under qualitative approach that emphasizes narrative construction or textual descriptive for the studied phenomenon. The method used in this research is descriptive method. According to Arikunto [9], the qualitative descriptive method describes and explains collected findings in order to gather actual and detailed information, identify problems, and make comparison or evaluation. This method furthermore help us to determine what other people do in dealing with the same problem and learn from their experiences in order to make plan and decision for the future. In this article, the descriptive qualitative method is used to describe physical condition of parks in big cities in Central Java and explain the elements and principles of designing landscape in particular. Surveys is used to collect data from certain places on natural setting. The data collected from observations, interviews, and study of literatures then are selected and arranged systematically based on the needs of the study. After that, the compiled data is analyzed with qualitative descriptive analysis method in certain steps of analysis.

The study is conducted in big cities across Central Java focused on parks as the object of observation which is typical for the city. The parks chosen as the sample study are selected according to their uniqueness, location, width, facilities, service areas, and the need of people around for public space. Six parks on six different cities are chosen as the sample of this study, as seen in table 1 below. These parks are inspected on their physical condition and comprehensiveness of their facilities as the representative of the parks in big cities in Central Java

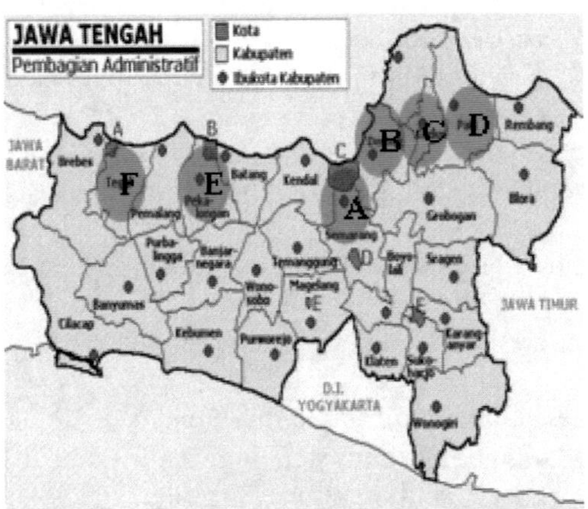

Information:

A = *Kota Semarang (Taman Menteri Supeno)*

B = *Kabupaten Demak (Alun-alun Simpang Enam)*

C = *Kabupaten Kudus (Alun-alun Simpang Tujuh)*

D = *Kabupaten Pati (Alun-alun Simpang Lima Pati)*

E = *Kota Pekalongan (Taman Monumen)*

F = *Kota Tegal (Taman Poci)*

Figure 1. Research Location. Source: Researcher, 2017

4. Results & Discussion

1.1 Analysis of Physical Condition of Parks in Big Cities in Central Java

Several parks in big cities in Central Java are also the icon for the city. The existence of the parks is one of government efforts to provide comfortabilty to meet the needs for public green space. Parks is not only as complementary feature of the city, but also serves to maintain coolness, climate, provide shades, rest area, and as a recreation site. Based on direct observation in the field, it is known that the physical condition for the parks are sampled as follows:

a) *Alun-alun Simpang Enam Demak*
   - Located *Jalan Sultan Fatah No. 41 Bintoro, Kecamatan Demak, Kabupaten Demak*
   - There are souvenirs vendor and culinary stalls.
   - The base is hard material and grass. There are also shade trees. The decoration tree cannot be used optimally so it cannot add the appeal of this place.
   - Custom large nameplate is built specially for this place and often photographed by visitors.
   - Activities that mainly happened are social activities, where people gathered in this place to chat, interact, enjoying themselves tourism, or in some cases exercising.

b) *Alun-alun Simpang 7 Kudus*

- Located in *Jalan Sunan Kudus*
- The existing vegetations are shade trees, decorative plants, and eyecatching bush.
- Equipped with support facilities, such as resting shelters, jogging tracks, flower gardens, park benches, garbage cans, and attractive garden lights.
- Pavement on the sidewalk is made as attractive as possible using granite stone temple and natural stone mosaic used to encircle the intersection.
- This square serves as a public space to hold car free day activities, festivals, sports, ceremonies, gathering places, tourist attractions, etc.

c) *Taman Monumen*
- Located in *Jalan Imam Bonjol, Bendan, Pekalongan.*
- Green open space is an active public park located in the center of Pekalongan City in area ± 0,69 Ha.
- Physically, the base uses pavement, surrounded by many trees but in light density because the trees are cypress and short shrubs.
- In addition, there are also some street furniture such as park bench, lamp posts, billboards (signed), flower pots, garbage bin and flagpole.
- During the day there are only a few street vendors inside the park not as many as in the afternoon and evening. In addition, only a few local people come to socialize or just play (relaxation).

d) *Taman Poci*
- Located in *Jalan Kolonel Sudiarto, Panggung, Tegal Timur.*
- There are a variety of beautiful plants and trees, as well as sculptures and cups that become icons for this park.
- Equipped with facilities such as parking areas, bathroom, praying room, and street vendors.
- Easy to reach by public transport and private vehicle because it is equipped with road signs as a signpost.
- In the park, there is a main building that resembles a large sailing ship and its top can be used to see the scenery around the park.
- Activities happen are leisure activities, afternoon walks, tours, playground for children.

e) *Alun-alun Simpang Lima Pati*
- Located in *Jalan Tombronegoro, Pati Kidul, Kecamatan Pati, Kabupaten Pati.*
- There are street vendors, children's playground, flagpole, dumpster, jogging track, garden lights, ornamental plants pots, gathering places, signage, sculpture.
- Equipped with shade vegetation, bush, and ornamental plants.
- The shape of the park is semi circular with a good drainage system to avoid puddle of rainwater.
- Pavement on the sidewalk is arranged in pattern with different colors and shapes.
- Activities appear in this park are leisure activities, tours, shopping, culinary, sports, and play for children.

f) *Taman Menteri Menteri Supeno*
- Located in *Jalan Menteri Supeno, Semarang.*
- This 5.000 m^2 park is located in the city center and surrounded by office buildings and equipped with seating, lamp post, billboards, bins, pedestrian paths, bathrooms, vegetation, parking lots, children's playground, vendors area, skateboard area, and a fountain with a statue that symbolizes the family planning program in the middle of the park.
- The number of trees and vegetation make the atmosphere around the park is cool. So, during the day residents often sit around or just simply hanging out, rest, sightseeing, and enjoying the view.
- In the afternoon and the evening, there are some street vendors selling various types of food. At night, this park is often used as a venue for event.

1.2 Analysis of The Model of Landscape Development in Big Cities in Central Java

The landscape of the city park is one of the element of urban space needed by the community. Therefore, the landscape model of urban parks as a public space must have good standards, especially in terms of design and facilities, and supporting elements. The provision of such facilities should be in line with the type of park and the needs of the community, to create a synergy between natural elements and the landscape generated by the diversity of vegetation, activities, with artificial elements provided as social and recreational facilities. The integration between these natural and artificial elements will create an attractive aesthetic visuals in every parks. Parks landscape models in observed areas are explained in Table 1 below:

Table 1. Urban Park Landscape Development Model in Big Cities in Central Java

No.	Name	Elements of Park Support	Landscape Design Principles
1	Alun-alun Simpang Enam Demak  	- Front view of this place is good enough indicated by the sculpture of the name of the place. - Support facilities are in good condition and able to provide comfortability for visitors. - Access to this place is also quite easy, it can be reached by various types of vehicles and are on a busy way passed by vehicle.	- There is a harmony between elements related to size, color, quantity, and appearance. - Hierarchy in this place is indicated by the differences in size, shape, and strategic location. - The rhythm and repetition of landscape elements are seen in one pattern with a balance of composition between the elements. - Exciting pavement and vegetation selection shows the contrasting elements of this place.
2	Alun-alun Simpang Tujuh Kudus  	- The front view of this place is very interesting marked with the sculpture of the place arranged in a unique way, so it looks interesting. - Behind the sculpture is a small garden equipped with different types of plants with different sizes, types and colors. - Overall, the supporting facilities are in good condition and designed as attractive as possible, thus making the surrounding environment looks beautiful. - Easy to reach because it is in the city center.	- Setup of supporting elements consist of the size, color, quantity, and appearance shows the existence of harmony among elements that exist. - The use of different shapes and sizes on the supporting elements indicates a hierarchy in this place. - The rhythm and repetition of landscape elements are seen in one pattern with a balance of composition between the elements. - Preparation of pavements with different shapes and types, as well as the use of different colors in plants indicate the presence of contrast elements.
3	Taman Monumen Pekalongan	- The front view of the park looks less attractive. Because the statue of the	- Setup of supporting elements gives minimal consideration for the size, color, quantity, and

	 	fighters who became a landmark for this garden looks dull, the plaque of war heroes name who are near the statue are damaged. - This park became a pride for the people of Pekalongan, dan because it was built to commemorate the struggle for independence, it become scared. - Some of park's supporting facilities are below maintained. - The arrangement of trees and plants in this park is quite good and provide coolness for the environment. - The park is easy to reach because it is on the main road.	appearance, so the harmony between elements are almost invisible. - The use of different shapes and sizes on the supporting elements indicates a hierarchy. - The rhythm and repetition of landscape elements are seen in one pattern with a balance of composition between the elements. - Contrast elements seen in the statue of the fighters who stood firmly in the middle of the hall or plaza in the park.
4	Taman Poci Tegal 	- The front view of this park is very interesting because the park is flanked by two replicas of objects that characterize the City of Tegal, the replica of the pot on the right and replicas of aci tofu on the left. - Support facilities are complete and in good condition. - The park is also equipped with good vegetation arrangement, so it gives the cool impression.	- The harmony between elements is quite visible from the use of size, color, quantity, and display of each element. - Hierarchy seen from different shapes and sizes of supporting elements. - The rhythm and repetition of landscape elements are seen in one pattern with a balance of composition between the elements. - Contrast elements seen in the use of colors and sizes on the replica of objects and nameplate.
5	Alun-alun Simpang Lima Pati 	- The front view of this park is less interesting because there is no marker that can be used as a landmark. - Support facilities are in good condition. - Pavement on sidewalks and flower pots has	- Existing elements look good. - Hierarchy is seen from different shapes and sizes of supporting elements. - The rhythm and repetition of landscape elements are seen in one pattern with a balance of composition between the elements.

		different color and shape compositions. - Easy to reach because it is in the city center.	- Contrast element are seen in the use of color and size on pavement pavement and flowerpot.
6	Taman Menteri Supeno Semarang  	- The front view of the park looks interesting with a large nameplate that has colored lights. - The park has complete support facilities for community activities. - The location of this park is strategically located in the city center, making it easy to reach. - Vegetation is enough to give a cool impression for the surrounding environment.	- Setup of supporting elements that consider the size, color, quantity, and appearance shows the existence of harmony among elements that exist. - Hierarchy is seen from different shapes and sizes of supporting elements. - Rhythm and repetition of landscape elements are seen in line elements in the form of roads and circular spaces. - Contrast element are seen in the arrangement of pavement garden floor. - The balance is visible from the shape of a triangle-shaped garden with an emphasis on the center of a garden with a fountain.

Source: Results Analysis, 2017.

Based on the analysis of parks as observed above, the assessment of this study can be seen in Table 2 below. The assessment is categorized using data nominal scale. Category used consisting of 3 = very good; 2 = good enough; and 1 = not good.

Table 2. Assesment of Urban Park Landscape Design Model in Big Cities in Central Java

No.	Name	Supporting Elements of Park	Landscape Design Principles	Landscape Elements	Total Score
1	Alun-alun Simpang Enam Demak	3	2	2	**7**
2	Alun-alun Simpang Tujuh Kudus	3	3	2	**8**
3	Taman Monumen Pekalongan	2	1	2	**5**
4	Taman Poci Tegal	3	2	2	**7**
5	Alun-alun Simpang Lima Pati	2	2	2	**6**
6	Taman Menteri Supeno Semarang	3	3	3	**9**

Source: Results Analysis, 2017.

The assessment of results of the research shows that model garden with highest score is a *Taman Menteri Supeno Semarang*. This is because *Taman Menteri Supeno* has complete supporting facilities which are located at the center of the city, using various elements of pavement, and having good vegetation arrangement giving cool atmosphere on the environment.

5. Acknowledgement

This research is supported by Direktorat Riset dan Pengabdian Masyarakat Direktorat Jenderal Penguatan Riset dan Pengembangan Kementerian Riset, Teknologi, dan Pendidikan Tinggi. The researcher also presents his sincere appreciation to Dean of Engineering Faculty of Diponegoro University. We like to give our gratitude to colleagues from Architecture Department, Engineering Faculty, Diponegoro University who provide insight and expert opinion in assisting this research, although they may not agree with all of the interpretations/ conclusion of this paper.

6. Conclusions

Based on studies and analysis that has been done, the researcher conclude that city park in big cities in Central Java serves well as public open space used by people to interact socially. Most of this park is in strategic location, so it is easy to reach. These park can be a tourist attraction, but some of them do not have facilities for disabilities people. The management and supervisor of a park also has low awareness. This can be seen through bad arrangement of parking lot and many broken facilities. Therefore, the city park landscape development model has to utilize potential and the unique culture of local residents to overcome any problems. The management of the park should be the collaboration of society and regional government.

7. References

[1] Hakim, Rustam dan Hardi Utomo. 2003. Komponen Perancangan Arsitektur Lansekap (Prinsip-Unsur dan Aplikasi Desain). Jakarta: Penerbit Bumi Aksara.
[2] Undang-undang Republik Indonesia Nomor 26 Tahun 2007 tentang Penataan Ruang.
[3] Darmawan, Edy. 2009. Ruang Publik Dalam Arsitektur Kota. Semarang: Badan Penerbit Universitas Diponegoro.
[4] Rob, Krier. 1979. Urban Space. London: Academy Editions.
[5] Restyanto, Yoga. 2011. Arsitektur Landscape Kawasan Wisata Outbound Bukit Tangkiling. Jurnal Perspektif Arsitektur. Volume 6 Nomor 2 Desember 2011.
[6] Danier, L. 2015. The Little Sustainable Landscape Book. Oxford: Global Canopy Programme.
[7] Carr, Stephen et al. 1992. Public Space. Australia: Press Syndicate of University of Cambridge.
[8] Susiloarifin, Hadi dan Nurhayati. 1994. Pemeliharaan Taman. Jakarta: Penebar Swadaya.
[9] Arikunto. 2006. Procedure Research: A Practical Approach. Jakarta: PT Rineka.

IOP Conf. Series: Earth and Environmental Science **152** (2018) 012034 doi:10.1088/1755-1315/152/1/012034

A Review of Thermal Environmental Quality in Residential Areas in Tropical Cities

Wienty Triyuly[1,2*], Sugeng Triyadi [1] and Surjamanto Wonorahardjo [1]

[1]Department of Architecture, SAPPK, Bandung Institute of Technology, Bandung, Indonesia
[2]Department of Architecture, Sriwijaya University, Palembang, Indonesia

*bunda_wienty@yahoo.com

Abstract. The development of residential areas in big cities is characterized by a vertical-horizontal growth in the center of the city and horizontal growth in the edge of the city. The growth of the city has changed Land Use and Land Cover (LULC) and significantly altered the thermal environmental quality and urban heat island (UHI). This paper discusses the effects of the change LULC in several cities to their thermal environmental quality due to the development of residential area. We compare the effects of the change of LULC that occurred in residential areas in tropical cities. This study involves some factors such as the character building of residential, land cover, sun, and wind orientation. The results of this study present a map of various issues on the change of LULC in residential areas in tropical cities.

1. Development of residential areas

The development in urban areas generates a massive urbanization that causes land use intensification and building density in urban areas. It is indicated by changing of land cover types and vegetation index [1]. The development in an urban area can be controlled by using land use planning regulation with consideration of urban structure, site selection, street layout and building density based on climatic adaptation and ecological principles [2].
The development is an attractiveness factor for urbanization and also for expansion residential areas. Residential areas become an important part of a city that influence directly to ambient of a city. Therefore the quality of residential areas gives a direct effect on thermal quality of the city. Housing needs in a city increase density of residential areas due to limited of land areas.
Limitation of area for residential generates a housing area expansion to other lands that previously are not recommended for residential area such as wetland. Many efforts to convert the wetland into residential area such as put some pillars as supports of houses that change the characteristics of the land. The most popular effort is landfill that converts wetland into dry areas. Characteristic of land area alter due to landfill for housing area. The changing of wet and dry land influence ecosystem quality of the city [3].

2. The impact of urban physical development on thermal environment quality

Urban development encourages the development of area became built-up area. The development of built-up area cause the changes in land cover, building structure material and reduction nature of surface [4]. Land cover and building structure material which use reflective properties cause increasing the value of surface albedo [5, 4], emissivity, and infrared emissivity [5]. Reduction nature of surface in the form of decreasing of vegetation affect evapotranspiration in the area [5]. The change of surface albedo, emissivity, infrared emissivity and evapotranspiration will affect the thermal environmental quality and microclimate [5].

Thermal environmental quality and microclimate an area are formed due to urban geometry [6, 7, 8, 9], construction materials [8] and urban texture [10]. Urban geometry is element configuration of the area that affects thermal environmental quality in the form street geometry, orientation [11], street axes and building height [6], sky view factor, local and neighborhood scale [9]. Urban geometry affects the thermal environmental quality and microclimate because this factors related to solar radiation exposure on the surface area which is influenced by street geometry and orientation, trees, wind speed and horizontal surface [11]. Urban geometry has an impact on the environment in the level mesoscale and microscale [12].

The urban texture is element configuration of surface characteristic that impacts air and surface temperature, solar irradiation on horizontal and vertical surfaces, wind speed and direction [13]. Urban texture related to construction material, form (smooth and roughly the surface) and color elements that affect the heat gains and albedo.

The change of microclimate in residential areas affect the intensity of Urban Heat Island (UHI) because most of the urban areas are the residential area. If microclimate of residential areas increase than the intensity of UHI will increase too. The increasing of UHI related with material used [14, 15], building configuration [14, 16], lack of vegetation and water bodies [17]. The increasing of UHI have a correlation with the increasing of urbanization, air pollution and anthropogenic heat sources [4].

3. The changes of land use and land cover (LULC)

Development of residential areas causes the changes in land use land cover. The change of land use in the form of open space, street and building, the change of land cover in the form of pavement, building, vegetation and water body. The changes in Land Use Land Cover (LULC) significantly alter the thermal environmental characteristics [18] with radius 50m–200m in the form hardscape and soft scape material [19].

The change of LULC affects thermal environmental quality due to the ability of the material to absorb and store the heat in the form of albedo [20, 5]. Asphalt roads and concrete with white aggregate, light-colored walls and the roof with low emissivity, vegetation and trees are the material with high albedo [21]. Material asphalt roads, concrete and material covered will absorb and store heat in the material [14, 15]. Asphalt roads and roof are the material with a large surface heat value in the daytime while the exterior wall material and the tree have a large surface heat value in the nighttime. Green areas have a low surface temperature in the daytime and high surface temperature in the nighttime while soil and building have a high surface temperature in the daytime and low surface temperature in the nighttime [22].

3.1. Open Space

Open space has some various material land cover with morphologic features and properties of the material forming which affect the thermal environment [8].

The used of reflective and constructive material affect the intensity of UHI because this material has a high heat capacity and conductivity so that can deliver the heat in the daytime and then absorb the heat in the material. Reflective material affect surface temperature and air temperature [23]. The land cover material of open space consists of pavement, vegetation, and water.

Open space uses pavement in the form of granite, concrete, and asphalt. The pavement of granite, concrete, and asphalt have a surface temperature which the value of the surface temperature asphalt higher than other pavement. Asphalt has surface temperature higher than the air temperature in the daytime [24].

Open space uses vegetation as material element in the form of trees, low vegetation, and grasses. The ability of trees as elements of microclimate controller depends on the shading performance of solar, longwave radiation and capability to produce latent heat [25, 26]. The ability of grass and low vegetation as elements of microclimate controller depends on the condition of groundwater because of the influence of the water surface more dominant than grass and low vegetation surface [8].

The land cover material in the form of water body has the ability to cool thermal environment. The location and the width of water body have a direct correlation with the intensity of Urban Cooling Island (UCI) and UCI efficiency [17], small water body has the intensity of UCI smaller than large water body in the same area. The water body can influence thermal environment quality in the form of river [27], pond [28], and paddy fields [29]. The ability of water body to cooling depends on the distance of water body [27] and the wind movement [28]. The influence of the water body can be reduced if the area near the water body used as a buffer zone, not wetland zone [30].

3.2. Street

The street is hardscape material which forms the street configuration and street canyon to wind movement in the area. Streets can be covered by asphalt, concrete and light-color material [14]. Street affects directly to the thermal environment with parameter ratio, orientation and the presence of trees as shading element [31]. Street with shading of vegetation and building have surface temperature lower than street without shading because the shading of vegetation and building can reduce the heat gain indirectly and decrease the surface temperature [32].

3.3. Building

The building is the element that has a volume (height and width) and the distance so that affect building density and building layout. Building density is influenced horizontal and vertical surface so building density has a linear correlation with albedo so effect directly to thermal environment quality [7] and UHI [33].

Building configuration in the form of building layout affect the movement of the wind flow [10] so building layout should consider wind and sun orientation. Wind orientation affects wind movement so the best configuration layout is the layout toward to the prevailing wind [26]. Sun orientation affect heat gain, the building with east-west orientation receive heat gain directly from the sunlight and the wall with north-south orientation receive heat gain indirectly from the surrounding [32].

The elements of a building that has an effect directly to the thermal environment quality are wall and roof. The capability of the wall and the roof to store and absorb the heat gain of sunlight depends on material and orientation of the building. The wall and roof can use reflective and constructive material as well as can use a cool element such the green wall, the green roof and cool roof.

The green wall (green envelope) using elements of vegetation on the building wall that function to reduce heat gain and cooling loads through shading, cooling transpiration and thermal insulation [34] with potential usage on the west side of the building [35]. The green roof is roof element which using the vegetation on the roof layer that function to reduce the heat gain to surrounding and heating effect to the building [36]. The cool roof is roof element which using high solar reflectance and high thermal remittance material that functions to absorb radiation of the sun so it can reduce the surface roof temperature [37].

4. Analysis

The residential area is a fast-growing area in the city so that the development of this area has a correlation significant with thermal environmental quality of the city. Thermal environmental quality in its relationship with the UHI depend on the surface material and its condition [28] so thermal environmental quality in residential areas in tropical cities depend on the surface and the characteristic tropical cities. The tropical cities have two seasons and receive directly the heat gain from sunlight all the day. This condition affects the solar radiation exposure on the surface area so this needs a planning of some element to control temperature The planning of topography and building geometry can be used to set sun angle so the surface area which is exposed to the heat from sunlight give small impact to day temperature [38].

Some researchers research surface and air temperature in tropical cities to find the elements that affect the thermal environmental quality of residential. The effect of the element in tropical cities is different with other cities because the sun and wind orientation influence thermal environmental quality significantly. Sun and wind orientation of tropical cities affect the capability of soil, building, vegetation and water body to absorb and store heat gain so that affects the air temperature and humidity of the tropical cities.

Table 1. The factor that affects residential thermal environmental quality in tropical cities

No	Researchers	Building	Wind	Factor								
				Pavement	Street	Building		Vegetation				
						Wall	Roof	Wall	Roof	Open Space	Park	Forest
1	Sharmin, Steemers, & Matzarakis, [39]	Vertical	v	v	v	v	v					
2	Ardalan et al [40]	Vertical						v	v	v		
3	Xue, Gou, Siu, & Lau [41]	Vertical				v			v		v	
4	Roth & Lim [42]	Vertical				v					v	
5	Didem, Stocker, Carruthers, & Hunt [43]	Vertical	v			v	v					
6	Ayanlade [44]	Horizontal										v
7	Li & Norford [45]	Vertical	v				v		v			
8	R, Pereira, & Karam [46]	Horizontal-Vertical										v
9	Sharmin et al [47]	Vertical	v		v	v	v			v		
10	Rajagopalan, Lim, & Jamei [48]	Vertical	v			v				v		
11	Wong & Yu [49]	Vertical								v	v	v

v=affect

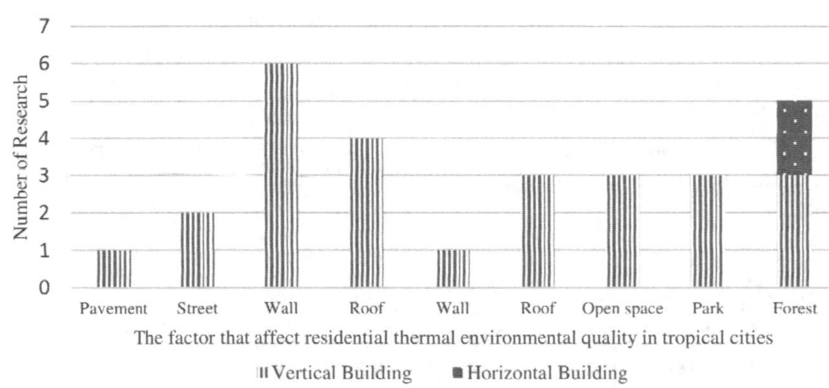

Figure 1. The number of residential thermal environmental quality research in tropical cities

4.1. Pavement and Street

Pavement and street are hardscape material that can absorb and store heat gain in the material. Pavement can use concrete, brick, granite, asphalt and etc while street can use asphalt, soil, and etc.

Pavement and street are influenced by texture and color of material. The dark color material has lower surface temperatures than light color material [8]. The texture and color of coating pavement material can cause the pavement be cool material that called by the cool pavement [50].

The use of concrete as material pavement and asphalt as street material can affect residential thermal environmental quality in tropical cities [39].

4.2. Building
Thermal environmental quality is influenced by building form and building distance with parameter H/W Ratio (building height/street width) and SVF (Sky View Factor) [39]. Building form and building distance set the building so that formed the character of building.

The character of building consists of horizontal and vertical building that influence residential thermal environmental quality in tropical cities. The most influential on the horizontal building is the roof while the most influential on the vertical building is wall and roof.

The influence of the building is determined by building surface material. The material of building will store and absorb heat gain in the vertical and horizontal surface in the form wall and roof. The use of material on the wall [48, 47, 41, 39, 43, 42] and roof building [47, 45, 39] can affect residential thermal environmental quality in tropical cities. The use of construction material (concrete and brick) can absorb and store heat gain in the daytime so that this material affects air temperature in the daytime and nighttime. The use of cool material and green material can absorb and reduce the heat gain so that the air temperature in the daytime can be controlled.

4.3. Vegetation
Vegetation can be used as wall material, roof material and land cover material. The use of vegetation as material on wall of building [40], roof of building [45, 40, 41], material in open space [49, 48, 47, 46, 40], park [49, 41, 42] and forest [49, 44, 45] can affect residential thermal environmental quality in tropical cities.

Vegetation is an active element that can absorb and reflect the solar radiation in large number. Vegetation has radiative temperature lower than other objects with same color. Reduction of vegetation can increase thermal environmental quality and affect the intensity of UHI [44].

4.4. Water Body
Water body as cooling element can affect thermal environmental quality. Temperature and humidity of the area will be declining if the area away from the water body [51] so water body affects microscale environment as evaporative cooling [28]. The scattered water affects thermal environmental quality better than the centralized water body [52] with water body orientation to prevailing wind [53].

Water body has a major impact on the thermal environmental quality in tropical cities because water body can reduce air temperature in the daytime and raise air temperature in the nighttime. The researcher rarely do research in this field whereas the influence of the water body in residential area in the tropical cities different from other areas and have specific characteristics.

5. Discussion
The development of residential area depends on housing need and the availability of the land. In the developing country, the development of residential area in the center of the city are dominated vertical-horizontal growth and the development of residential area in the edge of the city is dominated horizontal growth [32]. The growth of residential causes the changing land cover especially the change of open space become building area in the form of horizontal and vertical building. The change of land cover affects wind movement and surface area that exposed solar radiation. The horizontal building has the horizontal surface more dominant than the vertical surface so the absorption of solar irradiation is dominated by the horizontal surface. This condition cause thermal environmental quality in the residential area is influenced by the form and size of the building.

The growth of residential in the center of the city causes all areas used as a built-up area with the development of the area is dominated by pavement and building [39] so that reduce the space to vegetation and water body.

The development of horizontal residential in the center of the city is dominated by pavement and horizontal building so that this area becomes horizontal dense area [32]. The building was built near each other with small space for vegetation and water body. This condition causes this area be a dense area so that affects the wind movement in the area. Thermal environmental quality is influenced by small wind movement, pavement and building material (especially material roof).

The development of vertical residential in the center of the city is dominated by pavement, street, vertical building, vegetation and water body. The building was built vertically so that the area still has an empty land as the area of vegetation and water body. The distance between the vertical building causes wind movement can occur in the area although influenced by the height of the building. The vertical building causes the wall of building be the important elements that affect the thermal environmental quality because the wall of building affects the wind movement and have large surface if compared with other materials.

The development of horizontal residential on the edge of the city is dominated by pavement, street, horizontal building, vegetation and water body. The building can be built with a certain distance and the area can have space for vegetation and water body so that this area becomes horizontal rare area [32]. The distance between the building causes significant wind movement has occurred in the area. Thermal environmental quality is influenced by wind movement and material of pavement, street, building, vegetation and water body.

Thermal environmental quality is influenced by the ability of element material as the heating and cooling element. In the tropical cities, sunlight shine surface area every day throughout the year so that the element of this material will always affect the thermal environmental quality. Thermal environmental quality can be controlled by balancing the use of heating and cooling element so that the change of LULC does not affect the thermal environmental quality significantly. Horizontal residential in the center of the city can use light-colored pavement, roof material with low emissivity and green roof. Vertical residential in the center of the city can use light-colored pavement, wall and roof material with low emissivity, green wall (green envelope) and green roof. Horizontal residential in the edge of the city can use light-colored pavement with low emissivity, open space with vegetation, park with vegetation and water body as cooling element.

Table 2. The factor that affects thermal environmental quality in the city

Location	Form of Building	Wind Movement	Factor								
			Pave ment	Street	Building		Vegetation				Water Body
					Wall	Roof	Wall	Roof	Open Space	Park, etc	Pond, etc
In the center of the city	Horizontal	-	-	v	-	v	-	v	-	-	-
	Vertical	v	v	v	v	v	v	v	v	v	v
In the edge of the city	Horizontal	v	v	v	-	v	v	v	v	v	v

v = affect

In the tropical cities that have wetland area, the change of LULC should not change the characteristics of the area especially in the center of the city. Development of residential areas is not done with landfill because it will interfere ecology and increase surface and air temperature. Wetland area has the potential cooling element that can help to control the thermal environmental quality. The residential areas in the wetland have water body as cooling element to decrease the mean minimum temperature. The residential areas with the cooling element are the best condition because the cooling element can control the minimum cooling degree and maximum heating degrees [54].

The residential areas in wetland have UHI and UCI element that can affect the air temperature. This element can be used to balance the daytime and nighttime temperature. The temperature balance is done by using the heating from the elements that produce heat gains and the cooling from the elements that affect the humidity. The heating from the elements that produce heat gains is stored and then used to raise the nighttime air temperature while cooling element is used to provide the humidity of the daytime temperature. The existence of UCI element in the wetland in the tropical cities has direct effect significantly, the sun shines longer than another characteristic city so that affect evapotranspiration process. The evapotranspiration process from UCI element and the characteristic wind in the tropical cities (speed and direction wind) can provide maximum humidity for the surrounding.

6. Conclusions

The growth of the residential area in tropical cities consists of horizontal and vertical building in the center of city and in the edge of the city. The growth of the residential gives significant influence to urban heat island intensity that consists of several factors such as urban geometry, construction materials and urban texture. Development of residential areas in tropical cities affects thermal environmental and microclimate directly because of the changes LULC affect heating and cooling process during daytime and nighttime. The material element that affects thermal environmental quality is pavement, street, building, vegetation and water body.

The research of wetland residential in tropical cities still rarely so that this research can be done with the aim of the research to find out the influence of the water body against thermal environmental quality. Research can be emphasized on the horizontal research building in the center of the city. The existence of the wetland element in residential area in tropical cities can be used as UCI element. UCI element provides a possibility of balancing the day and night air temperature in a different way with air temperature in the dry land. The water body can influence humidity and store heat gain so that gives the cooling effect during the daytime and heating effect air temperature in the night time.

Acknowledgments

I wish to thank the Ministry of Research Technology and Higher Education Republic of Indonesia, Sriwijaya University and Bandung Institute of Technology for the opportunity to follow the education at Doctorate Architecture Bandung Institute of Technology. This publication is supported by the Dissertation Research Scheme from Faculty of Engineering, Sriwijaya University and the Research Scheme of Desentralisasi ITB PUPT Ristekdikti Indonesia 2017, contract No. 009/SP2H/LT/DRPM/IV/ 2017.

References

[1] S. L. Huang, S. H. Wang, and W. W. Budd, "Sprawl in Taipei's peri-urban zone: Responses to spatial planning and implications for adapting global environmental change," *Landsc. Urban Plan.*, vol. 90, no. 1–2, pp. 20–32, 2009.

[2] R. K. Herz, "Considering climatic factors for urban land use planning in the Sahelian zone," *Energy Build.*, vol. 11, no. 1–3, pp. 91–101, Mar. 1988.

[3] M. J. Antos, G. C. Ehmke, C. L. Tzaros, and M. A. Weston, "Unauthorised human use of an urban coastal wetland sanctuary: Current and future patterns," *Landsc. Urban Plan.*, vol. 80, no. 1–2, pp. 173–183, 2007.

[4] S. Bonafoni, G. Baldinelli, and P. Verducci, "Sustainable strategies for smart cities: Analysis of the town development effect on surface urban heat island through remote sensing methodologies," *Sustain. Cities Soc.*, 2016.

[5] H. Taha, "Urban climates and heat islands: albedo, evapotranspiration, and anthropogenic heat," *Energy Build.*, vol. 25, no. 2, pp. 99–103, 1997.

[6] E. L. Krüger, F. O. Minella, and F. Rasia, "Impact of urban geometry on outdoor thermal comfort and air quality from field measurements in Curitiba, Brazil," *Build. Environ.*, vol. 46, no. 3, pp. 621–634, Mar. 2011.
[7] X. Yang and Y. Li, "The impact of building density and building height heterogeneity on average urban albedo and street surface temperature," *Build. Environ.*, vol. 90, pp. 146–156, 2015.
[8] A. Chatzidimitriou and S. Yannas, "Microclimate development in open urban spaces: The influence of form and materials," *Energy Build.*, vol. 108, pp. 156–174, Dec. 2015.
[9] E. Jamei, P. Rajagopalan, M. Seyedmahmoudian, and Y. Jamei, "Review on the impact of urban geometry and pedestrian level greening on outdoor thermal comfort," *Renew. Sustain. Energy Rev.*, vol. 54, pp. 1002–1017, 2016.
[10] M. Ignatius, N. H. Wong, and S. K. Jusuf, "Urban microclimate analysis with consideration of local ambient temperature, external heat gain, urban ventilation, and outdoor thermal comfort in the tropics," *Sustain. Cities Soc.*, vol. 19, pp. 121–135, Dec. 2015.
[11] E. Andreou, "Thermal comfort in outdoor spaces and urban canyon microclimate," *Renew. Energy*, vol. 55, pp. 182–188, 2013.
[12] T. R. Oke, "Street design and urban canopy layer climate," *Energy Build.*, vol. 11, no. 1–3, pp. 103–113, Mar. 1988.
[13] M. Shahrestani, R. Yao, Z. Luo, E. Turkbeyler, and H. Davies, "A field study of urban microclimates in London," *Renew. Energy*, vol. 73, pp. 3–9, Jan. 2015.
[14] A. Dimoudi, A. Kantzioura, S. Zoras, C. Pallas, and P. Kosmopoulos, "Investigation of urban microclimate parameters in an urban center," *Energy Build.*, vol. 64, pp. 1–9, Sep. 2013.
[15] A. Cao, Q. Li, and Q. Meng, "Effects of Orientation of Urban Roads on the Local Thermal Environment in Guangzhou City," *Procedia Eng.*, vol. 121, pp. 2075–2082, 2015.
[16] M. H. Elnabawi, N. Hamza, and S. Dudek, "Numerical modelling evaluation for the microclimate of an outdoor urban form in Cairo, Egypt," *HBRC J.*, vol. 11, no. 2, pp. 246–251, Aug. 2015.
[17] R. Sun and L. Chen, "How can urban water bodies be designed for climate adaptation?," *Landsc. Urban Plan.*, vol. 105, no. 1–2, pp. 27–33, Mar. 2012.
[18] K. Raghavan, V. Mandla, and S. Franco, "Influence of urban areas on environment: Special reference to building materials and temperature anomalies using geospatial technology," *Sustain. Cities Soc.*, vol. 19, pp. 349–358, Dec. 2015.
[19] M. Mizuno, Y. Nakamura, H. Murakami, and S. Yamamoto, "Effects of land use on urban horizontal atmospheric temperature distributions," *Energy Build.*, vol. 15, no. 1, pp. 165–176, Jan. 1990.
[20] A. H. Rosenfeld, H. Akbari, S. Bretz, B. L. Fishman, D. M. Kurn, D. Sailor, and H. Taha, "Mitigation of urban heat islands: materials, utility programs, updates," *Energy Build.*, vol. 22, no. 3, pp. 255–265, Aug. 1995.
[21] A. Chudnovsky, E. Ben-Dor, and H. Saaroni, "Diurnal thermal behavior of selected urban objects using remote sensing measurements," *Energy Build.*, vol. 36, no. 11, pp. 1063–1074, Nov. 2004.
[22] L. Huang, D. Zhao, J. Wang, J. Zhu, and J. Li, "Scale impacts of land cover and vegetation corridors on urban thermal behavior in Nanjing, China," *Theor. Appl. Climatol.*, vol. 94, no. 3–4, pp. 241–257, 2008.
[23] J. Yang, Z.-H. Wang, and K. E. Kaloush, "Environmental impacts of reflective materials: Is high albedo a 'silver bullet' for mitigating urban heat island?," *Renew. Sustain. Energy Rev.*, vol. 47, pp. 830–843, 2015.
[24] R. Sadat, K. Binti, and N. Ujang, "Effect of pavement materials on surface temperatures in tropical environment," *Sustain. Cities Soc.*, vol. 22, pp. 94–103, 2016.
[25] X. Picot, "Thermal comfort in urban spaces: impact of vegetation growth: Case study: Piazza della Scienza, Milan, Italy," *Energy Build.*, vol. 36, no. 4, pp. 329–334, Apr. 2004.

[26] B. Hong and B. Lin, "Numerical studies of the outdoor wind environment and thermal comfort at pedestrian level in housing blocks with different building layout patterns and trees arrangement," *Renew. Energy*, vol. 73, pp. 18–27, 2015.
[27] R. Yao, Q. Luo, Z. Luo, L. Jiang, and Y. Yang, "An integrated study of urban microclimates in Chongqing, China: Historical weather data, transverse measurement and numerical simulation," *Sustain. Cities Soc.*, vol. 14, pp. 187–199, Feb. 2015.
[28] Y. Tominaga, Y. Sato, and S. Sadohara, "CFD simulations of the effect of evaporative cooling from water bodies in a micro-scale urban environment: Validation and application studies," *Sustain. Cities Soc.*, vol. 19, pp. 259–270, Dec. 2015.
[29] M. Yokohari, R. D. Brown, Y. Kato, and S. Yamamoto, "The cooling effect of paddy fields on summertime air temperature in residential Tokyo, Japan," *Landsc. Urban Plan.*, vol. 53, no. 1–4, pp. 17–27, 2001.
[30] Y. C. Chen, C. H. Tan, C. Wei, and Z. W. Su, "Cooling effect of rivers on metropolitan Taipei using remote sensing," *Int. J. Environ. Res. Public Health*, vol. 11, no. 2, pp. 1195–1210, 2014.
[31] M. W. Yahia and E. Johansson, "Influence of urban planning regulations on the microclimate in a hot dry climate: The example of Damascus, Syria," *J. Hous. Built Environ.*, vol. 28, no. 1, pp. 51–65, 2013.
[32] S. Wonorahardjo, "New Concepts in Districts Planning, Based on Heat Island Investigation," *Procedia - Soc. Behav. Sci.*, vol. 36, no. June 2011, pp. 235–242, 2012.
[33] H. Yan and L. Dong, "The impacts of land cover types on urban outdoor thermal environment: the case of Beijing , China," *Environ. Heal.*, vol. 13, no. 43, pp. 1–7, 2015.
[34] M.-T. Hoelscher, T. Nehls, B. Jänicke, and G. Wessolek, "Quantifying cooling effects of facade greening: Shading, transpiration and insulation," *Energy Build.*, 2015.
[35] R. Djedjig, E. Bozonnet, and R. Belarbi, "Analysis of thermal effects of vegetated envelopes: Integration of a validated model in a building energy simulation program," *Energy Build.*, vol. 86, pp. 93–103, Jan. 2015.
[36] P. La Roche and U. Berardi, "Comfort and energy savings with active green roofs," *Energy Build.*, vol. 82, pp. 492–504, Oct. 2014.
[37] M. Zinzi and S. Agnoli, "Cool and green roofs. An energy and comfort comparison between passive cooling and mitigation urban heat island techniques for residential buildings in the Mediterranean region," *Energy Build.*, vol. 55, pp. 66–76, Dec. 2012.
[38] J. Nichol and M. S. Wong, "Modeling urban environmental quality in a tropical city," *Landsc. Urban Plan.*, vol. 73, no. 1, pp. 49–58, 2005.
[39] T. Sharmin, K. Steemers, and A. Matzarakis, "Microclimatic modelling in assessing the impact of urban geometry on urban thermal environment," *Sustain. Cities Soc.*, vol. 34, no. February, pp. 293–308, 2017.
[40] A. Ardalan, M. Mirnezhad, A. Ghaffarianhoseini, A. Ghaffarianhoseini, H. Omrany, Z. Wang, and H. Akbari, "Urban heat island mitigation strategies : A state-of-the-art review on," vol. 62, pp. 131–145, 2017.
[41] F. Xue, Z. Gou, S. Siu, and Y. Lau, "Green open space in high-dense Asian cities: Site configurations, microclimates and users' perceptions," *Sustain. Cities Soc.*, 2017.
[42] M. Roth and V. H. Lim, "Evaluation of canopy-layer air and mean radiant temperature simulations by a microclimate model over a tropical residential neighbourhood," *Build. Environ.*, vol. 112, pp. 177–189, 2017.
[43] Y. Didem, J. Stocker, D. Carruthers, and J. Hunt, "A sensitivity study relating to neighbourhood-scale fast local urban climate modelling within the built environment," *Procedia Eng.*, vol. 198, pp. 589–599, 2017.
[44] A. Ayanlade, "Seasonality in the daytime and night-time intensity of land surface temperature in a tropical city area," *Sci. Total Environ.*, vol. 557–558, pp. 415–424, 2016.
[45] X. Li and L. K. Norford, "Evaluation of cool roof and vegetations in mitigating urban heat island in a tropical city , Singapore," *URBAN Clim.*, 2016.

[46] J. L. F. R, A. J. Pereira, and H. A. Karam, "Estimation of long term low resolution surface urban heat island intensities for tropical cities using MODIS remote sensing data," *UCLIM*, vol. 17, pp. 32–66, 2016.
[47] T. Sharmin, K. Steemers, and A. Matzarakis, "Analysis of microclimatic diversity and outdoor thermal comfort perceptions in the tropical megacity Dhaka, Bangladesh," *Build. Environ.*, vol. 94, pp. 734–750, 2015.
[48] P. Rajagopalan, K. C. Lim, and E. Jamei, "Urban heat island and wind flow characteristics of a tropical city," *Sol. Energy*, vol. 107, pp. 159–170, Sep. 2014.
[49] N. H. Wong and C. Yu, "Study of green areas and urban heat island in a tropical city," vol. 29, pp. 547–558, 2005.
[50] N. Anting, M. Fadhil, K. Iwao, M. Ponraj, K. Jungan, L. Yee, A. John, and L. Meng, "Experimental evaluation of thermal performance of cool pavement material using waste tiles in tropical climate," *Energy Build.*, vol. 142, pp. 211–219, 2017.
[51] E. Johansson and R. Emmanuel, "The influence of urban design on outdoor thermal comfort in the hot, humid city of Colombo, Sri Lanka," *Int. J. Biometeorol.*, vol. 51, no. 2, pp. 119–133, 2006.
[52] H. Jin, T. Shao, and R. Zhang, "Effect of water body forms on microclimate of residential district," *Energy Procedia*, vol. 134, pp. 256–265, 2017.
[53] N. Imam, M. Ichinose, E. Kumakura, S. Kardinal, K. Chigusa, and N. Hien, "Thermal environment assessment around bodies of water in urban canyons : A scale model study," vol. 34, no. June, pp. 79–89, 2017.
[54] R. A. Memon, D. Y. C. Leung, C.-H. Liu, and M. K. H. Leung, "Urban heat island and its effect on the cooling and heating demands in urban and suburban areas of Hong Kong," *Theor. Appl. Climatol.*, vol. 103, pp. 441–450, 2010.

IOP Conf. Series: Earth and Environmental Science **152** (2018) 012035 doi:10.1088/1755-1315/152/1/012035

Investigation on Consumers' Behaviour towards Energy Saving through Utilisation of Virtual SED (Smart Energy Displays) in Residential Building

Sandhika Adlisia Puspa Harani

MSc in Sustainable Energy and Entrepreneurship,
Department of Architecture and Built Environment, University of Nottingham

sandhika.adlisia@gmail.com

Abstract. The study is conducted by gathering data from interviews an in-home experiment, to examine the impacts of both virtual and physical SED toward user engagement. Business opportunity and benefits of virtual SED for stake holders are also discussed in this study. The research was conducted by interviewing method to respondens in Nottingham, UK. By comparing consumers' energy saving behaviour from physical and virtual SED users, virtual SED shows similar level of effectiveness as physical SED, but there is no evidence that the virtual versions are better than the physical ones in terms of reducing energy consumption. Nevertheless, virtual SED can be more beneficial for consumers who can get easier access. They also help educating users to be more concern about energy issue. Energy suppliers get benefits by having virtual versions of SED, in which they can reduce production and distribution costs, as well as diminishing waste from physical SED.

1. Introduction

The UK government has officially stated a framework contained the target to reduce 80% of the greenhouse gases (GHG) emissions by 2050 in every sector, including residential sector. In 2014, the residential sector accounted for 12% of the GHG emissions, which made it the fourth-highest GHG contributor to the environment [1], as shown in Figure 1.

Carbon dioxide was recorded to be the dominant substance in the emissions. In particular, heating played a significant role in the energy consumption, which reached 85% of the total consumption for this sector, including space heating, water heating, and cooking [2].

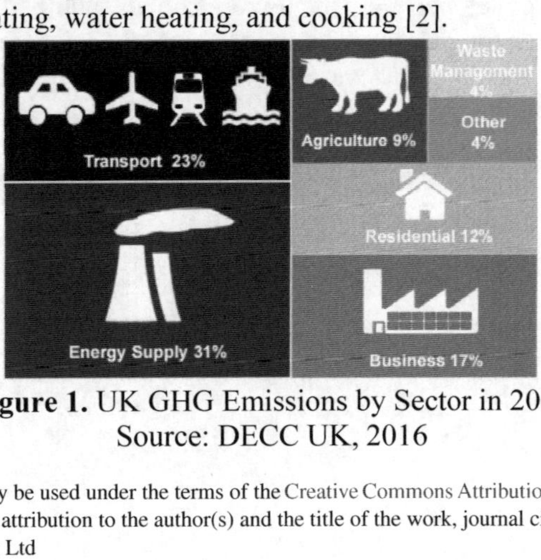

Figure 1. UK GHG Emissions by Sector in 2014
Source: DECC UK, 2016

One reason why residential sector consumed high amount of energy was because of the intangibility of energy itself. The form of energy, which was abstract, made people hard to visualise how much energy they spent every time. In this case, smart meters played role as devices that can record energy consumption regularly.

Smart meters are developed versions from conventional gas and electricity meters. The 'smart' devices collect consumption data and transmit the data to energy suppliers and home owners in near real-time. This makes manual recording no longer required. Besides providing energy consumption data, smart meters also give more accurate energy bills for consumers based on actual usage rather than estimated use [3].

The UK government has started 53 million of smart meter roll-out programme for residential and small businesses that should be completed by 2020 [2]. The programme is conducted through major energy suppliers in the UK, such as E.ON, British Gas, and SSE. Consumers are welcome to get free smart meter installation from the energy suppliers.

By having smart meters, home owners can monitor their energy consumption and improve their consumption behaviour. The improvement may result in annual savings, especially in electricity and gas, that is predicted to reach GBP 24 by 2020 per household and GBP 39 in 2030 [2].

Each smart meter will be accompanied by an in-home display, a device that allows users to monitor and access the information sent by the smart meter with ease. In-home display will also help users to visualise their energy consumption, so that they can change their paradigm about energy into a more dynamic, transparent, and controlled matter [4][5] .

Smart Energy Displays (SED) are one of physical in-home displays developed by E.ON for its customers. Consumers can monitor their energy consumption in currency (Poundsterling), kWh, or the amount of CO2 emission through the device.

Unfortunately, SED will be 'abandoned' gradually along with the users' routine activity due to lack of motivations in accessing the devices. This issue triggered further development of SED to be more accessible, attractive, and interactive in the form of virtual versions, such as mobile application and website. This innovation has great potential to be implemented worldwide, including in Indonesia, if virtual SED can be used effectively.

The study is conducted to assess the effectiveness of virtual SED, the benefits for E.ON as energy supplier, for consumers, for the UK government, and the impact of virtual SED on people's energy conservation behaviour, and to compare this with the effectiveness of physical SED. Potential of long term engagement between virtual SED and users are assessed as well..

2. Literature Review

2.1. Review on SED Technology and the Application

Currently, both physical and virtual in-home displays/SED products are widely available in the market. Some of these are developed by private companies, such as Owl, Current Cost, and Efergy, while others are distributed through energy supplier companies, which have become major players in the utility sector. Figure 2 shows eight examples of in-home displays products that are used in this study.

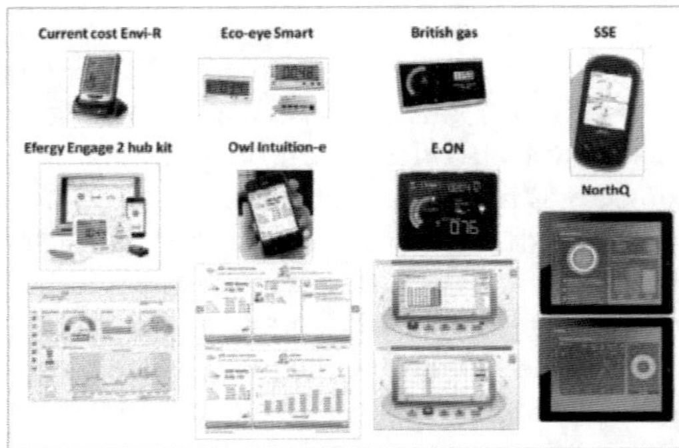

Figure 2. Examples of In-Home Displays Products

Four of the products come in physical devices, while the rest of them are complemented with virtual versions. Each product offers diverse sets of features.

Based on findings from Nguyen [6] and Dr. Lewis et al.[7], it has been proven that SED can reduce energy consumption from 5% to 15%, with an average reduction of 9%, within Europe and Great Britain. The savings are highly influenced by what is featured in SED. In Great Britain itself, the minimum typical saving for households is £147 over three years. It has been determined that having additional features meeting consumers' need increases the likelihood of reducing their energy bills.

2.2. *Related Research*

Several studies have found that there are four main elements to encourage users to use mobile applications or websites as virtual SED; namely, eco-visualisation, comparison, gamification, and social media. Intangible information such as energy consumption can be understood through eco-visualisation, which uses animations to illustrate levels of consumption, as non-numeric information is more likely to captivate users [8][9]. Meanwhile, the ability to compare levels of consumption and savings with those of other users can encourage further energy reduction [10][9]. Such comparison can provide a benchmark for users' success in decreasing consumption [11]. Users' consumption and savings can also be compared against those of similar household types, in order to obtain more validation [12]. Gamification, a term used for integrating game elements into a non-game aspect, combined with social media, can allow users to collaborate and socialise in order to accomplish their goals [13][14].

3. Methodology

In-home experiment for 10 days was conducted toward participants that had been selected randomly by installing a prototype of virtual SED, called NorthQ Home Manager starter kit, onto their electricity meter. The kit consisted of a Q-stick as the gateway and a Q-Power that received and transmitted information about energy consumption. The kit works when it is installed and connected to a network as illustrated in Figure 3. The data collected from the device could be accessed through a mobile application and website.

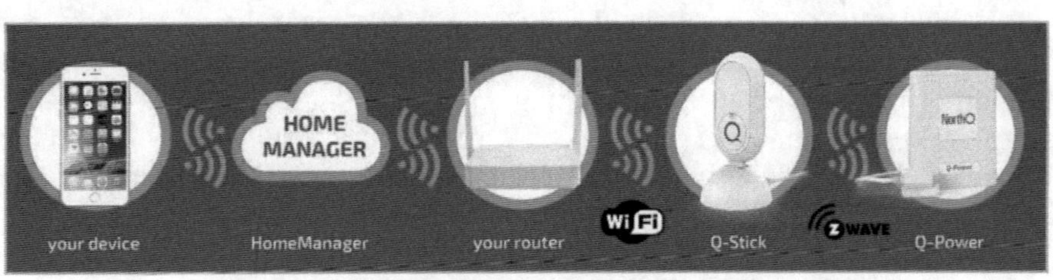

Figure 3. How NorthQ Electricity Starter Kit Works
Source: NorthQ Home Manager, 2016

IOP Conf. Series: Earth and Environmental Science **152** (2018) 012035 doi:10.1088/1755-1315/152/1/012035

The experiment was held to examine the amount of their energy consumption and their savings compared to their previous consumption (without virtual SED). Furthermore, interviews with the participants were held to gather feedback about user interface, SED visualisation, and SED contents.

2 sets of groups were involved in data collection through an in-home experiment. Group 1 was a group of small family (3 persons, age 25-34 years old and 0-5 years old) and Group 2 was consisted of students (4 persons, age 18-24 years old) that never used any form of SED before.

Several methods are used to analyse the results, those are comparative analysis as well as cause and effect analysis. Comparative analysis is used to compare the benefit, impact, and effectiveness between physical and virtual SED. Cause and effect analysis as well as SWOT analysis are conducted to get a better understanding of what works well and what needs to be improved.

4. Result and Discussion

Information regarding the impacts of mobile application usage, the intensity of interaction, and user preference are obtained from the experiment involving both groups of participants.

4.1. The Experiment of Virtual SED Towards Energy Saving Behaviour

Some findings have been found based on 10 days experiment toward the groups. They had been interviewed before and after the experiment conducted to gather information about their perception about virtual SED.

Before the test, both groups admitted that they not really paid attention how much energy they consumed. However, they argued that they had applied energy saving behaviour in daily life, such as optimising household appliances, adjusting their schedule based on energy tariff, and shutting down appliance when it was not used.

The family group felt optimistic they would be more motivated to save energy in longer term with the help of virtual SED. On the other hand, the student participants thought that virtual SED would indeed help increasing their energy awareness, but the device would not play a significant role in reducing energy consumption. The reason behind their skeptical thought was because energy, especially electricity, was primary needs and it was difficult to replace, so it would be hard to conserve the energy.

Final interview was held after 10 days of experiment. The features and performance of NorthQ Home Manager as virtual SED device were evaluated by both groups of participants. In general, both groups agreed that it took around 1-2 days to access and operate the application proficiently. The information was presented clearly through graphs and diagrams, so it could be understood easily. The visualisation was displayed quite attractive to motivate them accessing the application.

After using the virtual SED, all participants became more aware about how much energy they consumed which could be monitored regularly through its mobile application and website. However, there was no significant change towards their energy saving behaviour and their consumption reduction compared to the previous situation without SED.

The family group felt that it was extremely important to have a virtual SED device, so that they could make an effective strategy to get further savings. The same response was provided by the students group who thought that they could understand their consumption with the support of the device.

The Daily Consumption overview in the menu was the most attractive feature for the family participant. The feature showed electricity consumption overview for today, yesterday, as well as average electricity usage in kWh or Poundsterling. By looking at the information displayed in graphs, the participant could analyse what activities consumed most energy. Meanwhile, the students group prefered the compare feature as well as the historical data view. The features allowed them to monitor and compare their consumption in more detail and for specific times. Based on the comparison, they would be pleased if they managed to consume less energy compared to yesterday and would try to reduce the consumption if it appeared higher than yesterday. Those features are presented in Figure 4.

IOP Conf. Series: Earth and Environmental Science **152** (2018) 012035 doi:10.1088/1755-1315/152/1/012035

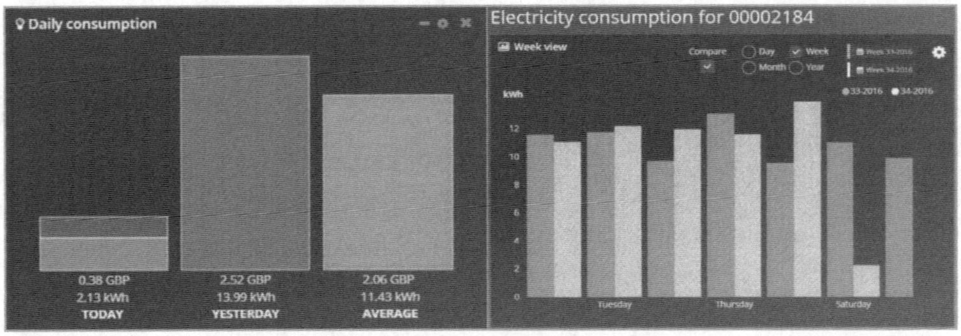

Figure 4. Daily Consumption Feature (left) and Compare Feature (right) of NorthQ Home Manager Apps

Reducing energy bills became the main reason for all the participants to use the SED. Because of this reason, they tended to see the consumption in currency rather than kWh, so they could easily imagine how much they spent on electricity. The device was able to attract all the participants to interact with it regularly by having favourable layout composition and colour schemes. Nevertheless, the students group felt that it took users' initiative to check the application and it would be good if there was an alert feature to notify users. This issue would affect participants' willingness to use the application in long term. If there was no update and improvement of the product, they might stop using the device gradually when their objectives were achieved and they managed to reduce a considerable amount of energy. Moreover, all participants agreed that virtual SED would be more beneficial if it provides additional information about energy consumed by household appliances. They also thought that physical SED would be useful to complement mobile application or website when their cell phones were in used.

5. Business Opportunity

According to a survey [15], 56% of mobile leaders testified that it takes at least 7 months to develop an application, while 18% of them said that it costs approximately $500,000 to more than $1,000,000 per application as shown in Figure 5.

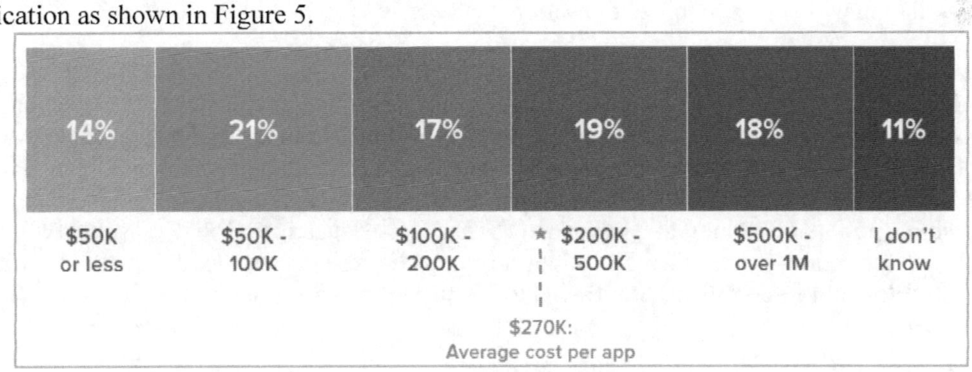

Figure 5. Costs of a Mobile Application Development (Formotus, 2016)

Since smart meters should be installed in around 53 million homes and small businesses by 2020 and the market share of E.ON is 14%, it is estimated that the company should deploy smart meters to around 8.15 million customers.

If it is assumed that a moderate physical SED will have £35 of production cost, E.ON should spend approximately £285 million to produce similar physical SED. Compared to an application, its development costs an average of $270,000 or around £206,000 for mid-range applications. This will be huge savings for E.ON as a company.

Not only it can give financial savings, virtual SED have more benefits. For example, the ease of products distribution, less production of waste, large market potential, and the growing trend of smart home technology will support the product development.

On the other hand, energy suppliers must be responsive in addressing the shortcomings of existing products and problems that may arise in the future, such as the accuracy of products in displaying information and competition that will arise with companies engaged in smart home technology.

Wide market segmentation may become an obstacle, in which virtual SED should be easily accessed and accepted by society from various background

6. Conclusion

Currently, most of the energy suppliers in the UK have developed their SED products to comply with the government target, which is to finish the installation of over 53 million smart meters and SED in the UK by 2020. Not only energy suppliers, but several other companies engaged in the business of smart home technology have manufactured their products that makes the competition tougher.

Moreover, the motivation of using an SED is decreased after a certain amount of time due to the lack of user interest. However, more attractive visualisation, a favourable user interface, comparison feature, and additional features increase the engagement on SED utilisation. This will lead to higher energy savings.

In addition, in terms of the energy awareness and energy consumption reduction, virtual SED may be as effective as physical SED, but there is no evidence shown that virtual SED is better compared with the physical versions. Overall, every respondent, agree that their energy awareness and energy saving behaviour may be improved by utilising the device. Some participants argued that the improvement was mainly influenced by the real-time energy overview feature, which can assist them to understand their energy usage.

Despite that, virtual SED offer more benefits, such as ease of access and more thorough information can be presented in a single view, as well as the virtual versions may eliminate the waste that comes from physical SED. Furthermore, virtual SED development will also benefit companies from savings in production and distribution costs. The cost can be reduced further by developing an application, rather than manufacturing the physical products. For the government, smart meters and SED installation will help to stabilize the national energy security.

7. Acknowledgement

All praise to Allah, the Almighty, on whom I seek guidance to live and to finish this research.

I would like to express my gratitude to Professor Mark Gillot, Professor Simon Mosey, Dr. Yuehong Su for their supervision. Also to thank Mr. Simon Church and Ms. Sian Green, The E.ON's Connected Homes Research Team. Furthermore, I would like to thank Indonesian Education Scholarship from the Indonesian Endowment Fund for Education (LPDP), Ministry of Finance Republic of Indonesia, for the financial support during the research in the University of Nottingham, UK. Lastly, I would like to thank Mr. Dewanto Harjunowibowo for supporting me in publishing this paper.

References

[1] Department of Energy and Climate Change (DECC) UK, Estimates of Heat Use in the United Kingdom in 2013, Department of Energy and Climate Change UK: London, 2014.
Department of Energy and Climate Change (DECC) UK, UK National Energy Efficiency Action Plan, London: Department of Energy and Climate Change UK, 2014.
[2] Department of Energy and Climate Change (DECC) UK, 2014 UK Greenhouse Gas Emissions, Final Figures, London: Department of Energy and Climate Change UK, 2016.

[3] Department for Business, Energy & Industrial Strategy, "Smart Meters: A Guide," 8 October 2013. [Online]. Available: https://www.gov.uk/guidance/smart-meters-how-they-work. [Accessed 31 July 2016].

[4] T. Hargreaves, M. Nye and J. Burgess, "Keeping energy visible? Exploring how householders interact with feedback from smart energy monitors in long term," Energy Policy, vol. 52, pp. 126-134, 2013.

[5] Faruqui, S. Sergici and A. Sharif, "The impact of informational feedback on energy consumption—A survey of the experimental evidence," Energy, vol. 35, p. 1598–1608, 2010.

[6] S. P. Nguyen, "Mobile Application for Household Energy Consumption Feedback Using Smart Meters: Increasing Energy Awareness, Encouraging Energy Savings and Avoiding Energy Peaks," in International Conference on Collaboration Technologies and Systems, Minneapolis, 2014.

[7] P. Lewis, A. Bogacka, R. Grigoriou and S. Xu, "Assessing the Use and Value of Energy Monitors in Great Britain," VaasaETT, Helsinki, 2014

[8] T. Kim, H. Hong, B. Magerko and Coralog, "Use-aware visualisation connecting human micro-activities to environmental change," in In CHI: Extended Abstracts on Human Factors in Computing Systems, Boston, 2009.

[9] T. Holmes, "Dialogical Encounters: Art and Energy Awareness via Eco-Visualisation.," ACM Magazine for Students - Green Technologies: From Pollution to Pixels, vol. 17, no. 4, pp. 52-55, 2011.

[10] J. Simon, M. Jahn and A. Al-Akkad, "Saving Energy at Work: The Design of a Pervasive Game for Office Spaces," in 11th International Conference on Mobile and Ubiquitous Multimedia (MUM), Ulm, 2012.

[11] P. Petkov, M. Foth, F. Kobler and H. Krcmar, "Motivating Domestic Energy Conservation through Comparative, Community-Based Feedback in Mobile and Social Media," in 5th International Conference on Communities & Technologies, Brisbane, 2011.

[12] J. Kjeldskov, M. B. Skov, J. Paay and R. Pathmanathan, "Using Mobile Phones to Support Sustainability: A Field Study of Residential Electricity Consumption," in CHI, Austin, 2012.

[13] A. Gustaffson, M. Bang and M. Svahn, "Power Explorer - A Casual Game Style for Encouraging Long Term Behaviour Change among Teenagers," in International Conference on Advances in Computer Entertainment Technology, New York, 2004.

[14] L. e. a. Gamberini, "Saving is Fun: Designing a Persuasive Game for Power Conservation," in International Conference on Advances in Computer Entertainment Technology, Lisbon, 2011.

[15] Formotus, "Figuring the Costs of Custom Mobile Business App Development," Formotus, 2016. [Online]. Available: http://www.formotus.com/14018/blog-mobility/figuring-the-costs-of-custom-mobile-business-app-development#prettyPhoto. [Accessed 1 August 2016].

IOP Conf. Series: Earth and Environmental Science **152** (2018) 012036 doi:10.1088/1755-1315/152/1/012036

The Study of Thermal Comfort in Transforming Residential Area in Bandung using ENVI-met Software. Case Study: Progo Street

Tubagus M Aziz SOELAIMAN[1*], Woerjantari K SOEDARSONO[1], and M Donny KOERNIAWAN[1]

[1]Architecture Study Programme, Institute Technology of Bandung, Indonesia

*aziz.soelaiman@gmail.com

Abstract. Bandung has a high potential in attracting tourists. This potential impact on building function near tourist attraction that can transform residential uses into commercial uses. Progo Street and its surrounding area used as the case study, which is close to Gedung Sate and Riau Street as tourist destinations in Bandung. Moreover, this transformation is also reinforced by the spatial planning policies in Bandung, known as RTRW and RDTR, said that this area will be fully non-residential area. This condition in some cases could affect thermal comfort. This paper provides the changes of thermal comfort phenomenon that occurs using EnviMet software. The study compares Predicted Mean Voted (PMV) as thermal comfort indicator between existing and Bandung detailed spatial plan (RDTR) condition. The result shows that the PMV value of current condition is higher than future planning, nonetheless the planned area will be changed into higher non-residential buildings and less greeneries. Some environmental factors that are used to calculate PMV such as air temperature, mean radiant temperature, humidity, and wind speed are also examined to find out what makes the plan more comfortable than the existing. Simulations using ENVI-met software could be considered in making more objective planning policy in the future.

1. Intruduction

Bandung city development as one of tourist destinations in Indonesia has encouraged the provision of additional facilities to accommodate the tourists or visitors, both from other cities and from abroad. Data from the Central Bureau of Statistics (BPS) shows that there is an increase number of tourists visiting Bandung and has reached its peak in 2011 with more than 6.7 million people. This phenomenon encourages the transformation of building functions or land uses from residential uses into commercial uses in some strategic corridors. Irregular city growth could make each area within the city developes based on the economic potential that can generate the area. As a result, space for development is getting bigger and wider to accommodate the needs of the visitors. This growth will affect the microclimate of the region, for example more pavement areas for car parking in commercial use will give an increase temperature in the area. The phenomena of Urban Heat Island (UHI) is a result of city development that modify the regional atmosphere situation [1] and not considering the capacity of the environment. To overcome its negative impact, it is necessary to study the area on achieving thermal comfort, which can be affected by several factors such as air temperature (Ta), humidity, level of radiation (Tmrt), and wind speed.

This study proposes an engineering approach that discovers the level of thermal comfort in a transformed residential area in Bandung, so it can be the information base on how the spatial plan can be used to make better space.

2. Case Study

Area between Gedung Sate building as one of tourist destinations in Bandung and LL RE Martadinata Street (known as Riau Street) is used as case study area, particularly Progo Street and surrounding area.

Progo Street is used to be a residential district in the colonial era, which has similar architectural style for all buildings along the corridor – one floor villa type with green terrace and innercourt. It is the second layer street after Riau Street, which serve as primary collector road, and parallel with Cimandiri Street in the North of the case study area. These streets have the same orientation, West-East, but different land uses and road width.

Figure 1. Current Condition of Progo Street, mixed of residential plots and commercial uses along lush corridor.

Nowadays, Progo Street and its surrounding area has gradually transformed. Some of the land uses have changed from residential to commercial, building height become higher, façade design began vary, and green terraces become pavements for parking lots. Although the area partly altered into commercial uses and become more public, the infrastructure is not supported yet with pedestrian path along the streets. Moreover, the streets are more crowded with vehicles that parking along the street, since the parking space inside the lot is limited. However, this area is mostly still has lush trees that give shades and comfort.

2.1. Planned

To find out the comparison on how comfort the area of the case study, this research uses the Detailed Spatial Plan (draft) of Bandung City 2015, known as RDTR, for simulation. By simulating the spatial plan, elements or factors for consideration in making ke plan in the future can be found.

Figure 2. Spatial Plan Map of Cibeunying District with enlargmenet map of case study area, from Draft Detailed Spatial Plan (Rencana Detail Tata Ruang - RDTR) of Bandung City 2015.

In RDTR, it can be seen that the case study area will be mostly commercial uses, facilities, and offices. No more residential uses within the area.

From the spatial plan, information about Building Coverage Ratio (BCR), Floor Area Ratio (FAR), Land Uses, and Setback are taken from the document for simulation.

2.2. *Development Changes*

There are several differences between the existing condition of the study site and the projected/detailed spatial plan. The most significant change is the building configuration and intensity. Based on Draft Detailed Spatial Plan document, the buildings along Progo Street will be higher than current condition, with average about 3 floor height (based on calculation of Building Coverage Ratio and Floor Area Ratio taken from Detailed Spatial Planning of Bandung 2015). Moreover, the building coverage is maximized on the first floor without expanded over the setback.

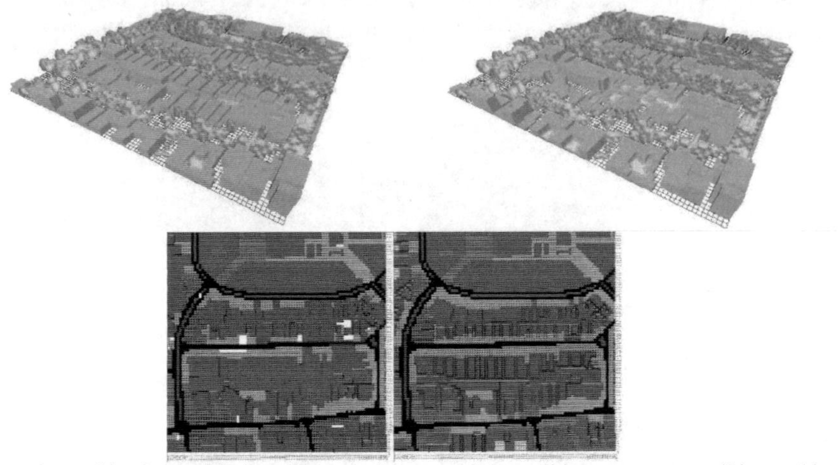

Figure 3. 3D Illustration and land material plan of case study area with 2 conditions: current condition and based on future plan (RDTR), created using EnviMET software.

Second change that can be seen from Figure 3 is the material in front of the buildings, which is changed into pavement (light grey) and concrete (dark grey). This is to accommodate the parking area in front of the commercial uses.

In this study, the trees are kept the same for both plan, the building material is also assumed to be the same since the simulation is taken in wider or urban context.

3. Thermal Comfort Assessment Using Envi-Met Software

In several research or studies, thermal comfort assessment is usually associated to ENVI-met software. One of the studies has been conducted by Kurniawan and Gao [2], where the thermal comfort of urban open spaces in tropical city (Jakarta) was assessed using this software. ENVI-met Simulation Software itself is a three dimensional computer model that can evaluates environmental conditions, especially related to thermal condition in urban context [3]. Microclimate and air quality in open spaces and urban structures will be calculated with this software, consisting of dozens of dynamic subsystems ranging from atmospheric dynamics, over soil physics, vegetation response, to building indoor climate. Among all environmental factors that analysed in this software, Predicted Mean Voted (PMV) will be the main factor used as thermal comfort index.

3.1. *PMV*

Predicted Mean Voted (PMV) can be obtained from ENVI-met simulation result and used as thermal comfort index. It is one of the most widely used thermal comfort indext nowadays. In ENVI-met software, PMV is calculated from personal factors (metabolic rate (met) and clothing insulation (clo)) and environmental factors (air temperature (Ta), radiant temperature (Tmrt), air velocity, and relative humidity).

The thermal scale for PMV based on Fanger's thermal comfort model (1972), range from Cold (-3) to Hot (3). This comfort sensation scale is also adopted as an ISO standard, particularly in ISO 7730 [4]. If the result is out of the range, the area will be considered as uncomfort zone and need to conduct further study to make the area more comfort.

Figure 4. Images of the selected area: (1) aerial satellite view picture; (2) CAD map of building footprints including trees and their type of height; (3) Input file images for EnviMet simulation, based on the aerial satellite view pictures. Source: Google Maps, GIS Map, and EnviMet

4. METHODOLOGY

As shown in Figure 4, the selected area in Progo Street and its surroundings is traced from aerial satellite view (Google maps) and defined through CAD map. The map then digitized on ENVI-met software with 90 x 90 grids of 5 m resolution, so the model represented 450 x 450 m2 of the selected area for study. The buildings are assumed have the same material but different heights, in accordance with the existing and future plan. Meanwhile the ground surface is mostly covered by asphalt road, except the area in front of the buildings within the plots. The trees are also mapped with various heights and diameters.

The simulation conducted on June 21st, one of the typical hottest day in Indonesia, between 1am and 12am (24 hours).

As can be seen in Table 1, the initial condition of simulation into 4 parts: basic settings, meteorological settings, soil temperature, and PMV settings. The basic settings contain information regarding the time and location taken for simulation, while meteorological basic settings represent the real condition of the existing location.

To calculate the PMV model, or known as biomet model in ENVI-met, some information about personal parameters should be set first, as the energy balance of human body is related to the personal comfort to surrounding environment.

Table 1. The Initial Condition of Simulation

Basic Settings	
Boundary Model	450x450 m
Grids	90x90
Simulation Date (DD.MM.YYYY)	21.06.2017
Start Simulation at Time (HH:MM:SS)	06:00:00
Total Simulation Time in Hours	23
Save Model State each 60 min	60

[Meteorology] Basic Settings	
Wind Speed in 10 m ab. Ground [m/s]	3.12
Wind Direction (0=N,.90=E,.180=S,.270=W)	200
Roughness Length z0 at Reference Point	0.01

IOP Conf. Series: Earth and Environmental Science **152** (2018) 012036 doi:10.1088/1755-1315/152/1/012036

Initial Temperature Atmosphere [^{0}C]	27
Specific Humidity in 2500 m [g Water/kg air]	7
Relative Humidity in 2m [%]	84

[SOILDATA] Settings for Soil	
Initial Temperature Upper Layer (0-20cm) [K]	293
Initial Temperature Middle Layer (20-50cm) [K]	293
Initial Temperature Deep Layer (50-200cm) [K]	293
Relative Humidity Upper Layer (0-20cm) [%]	50
Relative Humidity Middle Layer (20-50cm) [%]	60
Relative Humidity Deep Layer (50-200cm) [%]	60

[PMV] Settings for PMV-Calculation	
Weight [kg]	60
Height [m]	1.75
Surface Area [m^2}	1.74
Walking Speed [m/s]	0.9
Static Clothing Insulation	0.75
Total Metabolic rate [W]	156.24

5. Results

As the main factor for thermal comfort index that considered in the study, the distribution map of PMV values for both existing condition and based on spatial plan are presented in Figure 5. It can be seen that at first glance there is no significant changes, even though the building layout considerably changed. To some points, the decline of PMV value occurred from above 4 to lower than 4. To be more exact, Figure 6 presents the comparison of PMV value composition between two conditions. It shows that in general the future plan has lower PMV index (more comfortable) than current condition. Nevertheless, overall result in both conditions are not in an ideal comfort zone as most of the area have value over than 3 – more than ideal thermal comfort range.

To find out the cause of the lower value, the environmental factors of thermal comfort have been analyzed through the same simulation software. In Figure 7, it shows the air temperature of current condition in several area are higher than in Detailed Spatial Plan, particularly on Progo Street and Riau Street, with 1-20 C difference. This trend is also proved in Figure 8, where air temperature in existing condition mostly higher than in Detailed Spatial Plan in a day (23 hours).

Figure 9 shows the simulation result of mean radiant temperature (Tmrt). In general, there is no significant changes of the Tmrt, but the temperature range in Detailed Spatial Plan (36.1-66.6oC) is lower than in current condition (37.3-67oC). More clearly can be seen in Figure 10, the radiant temperature level between both conditions has a gap during the day (10am-2pm).

Meanwhile from Figure 11 and Figure 12, relative humidity in Progo Street and its surroundings generally decreased. During the day, the humidity for both conditions are comparatively the same, but at night the future plan has less humid compared to the existing condition.

Wind speed as the last environmental factor also simulated with the result in Figure 13. Similar trend with the other factors, wind speed in Progo Street and its surroundings will be decreased in the future, with 1 m/s in average. The gap is clearly wide, although the future plan has higher buildings than now.

Although the future plan has higher buildings, more building area, and less green surfaces, they do not affect the thermal comfort. However, the future plan is still need to be reviewed since the overall PMV value is still over the thermal comfort standard.

IOP Conf. Series: Earth and Environmental Science **152** (2018) 012036 doi:10.1088/1755-1315/152/1/012036

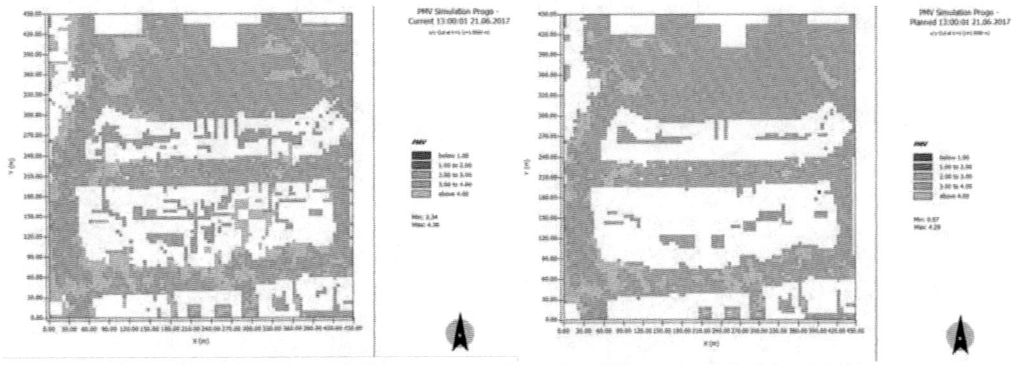

Figure 5. PMV Simulation for current condition condition and Detailed Spatial Plan

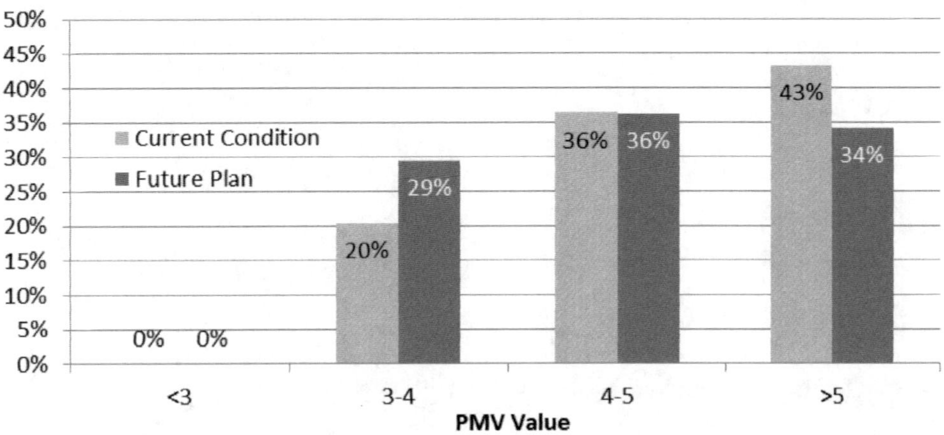

Figure 6. Comparison of PMV Value on Progo Street and surrounding Area (at 1pm)

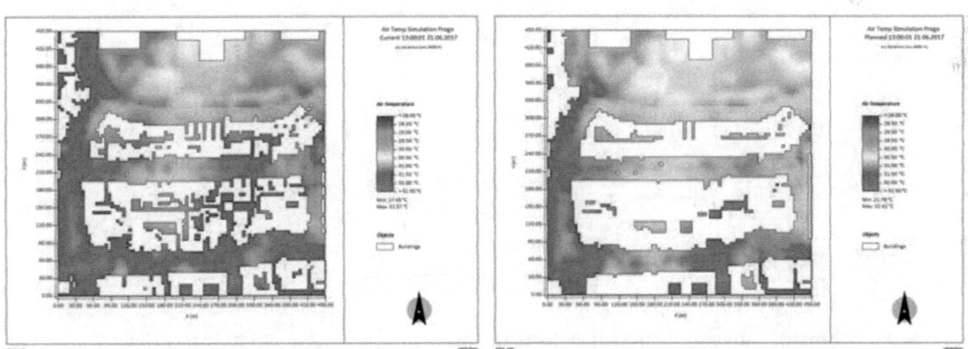

Figure 7. Air Temperature simulation result for current condition and Detailed Spatial Plan

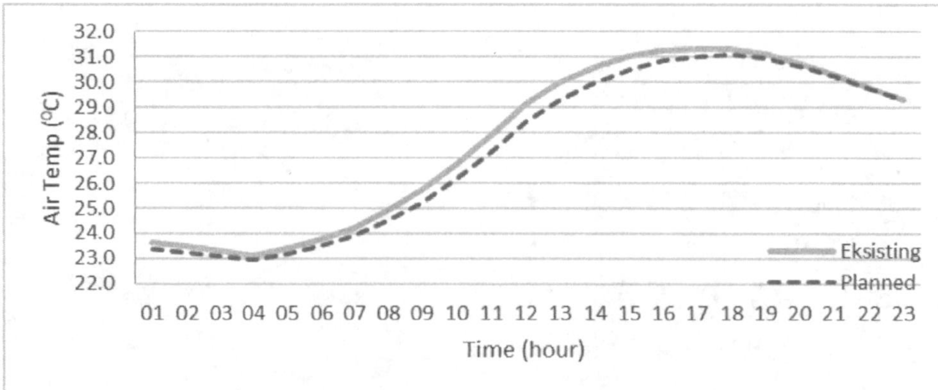

Figure 8. Diurnal air temperature change at human height level (1.5m) on a typical hottest day, from 1 am to 11 pm on June 21st

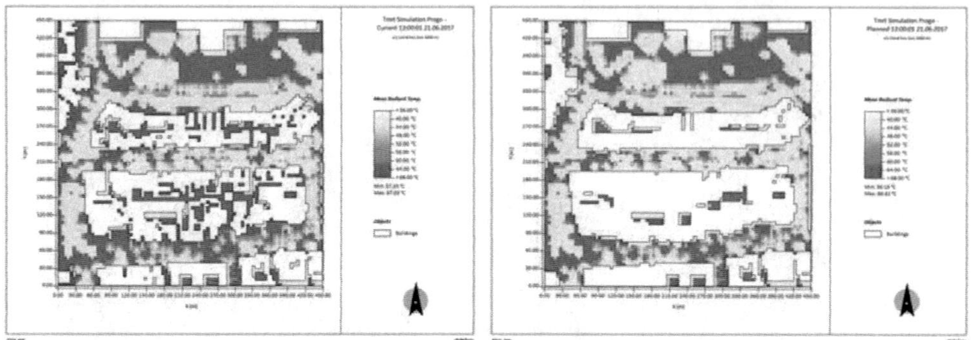

Figure 9. Mean Radiant Temperature (Tmrt) simulation based for current condition

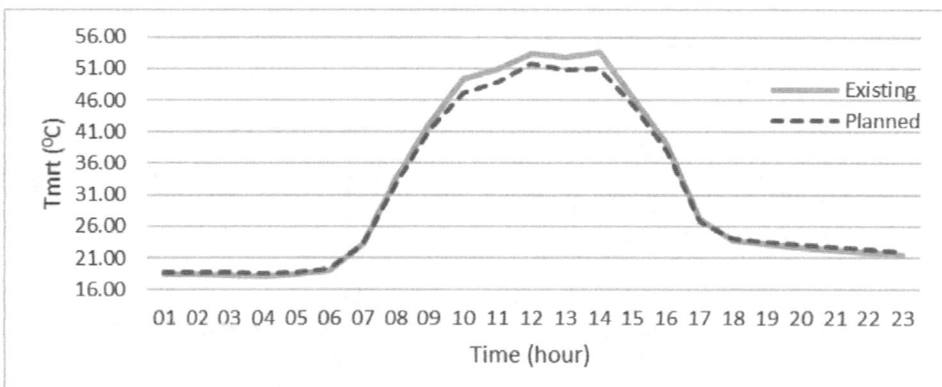

Figure 10. Diurnal mean radiant temperature change, from 1 am to 11 pm on June 21st

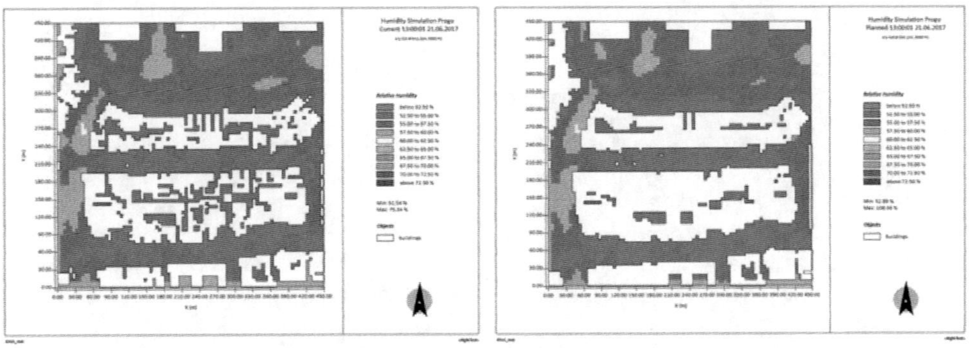

Figure 11. Relative Humidity simulation for current condition and Detailed Spatial Plan

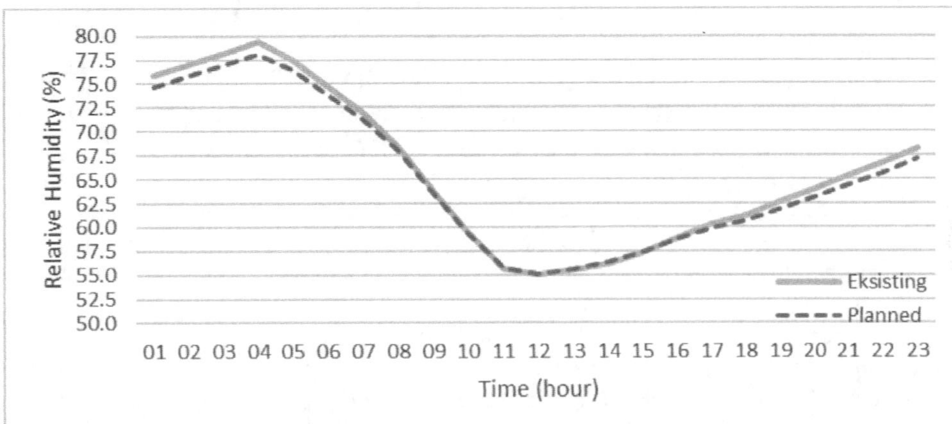

Figure 12. Diurnal relative humidity at human height level (1.5m) on a typical hottest day, from 1 am to 11 pm on June 21st

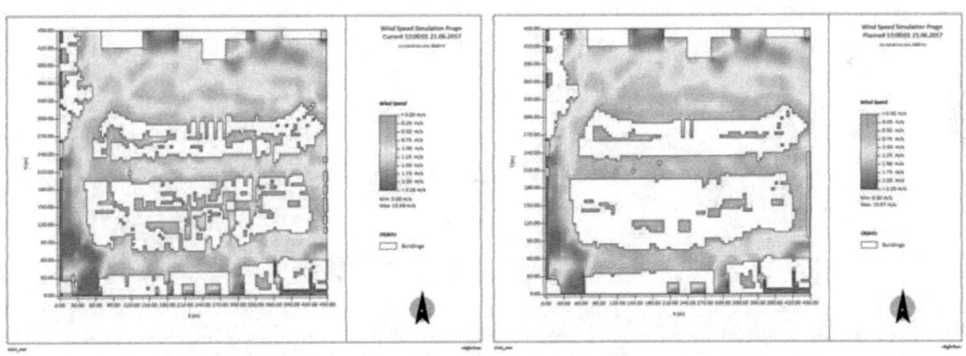

Figure 13. Wind Speed simulation for current condition and Detailed Spatial Plan

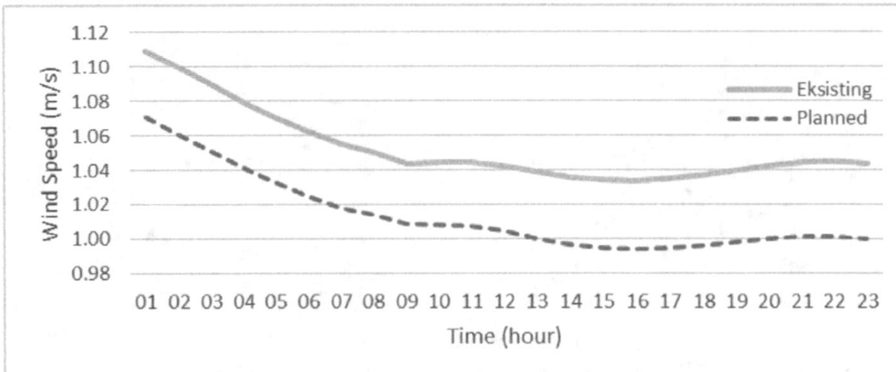

Figure 14. Diurnal wind speed change, from 1 am to 11 pm on June 21st

6. Conclusions

Overall, the thermal comfort index values (indicated by PMV value) on both existing and planned conditions of Progo Street and surrounding area are over the standard index for hot humid climate countries. Although there are a lot of lush trees and relatively low height buildings, the thermal comfort is still not reached yet.

Compared to each other, the existing condition has higher thermal comfort index values than the projected condition. Based on this research and literature study, it is clearly seen that several factors such as building heights, building area or configuration, and terrace materials have an impact to the thermal comfort.

This research has collected limited data to find out more exact and objective causes of the high thermal comfort index. If more accurately variables and causes can be gathered through other research methods, factors for considerations on making spatial planning policy can be determined. Therefore, further research is needed.

7. Acknowledgements

We would like to express a sincerely gratitude to Institute of Technology Bandung (ITB) for providing the grant through Research, Community Service, and Innovation Program, known by Program Penelitian, Pengabdian Masyarakat, dan Inovasi (P3MI) that organised by LPPM ITB.

The research process is also conducted together with Hersanti from Landscape Architecture programme, Marisa and Aryo from Urban Design programme, and David from Centre of Urban Design Studies as the assistants of researcher.

8. References

[1] Kuttler, W. 2010. Urban climate, Part I. Gefahrstoffe – Reinhaltung der Luft, 70(7/8), pp 329–340.

[2] Koerniawan, M. D., Gao, W. 2014. *The Simulation Study of Thermal Comfort in Urban Open Space of Commercial Area Using EnviMet Software*, a conference paper, January 2015.

[3] Wang, Y., Akbari, H. 2016. *The Effects of street tree planting on Urban Heat Island mitigation in Montreat*. Journal of Sustainable Cities and Society, 27: pp 122-128

[4] ISO 7730. 1994. Moderate Thermal Environments – Determination of the PMV and PPD Indices and Specification of the Conditions for Thermal Comfort. International Standard.

[5] Fanger, P.O. 1972. *Thermal comfort. Analysis and applications in environmental engineering*. New York: McGraw-Hill

[6] Bandung Local Regulation no. 10 of 2015 about Detailed Spatial Planning of Bandung 2015-2035. Spatial Planning Department of Bandung.

[7] Central Bureau of Statistics (BPS). 2017. *Jumlah Wisatawan Mancanegara dan Domestik di Kota Bandung, 2016*. Cultural and Tourism Department of Bandung.

IOP Conf. Series: Earth and Environmental Science **152** (2018) 012037 doi:10.1088/1755-1315/152/1/012037

Sustainable Resilience of Vulnerable Urban Kampong Fisherman Settlement in Dadap, Indonesia

Assoc. Prof. DR. Ir. Budi Prayitno, M. Eng.

Department of Architecture and Planning, Faculty of Engineering, Universitas Gadjah Mada, Yogyakarta, Indonesia

budiprayitno_ugm@yahoo.com

Abstract. The congested space in a vulnerable urban fisherman kampong settlement is triggered by the inability to compete and the weak empowerment of the kampong's sustainable assets. This situation necessitates a paradigm shifting from the competition between the interests of public and private sector into a harmonious coopetition. By using a case study in Dadap, this study aimed to map the vulnerability, resilience, and sustainability of the vulnerable kampong settlement as a basis for constructing a concept for the kampong's sustainable resilience reinforcement. The research methods consisted of a descriptive qualitative method to analyse the level of the kampong's sustainable resilience and space syntax to simulate the system of the kampong's spatial structure. The results of this research show that reinforcing the resilience of living space and livelihood system of a vulnerable space of an urban kampong fisherman settlement can be done through constructing the triggers in the form of community-based economic empowerment facilities and sustainable coopetition approach through empowering the kampong's sustainable assets with the concept of co-working, co-benefit and co-existence. Regarding the reinforcement of the kampong's structural space system, it can be done through changing its configuration system from intra-space of inner area system to inter-space of area-wide system.

1. Introduction

The existence of fisherman settlement in Indonesia is identical with the image of poverty and slum. Kampong Dadap, which is located in Kosambi Subdistrict (kecamatan), Tangerang Regency, Banten Province, was initially a fishing village settlement. Over time, it became a densely populated urban kampong filled with economic, social, and environmental problems. Daily tidal flooding that strikes the kampong is one of the reasons why it is environmentally uninhabitable.

Figure 1. Tidal flood in fisherman Kampong Dadap Settlement.

Moreover, the prostitution business in Dadap Cheng In adds to the complexity of the social and economic problems. In the beginning Dadap was known as seafood culinary region and the majority of its residents are fisherman. It covers an area of 15 hectares with a population of 12.743; most of them occupy government-owned lands. Kali Perancis, which was a Soekarno-Hatta Airport drainage channel that flowed into the sea, is located here. In the course of time, the river experienced high levels of pollution due to the emergence of settlements along its banks. The pollution comes from domestic and industrial wastes that carry sediment and other waste materials; thus, they damage the estuary ecosystem.

Dadap's built-up area reaches 85 percent and most of it consists of houses that cause a very high density and lack of public open space. The tidal flooding that hit this dense settlement along with lack of proper settlement infrastructure makes an unhealthy environment.

Figure 2. Tidal flood in Dadap school area.

Even worse, all the residents use the river and ditches to dispose all of the waste from household and small home-based industry. To fulfill their needs for clean water, a small proportion them use urban drinking water services while most of the others buy water from vendors. These social, economic, and environmental aspects are highly vulnerable to cause crimes and massive environmental issues.

The vulnerability of such settlements is generally generated from the residents' failure to access their livelihood assets and therefore results in poverty [1]. And this is a part of urbanization process [2][3].

The complexity in this informal settlement is a highly vulnerable problem. On the other hand, Dadap is a very strategic area or otherwise known as 'Golden Site'. There had even been a plan to develop it into Dadap International City central area. The idea was based on the strategic value of the area which is adjacent to Soekarno-Hatta International Airport and Pantai Indah Kapuk, a luxurious commercial and residential mega-project. The plan that was initially based on the area's strategic and commercial potentials was rejected by the residents. Thus, by using social space approach through Sustainable Urban Renewal concept it was decided to maintain the area as a fisherman settlement. This concept was performed using Creative Co-Planning and Management approach. The social space approach in an informal settlement arrangement will be largely determined by the level of inclusiveness in the region design process [4]. The strategic and commercial potentials of this area have to synergize with the solution to the vulnerability of environmental degradation in Dadap.

The urban space of the synergy between spatial, economic, social, and environmental assets and the solution to the vulnerability of those assets needs to be re-framed in the hybrid space context [5].

2. RESEARCH METHODS

This study uses qualitative descriptive method in analyzing the vulnerability, resilience, and sustainability aspects; whereas in analyzing the spatial structure of the settlement the study uses spatial observation through space syntax simulation. The observation location is the fisherman settlement in Dadap, Tangerang Regency. It is adjacent to the reclaimed settlement area, i.e., Pantai Indah Kapuk and Soekarno-Hatta International Airport.

The spatial simulation approach used in this study is space configuration with the integration and intelligibility values of the area. A high integration value provides an overview of how the Dadap area connects with its surrounding areas in meso-scale (intra-space) and macro-scale (inter-space). Intelligibility value is a space clarity value, determined by looking at two variables which are integration and connectivity values. The two variables have cognitive and spatial-physical relationship. Space proximity in a regional system could be discovered by knowing the intelligibility value. Measurement of the integration and intelligibility value reveals the accessibility rate of research area. Therefore, it is easier to observe the movements from, to, or pass the research area.

3. VULNERABLE SPACE

The vulnerability of a space is always associated with access and infrastructure. As an informal settlement, Dadap's condition is largely determined by urban livelihood aspects such as location, accessibility, mobility, and linkage. The spatial expansion process that commonly occurs in an urban space is highly dependent on the integration and the informal linkage between activities and economic generators [1]. Spatial assets within the urban space linkage system are influenced by the radius range of urban economic network system. Dadap, an urban kampong fisherman settlement, was one of the fishing port centers and it is now defeated by a fishing port in Muara Kamal whose economic activities are much stronger. As a result, the economic activities in Dadap shifted into services activities that are at risk of social and public health vulnerability, namely the prostitution business and other economic activities that support it. What remains of the fisherman village is an abandoned dock for some fisherman boats and the green mussel cultivation using keramba, a floating fish nets or cages.

Figure 3. Abandoned fisherman boats in Dadap

The social and economic resilience to the vulnerable situation could be accomplished through an adaptation process which utilizes the remaining assets [6][7]. In this case, the remaining asset is the green mussel cultivation using keramba.

The slum in Dadap is the result of weak government intervention in local planning, urban health facilities, and lack of livelihood potential for the residents; thus, they lead to the massive slum phase or otherwise known as attaching, the third phase of three phases of informal settlement after settling and inserting [8][9][10][11].

3.1. Network Analysis

On a macro scale, Dadap is flanked by several fast-growing facilities namely the development of reclaimed islands, housing for the high-income group, and the expansion of Soekarno-Hatta airport. The area is a strategic node located between Tangerang Regency and Jakarta Metropolitan City. The interface zone acts as a driving force that is often analogous to a machine [12]. Circulation and intensity of activities that constitute a network could be managed through spatial simulations and assessed from connectivity, integration, and intelligibility of the space.

The road network in Dadap consists of two types; neighborhood streets and kampong streets such as alleys. The alleys follow the irregular pattern of the housing. The road network pattern is influenced by the configuration of neighborhood streets along the Kali Perancis banks and the public facilities distribution pattern in Dadap. The surrounding area of Dadap is a mix of residential and industrial area. The industrial area is a big plot, and it is a private zone that follows the grid pattern. Spatial analysis through space syntax method shows the intra network-based value is moderate with the average of 0.36 from the range of 0.31-0.98. The intelligibility value, which is a relationship between the integration value and connectivity, of Dadap is R2=0.03. That means the settlement has low intelligibility value in the spatial relationship. On the contrary, inter network-based analysis shows a quite high value with the average of 1.77 from the range of 0.3-3.7. The relationship between the

region's integration and connectivity or the intelligibility value scores R2=0.4, which means the region has a medium intelligibility value.
The different results of the two simulations above are the outcome of Dadap's weak integration with its surrounding area. This region is known as a pass-through only area, which is an area that is only passed through by people, not as the main destination. This situation causes the settlement to have its own intra-system which is very vulnerable to be resilient due to lack of livelihood asset potentials, urban services shortage, and weak government intervention.

3.2. Space as a competition domain
It could be seen from the space syntax simulation that Dadap is experiencing congested domain. Its pattern of intra-network space that filled with informality characters, degraded environmental quality and livelihoods shows a highly vulnerable sustainability system. The trigger of the vulnerability is the lack of access to livelihoods [13][14]. The decline of Dadap's fishing port functions due to competitiveness loss against Muara Kamal fishing port causes the fishing port livelihoods access to become very weak. The residents' response to this condition was informal and unplanned. As a result, they are environmentally, economically, and socially marginalized; thus, they have no access to development agencies or sources. The failure to adapt in this vulnerable space is a consequence of access, livelihoods and space shortages [8][15][16].

Without a strong intervention and adequate resources, the congestion of various marginalized space adaptation processes would result in social, economic, and environmental degradation. In other words, the sustainability value of the settlement in Dadap would experience degradation because it is economically and spatially defeated. On the other hand, the reclamation development and airport expansion are spatially supported by various abundant resources. The contrast is very apparent even though Dadap is located in the interface area between the two mega-development projects.

4. SUSTAINABLE RESILIENCE
Regional resilience is the ability of a region to survive and adapt to exogenous disturbances that threaten the sustainability of its function, space structure, and identity [17][7][18]. Therefore, mitigation strategies against those disturbances are needed. They have to be capable of responding to environmental, economic, and social changes as well as being future-oriented. As a waterfront settlement, Dadap must be able to adapt to the climate and tidal change. Similarly, the dynamics of socio-economic life and the fisherman livelihoods, which are vulnerable to the failure to access livelihood resources, have to be capable of creating diverse activities that are adaptive to the socio-economic dynamics of fisheries.

4.1. Co-exist, Co-working, and Co-benefit
Sustainable space design principle, which is based on social; economic; and environmental pillars, has to be adopted into a resilient and adaptive system. From a social point of view, the residents of fisherman settlement have to harmonically synergize with the life of commercial settlement for the high-income group. Public open space is a medium that has the potential to synergize these two different conditions [19]. Co-existential theory of informal and formal sectors is widely developed in urban areas [5]. Environmental coexistence could be interpreted as the ecological relationship between land and water [20]. Social co-working is, for example, a partnership that involves various stakeholders such as fisherman community, commercial settlement community, private sectors, and government [20]. Economic co-benefit means the synergy of economic activities between community-based livelihoods and commercial-based enterprises [20]. This sustainable resilience has to be spatially structured in order to synergize the social, economic, and natural potentials in macro spatial context to face the development of the airport and reclaimed land.

The initial approach was relocating the residents to vertical housings in the same location, but it was rejected by Dadap fishing community. The rejection was underpinned by the lack of secure tenure and livelihood certainty. For them, the acceptable solution was tenure and livelihood security by giving cheap vertical housings and building an integrated culinary tourism center. These two facilities would trigger the fisherman community's willingness to work together with the government and private sectors in organizing the area by using sustainable urban renewal.

Figure 4. Proposed Dadap Urban Kampong Renewal.

Environmental sustainability and resilience are realized with the concept of floating culinary facilities, pillar housings, and polder system engineering. Social sustainability and resilience are, for example, an interaction between the management of oceanic culinary tourism center and Dadap residents with the residents of Pantai Indah Kapuk as their target market. Economic sustainability and resilience, for instance, a management partnership using shared values approach between community-based businesses and private enterprises

Figure 5. Dadap oceanic culinary complex building (using floating construction technology).

4.2. From Competition to Coopetition

As a congested space, which is due to being spatially and economically defeated by the rapid development of Soekarno-Hatta airport and the reclamation for commercial area construction, Dadap needs to be resilient. A sustainable approach that could be performed is harmonious coopetition. The waterfront urban space could participate in a coopetition by means of diversification through job creation and economic regeneration [21]. The competition in waterfront space is a model of urban waterfront regeneration where the shape of mix used and hybrid function become the concept of architectural development.

Waterfront space constructs a place that based on local identity. The architectural identities of the commercial residential area Pantai Indah Kapuk and the expansion of Soekarno-Hatta airport have to be capable of harmonious coopetition and competition that results in production and place identity and thus become their brand marketing. Dadap's new identity needs to synergize with the Pantai Indah Kapuk's identity, which is an integrated luxurious residential area, and also with the modern identity of Soekarno-Hatta airport expansion. The integration of those identities does not always mean a similarity in their appearances, but rather a co-presence relationship between different groups [12].

Coopetition in spatial structures is determined by the degree of connectivity and spatial integration as shown by its intelligibility value. An architectural work will be more integrated if it successfully constructs a place and architecture imagery harmoniously [21]. Similarly, spatial and architectural transformation in waterfront space is also a branded form of place making [22]. Therefore, a sustainable urban renewal in Dadap, a transformation from slum fisherman settlement to a new residential area, needs to be developed by using appropriate spatial structure and local architectural imagery.

IOP Conf. Series: Earth and Environmental Science **152** (2018) 012037 doi:10.1088/1755-1315/152/1/012037

Figure 6. Proposed model of sustainable fisherman low-rise multistory and pillar detached housing in Dadap (using photovoltaics panels and rand rain water harvestness technology).

5. INTER-SPACE REFRAMING DESIGN

Urban spatial structure is a determinant of urban movement for both pedestrian and vehicular, whose functions change rapidly [12]. Pedestrian and vehicular spatial patterns form movement and social interaction patterns [23][24][25]. These spatial patterns are spatial assets for the establishment of network, social, economy, and environment [10]. Spatial vulnerability due to settlement and environment quality degradation leads to weak networks, and neighboring relationships as well as lack of opportunities to improve livelihoods.

5.1. Place and Network

A place is a space whose life within it is largely determined by the effectiveness of the network system that forms it. As a place, Dadap experiences quality degradation in its space and its network. This condition could be seen from its low intelligibility value from the space syntax simulation. Therefore, it is necessary to strengthen its intelligibility value; from a region that has a weak relationship with its surrounding area and a pass-through only area to be the main destination area.

By strengthening the inter-space network and the establishment of regional activation generator in the form of an integrated culinary tourism center, it is expected that they could increase Dadap's intelligibility value, from a pass-through only area to be the main destination area.

5.2. Reframing Space

To transform a region from a pass-through only area into a destination area, reframing space structure is required by strengthening its connectivity and integration network.

Space syntax simulation shows that there is a 0.17 increase in the existing value. The intra-space (local integration) intelligibility value also rises from 0.41 to 0.58, and from the average integration value of 1.77 to 1.81.

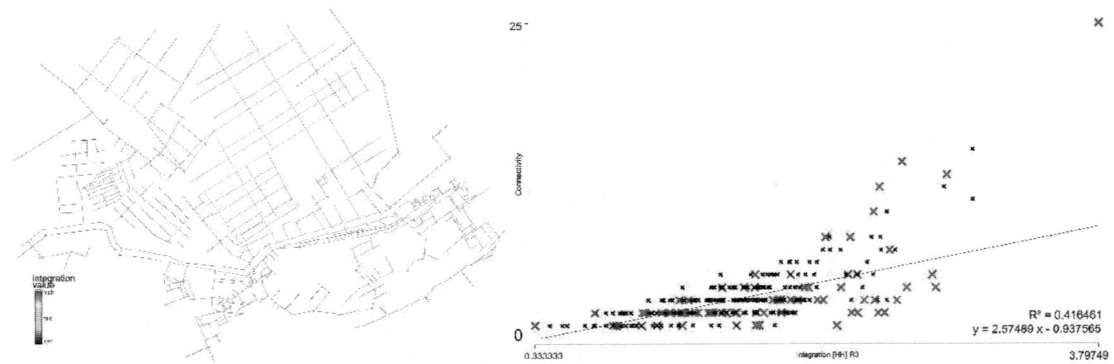

Figure 7. Existing local integration value.

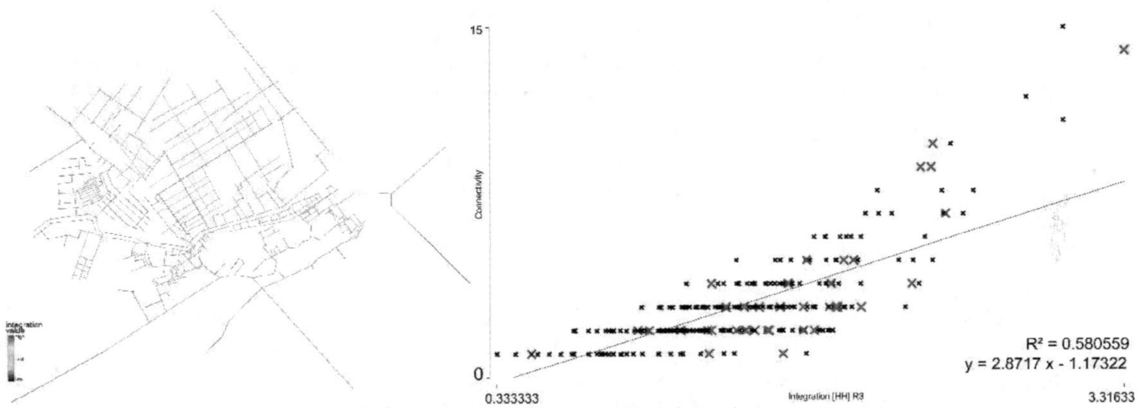

Figure 8. Proposed local integration value.

The inter-space (global integration) intelligibility value, which is the `relationship of settlement in Dadap with its surrounding area, also experiences improvement from 0.038 to 0.045, but the average integration value falls from 0.36 to 0.34.

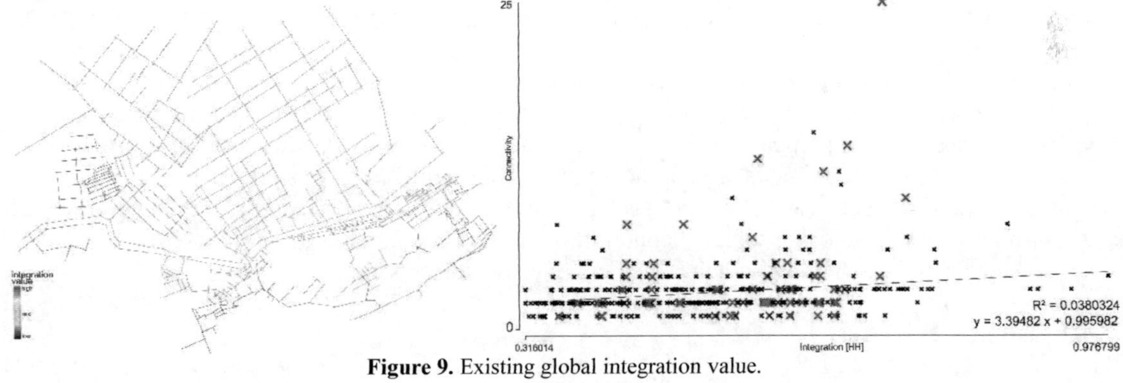

Figure 9. Existing global integration value.

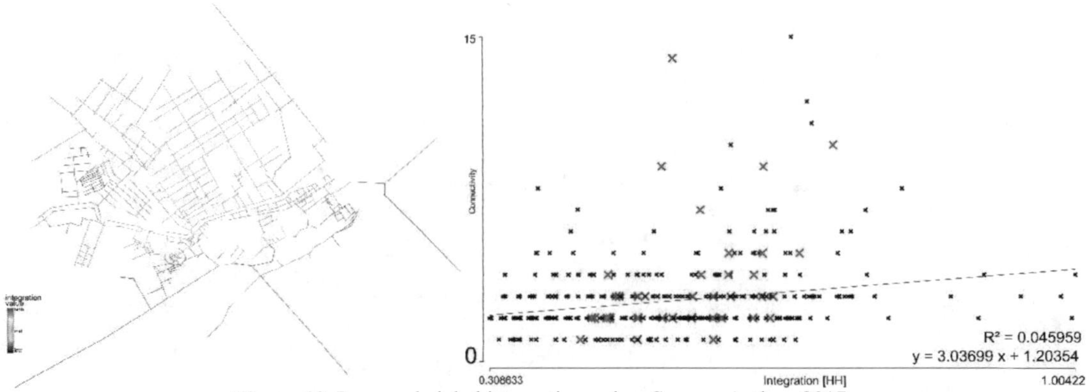

Figure 10. Proposed global integration value. Source: Author, 2017

The comparison between local and global context, both in terms of the integration and intelligibility value, shows that there is an improvement of regional value but a decline in global value of the region. This result implies that Dadap could be transformed into a destination area and not as a passed-through only area anymore. The simulation results reveal that the structural value of a region would rise after its network is improved by connecting it with the surrounding area. In addition, the improvement also results in the decline of the region's global value.

6. CONCLUSION

Fisherman urban kampong settlement is vulnerable because it lost in competing against the strong development of its surrounding area; hence, this leads to the degradation of its quality and environment. The fisherman settlement that is always vulnerable to the dynamics of environmental change must be able to adapt ecologically. Therefore, the resilience concept which is required in order to improve the adaptability and strengthening values must be based on social, economic, and environmental aspects. The paradigm shift in the way of looking at a competition, from something that would bring negative impacts to one that would give positive outcomes, is known as coopetition. Coopetition is an act that synergizes competition and coopetition. It could be adopted into the sustainable resilience concept by using social co-working, economic co-benefit, and co-existence in the ecological environment.

The continuous strengthening of the resilience value has to be structured based on the concept of inter-area and reinforced with trigger generators as the regional attractions. Efforts to redefine the function of fisherman settlement areas, reconnect the spatial network structure, and revitalize the function of community livelihoods are embodied in the concept of inter-space reframing design.

The discussion of space resilience and sustainability theory discovers that the resilience level of a space is highly determined by social, economic, and environmental aspects. Those aspects are the pillars of sustainability which will decide the resilience level of a space.

7. REFERENCE

[1] Chambers, R. (1995). *Poverty and Livelihoods: Whose Reality Count?* Journal of Environment and Urbanization. Volume 03 Number 03 pp. 173-204
[2] Ellin, N. (2006). *Integral Urbanism. Routledge*, Taylor & Fracis Group. New York
[3] Rakodi, C and Lloyd-Jones, T. (2002). *Urban Livelihoods: A People-centered Approach to Reducing Poverty*. Earthscan Publication Limited. London
[4] Lloyd-Jones, T. (2000). *Compact City Policies for Megacities*, in Jenks, M and Burgess, R (eds) The Compact City in Developing Countries. E&FN Spon. London. pp. 37-53
[5] Cho, I., Heng, K., and Trivic, Z. (2015). *Re-framing Urban Space: Urban Design for Emerging Hybrid and High-Density Conditions*. New York: Routledge
[6] Walker, B., Holling, C. S., Carpenter, S. R., and Kinzig, A. (2004) *Resilience, Adaptability and*

IOP Conf. Series: Earth and Environmental Science **152** (2018) 012037 doi:10.1088/1755-1315/152/1/012037

Transformability in Social–Ecological Systems. Ecol. Soc. 9,5.

[7] Walker, B., Gunderson, L., Kinzig, A., Folke, C., Carpenter, S., and Schultz, L. (2006). *A Handful Of Heuristics And Some Propositions For Understanding Resilience In Social-Ecological Systems*. Ecology and society, 11(1).

[8] Prayitno, B. (2014). *Skema Inovatif Penanganan Permukiman Kumuh. Jogja* : Gama Press , UGM.

[9] Dovey, K. and King, R. (2012). *Informal Urbanism and the Taste for Slums, Tourism Geographies*, 14(2), pp 275-293.

[10] Dovey, K. (2012). *Informal Settlement and Complex adaptive Assemblage,* International Development Planning Review, 34 (3). 371-90.

[11] Dovey, K. and King, R. (2011). "*Forms of Informality : Morphology and Visibility of Informal Settlements,*" Built Environment, 37, no 1 pp 11-29.

[12] Hillier, B. (1996). *Space is the machine*, Cambridge: Cambridge University Press.

[13] Tipple, A. G. (2000). *Extending Themselves: User-Initiated Transformations Of Government-Built Housing In Developing Countries*. Liverpool: Liverpool University Press.

[14] Napier, M. (2007). *Informal settlement integration, the environment and sustainable livelihoods in sub-Saharan Africa*. Council for Scientific & Industrial Research in South Africa.

[15] Kamalipour, H. (2016). *Forms of informality and adaptations in informal settlements*, 10(3), 60–75.

[16] Rothenberg, A.D., Gaduh, A., Burger, N.E., Chazali, C., Tjandraningsih, I., Radikun, R., Sutera, C., and Welant, S. *Rethinking Indonesia's Informal Sector*. World Development, 80, 96–113. http://doi.org/10.1016/j.worlddev.2015.11.005

[17] Hodgson, D., McDonald, J. L., and Hosken, D.J. (2015). *What do you mean, 'resilient'?*. Trends in ecology & evolution, 30(9), 503-506.

[18] Cutter S.L .(2008*). "A Place-Based Model for Understanding Community Resilience to Natural Disasters."* Global Environmental Change 18: 598-606

[19] Madanipour, A. (1999). *Why are the Design and Development of Public Space Significant for Cities?,* Environment and Planning B : Planning and Design, Vol. 26 pp 879-891.

[20] Prayitno, B. (2017). *Co-habitation space: A Model for Urban Informal Settlement Consolidation for the Heritage City of Yogyakarta, Indonesia*. Journal of Asian Architecture and Building Engineering. Vol 16, no 3.

[21] Crilley, D. (1993) *'Architecture as Metaphor'*. in: Kearns, &. Philo, Selling Places, pp.231-252.

[22] Hillier, B. and Netto, V. (2002) "*Society seen through the prism of space: outline of a theory of society and space,*" Urban Design International, pp. 181-203.

[23] McLennan, J. (2004). *The philosophy of sustainable design*, New York: Bookbaby Publisher.